Lecture Notes in Computer Science 6305

Commenced Publication in 1973
Founding and Former Series Editors:
Gerhard Goos, Juris Hartmanis, and Jan van Leeuwen

Rainer Keller   Edgar Gabriel
Michael Resch   Jack Dongarra (Eds.)

Recent Advances in the Message Passing Interface

17th European MPI Users' Group Meeting, EuroMPI 2010
Stuttgart, Germany, September 12-15, 2010
Proceedings

 Springer

Volume Editors

Rainer Keller
Michael Resch
High Performance Computing Center Stuttgart (HLRS)
70569 Stuttart, Germany
E-mail: {keller, resch}@hlrs.de

Edgar Gabriel
University of Houston, Parallel Software Technologies Laboratory
Houston, TX 77204, USA
E-mail: gabriel@cs.uh.edu

Jack Dongarra
University of Tennessee
Department of Electrical Engineering and Computer Science
Knoxville,TN 37996, USA
E-mail: dongarra@cs.utk.edu

Library of Congress Control Number: 2010933611

CR Subject Classification (1998): C.2.4, D.1.3, C.2.5, F.2, C.4, D.2, D.4,

LNCS Sublibrary: SL 2 – Programming and Software Engineering

ISSN   0302-9743
ISBN-10   3-642-15645-2 Springer Berlin Heidelberg New York
ISBN-13   978-3-642-15645-8 Springer Berlin Heidelberg New York

springer.com

© Springer-Verlag Berlin Heidelberg 2010
Printed in Germany

Typesetting: Camera-ready by author, data conversion by Scientific Publishing Services, Chennai, India
Printed on acid-free paper   06/3180

Preface

Parallel Computing is at the verge of a new era. Multi-core processors make parallel computing a fundamental skill required by all computer scientists. At the same time, high-end systems have surpassed the Petaflop barrier, and significant efforts are devoted to the development of hardware and software technologies for the next-generation Exascale systems. To reach this next stage, processor architectures, high-speed interconnects and programming models will go through dramatic changes. The Message Passing Interface (MPI) has been the most widespread programming model for parallel systems of today. A key questions of upcoming Exascale systems is whether and how MPI has to evolve in order to meet the performance and productivity demands of Exascale systems.

EuroMPI is the successor of the EuroPVM/MPI series, a flagship conference for this community, established as the premier international forum for researchers, users and vendors to present their latest advances in MPI and message passing system in general. The 17th European MPI users group meeting was held in Stuttgart during September 12-15, 2010. The conference was organized by the High Performance Computing Center Stuttgart at the University of Stuttgart. The previous conferences were held in Espoo (2009), Dublin (2008), Paris (2007), Bonn (2006), Sorrento (2005), Budapest (2004), Venice (2003), Linz (2002), Santorini (2001), Balatonfured (2000), Barcelona (1999), Liverpool (1998), Krakow (1997), Munich (1996), Lyon (1995) and Rome (1994).

The main topics of the conference were message-passing systems – especially MPI, performance, scalability and reliability issues on very large scale systems.

The Program Committee invited five outstanding researchers to present lectures on different aspects of high-performance computing and message passing systems: William Gropp presented "Does MPI Make Sense for Exascale Systems?," Jan Westerholm presented "Observations on MPI Usage in Large-Scale Simulation Programs," Jack Dongarra presented "Challenges of Extreme Scale Computing," Jesus Labarta presented "Detail at Scale in Performance Analysis," and Rolf Hempel presented "Interactive Visualization of Large Simulation Datasets."

The conference also included a full-day tutorial on "Advanced Performance Tuning of MPI Applications at High Scale with Vampir" by Wolfgang Nagel and Matthias Müller, and the 9th edition of the special session: "ParSim 2010: Current Trends in Numerical Simulation for Parallel Engineering Environments."

In all, 41 full papers were submitted to EuroMPI, out of which 28 were selected for presentation at the conference, along with 5 posters. Each paper had between three and five reviews, guaranteeing that only high-quality papers were accepted for the conference. The program provided a balanced and interesting view on current developments and trends in message passing. Three papers were selected as outstanding contributions to EuroMPI 2010 and were presented at a plenary session:

- "Load Balancing Regular Meshes on SMPs with MPI," by Vivek Kale and William Gropp,
- "Adaptive MPI Multirail Tuning for Non-uniform Input/Output Access," by Stéphanie Moreaud, Brice Goglin and Raymond Namyst
- "Using Triggered Operations to Offload Collective Communication Operations," by Scott Hemmert, Brian Barrett and Keith Underwood.

The Program and General Chairs would like to sincerely thank everybody who contributed to making EuroMPI 2010 a success, by submitting papers, providing a review, by participating or sponsoring the event.

September 2010

Michael Resch
Rainer Keller
Edgar Gabriel
Jack J. Dongarra

Organization

EuroMPI 2010 was organized by the High-Performance Computing Center Stuttgart (HLRS), of the University of Stuttgart, Germany, in association with the Innovative Computing Laboratory (ICL) of the University of Tennessee, Knoxville.

Program Committee Chairs

General Chair	Jack J. Dongarra, (UTK)
Program Committee Chair	Michael Resch (HLRS)
Program Committee Co-chairs	Rainer Keller (HLRS / ORNL)
	Edgar Gabriel (University of Houston)
Local Organizing Committee Chair	Stefan Wesner (HLRS)
Tutorials	Wolfgang Nagel (ZIH)
	Matthias Müller (ZIH)

Program Committee

Richard Barrett	Sandia! National Laboratories, USA
Gil Bloch	Mellanox, Israel
Ron Brightwell	Sandia National Laboratories, USA
George Bosilca	University of Tennesse, USA
Franck Cappello	INRIA, France
Barbara Chapman	University of Houston, USA
Jean-Christophe Desplat	ICHEC, Ireland
Yiannis Cotronis	University of Athens, Greece
Frederic Desprez	INRIA, France
Erik D'Hollander	Ghent University, Belgium
Jack J. Dongarra	University of Tennessee, USA
Edgar Gabriel	University of Houston, USA
Javier Garcia Blas	Universidad Carlos III de Madrid, Spain
Al Geist	Oak Ridge National Laboratory, USA
Michael Gerndt	Technische Universität München, Germany
Ganesh Gopalakrishnan	University of Utah, USA
Sergei Gorlatch	Universität Münster, Germany
Andrzej Goscinski	Deakin University, Australia
Richard L. Graham	Oak Ridge National Laboratory, USA
William Gropp	University of Illinois Urbana-Champaign, USA
Thomas Herault	INRIA/LRI, France
Josh Hursey	Indiana University, USA
Torsten Hoefler	Indiana University, USA

Yutaka Ishikawa	University of Tokyo, Japan
Tahar Kechadi	University College Dublin, Ireland
Rainer Keller	HLRS, Universität Stuttgart, Germany / ORNL, USA
Stefan Lankes	RWTH Aachen University, Germany
Alexey Lastovetsky	University College Dublin, Ireland
Andrew Lumsdaine	Indiana University, USA
Ewing Rusty Lusk	Argonne National Laboratory, USA
Tomas Margalef	Universitat Autonoma de Barcelona, Spain
Jean-Francois Mehaut	IMAG, France
Bernd Mohr	Forschungszentrum Jülich, Germany
Raymond Namyst	University of Bordeaux, France
Rolf Rabenseifner	HLRS, Universität Stuttgart, Germany
Michael Resch	HLRS, Universität Stuttgart, Germany
Casiano Rodriquez-Leon	Universidad de la Laguna, Spain
Robert Ross	Argonne National Laboratory, USA
Martin Schulz	Lawrence Livermore National Laboratory, USA
Stephen F. Siegel	University of Deleware, USA
Bronis R. de Supinski	Lawrence Livermore National Laboratory, USA
Jeffrey Squyres	Cisco, Inc., USA
Rajeev Thakur	Argonne National Laboratory, USA
Jesper Larsson Träff	University of Vienna, Austria
Carsten Trinitis	Technische Universität München, Germany
Jan Westerholm	Abo Akademi University, Finland
Roland Wismüller	Universität Siegen, Germany
Joachim Worringen	International Algorithmic Trading GmbH, Germany

External Reviewers

Mats Aspnäs
Michael Browne
Alejandro Calderon
Anthony Chan
Wei-Fan Chiang
Gilles Civario
Carsten Clauss
Javier Cuenca
Robert Dew
Kiril Georgiev
Brice Goglin
David Goodell
Haowei Huang
Lei Huang
Florin Isaila

Emmanuel Jeannot
Christos Kartsaklis
Philipp Kegel
Jayesh Krishna
Pierre Lemarinier
Guodong Li
Diego Rodriguez Martinez
Alastair McKinstry
Dominik Meiländer
Gara Miranda-Valladares
Jan Pablo Reble
Justin Rough
Lucas Mello Schnorr
Adam Wong

Sponsors

We would like to thank the following companies for their kind support:

- Cisco
- Cray
- IBM
- Microsoft
- NEC

Table of Contents

Collective Operations

Applications

MPI Internals (I)

Fault Tolerance

Best Paper Awards

MPI Internals (II)

Poster Abstracts

A Scalable MPI_Comm_split Algorithm for Exascale Computing

Paul Sack and William Gropp

University of Illinois at Urbana-Champaign

Abstract. Existing algorithms for creating communicators in MPI programs will not scale well to future exascale supercomputers containing millions of cores. In this work, we present a novel communicator-creation algorithm that does scale well into millions of processes using three techniques: replacing the sorting at the end of MPI_Comm_split with merging as the color and key table is built, sorting the color and key table in parallel, and using a distributed table to store the output communicator data rather than a replicated table. This reduces the time cost of MPI_Comm_split in the worst case we consider from 22 seconds to 0.37 second. Existing algorithms build a table with as many entries as processes, using vast amounts of memory. Our algorithm uses a small, fixed amount of memory per communicator after MPI_Comm_split has finished and uses a fraction of the memory used by the conventional algorithm for temporary storage during the execution of MPI_Comm_split.

1 Introduction

The Message Passing Interface Forum began developing the message-passing interface (MPI) standard in early 1993. MPI defines a communication interface for message-passing systems that includes support for point-to-point messaging, collective operations, and communication-group management. The crux of communication-group management is the *communicator*: a context in which a group of processes can exchange messages.

In 1993, the fastest supercomputer in the Top 500 was a 1024-processor Thinking Machines system that could sustain nearly 60 Gigaflops [6]. In the June 1997 list, the 7264-processor Intel ASCI Red system broke the Teraflop barrier. In early 2010, the fastest supercomputer was a Cray system with 224,000 cores at 1.8 Petaflops, and a Blue Gene system ranked fourth has nearly 300,000 cores achieves nearly 1 Petaflop. If scaling trends continue, supercomputers with *millions* of cores will achieve Exaflop performance. Current methods for managing communicators in MPI do not perform well at these scales – in space or in time – and programmers cannot program their way around using communicators.

In this work, we first examine the state of the art in section 2 and identify the communicator-creation functions that are not inherently unscalable. In section 3, we propose novel communicator-creation algorithms that do scale to million-core supercomputers. In section 4, we detail our evaluation methodology and present our results. In section 5, we discuss future work. We conclude in section 6.

R. Keller et al. (Eds.): EuroMPI 2010, LNCS 6305, pp. 1–10, 2010.

2   Background

MPI programs start with one communicator, MPI_COMM_WORLD, that contains every process in the program. It is common to form smaller communicators containing a subset of all the processes. MPI provides two ways to do this:

- MPI_Comm_create: processes must enumerate all the members of the new communicator.
- MPI_Comm_split: processes specify a color; processes whose colors match become the members of new communicators.

MPI_Comm_create requires as input a table specifying membership in the new communicator for every process in the old communicator. Only one communicator is created per call (in MPI 2.1), and ranks can not be reordered. Each process in the old communicator must call this function with the same table, whether or not the process is included in the new communicator.

This algorithm scales poorly as the memory and computation costs scale linearly with the size of of the input communicator.

MPI_Comm_split requires only a color and a key as input. (The key is used to reorder ranks in new communicators.) It is more versatile, since it can create many communicators in one call, can reorder ranks, and does not require the programmer to build a large table to specify communicator membership. The ranks in the new communicator are assigned by sorting the keys and using the ranks in the old communicator as a tie-breaker.

We examined the implementation of MPI_Comm_split in two widely-used open-source MPI implementations: MPICH [4,5] and OpenMPI [3], which is derived from LAM-MPI. Unfortunately, MPI_Comm_split scales equally poorly in these current MPI implementations, which operate as follows:

1. Each process builds a table containing the colors and keys of all the other processes using MPI_Allgather.
2. Each process scans the table for processes whose colors match.
3. Each process sorts the keys of processes whose colors match.

MPICH makes use of "recursive doubling" to build the table [10]. In the first step, each process exchanges its color and key information with the process whose rank differs only in the last bit. In the second step, each process exchanges its color and key table (now containing two entries) with the process whose rank only differs in the second-last bit. This continues, and in the final step, each process exchanges a table containing $P/2$ entries with the process whose rank differs only in the first bit.

Open MPI uses an operation similar to recursive doubling to build the table. In the first step, process p_i send data to process $p_{i+1 \bmod P}$ and receives data from $p_{i-1 \bmod P}$. In the following steps, process p_i sends data to process $p_{i+2 \bmod P}$, then $p_{i+4 \bmod P}$, and so on, while receiving data from $p_{i-2 \bmod P}$, then $p_{i-4 \bmod P}$. We assume the use of recursive doubling in the rest of this paper, but it makes little difference in the analysis which is used.

In both implementations, after the tables are built, the tables are scanned for entries with matching colors, and then those entries are sorted by key.

This algorithm incurs a memory and communication cost of $\Theta(n)$, and a computation cost dominated by the sorting time: $O(n \lg n)$. Only the entries in the table whose colors match are sorted, so for smaller output communicators the computation cost can be much less.

As we will show, the sorting phase of MPI_Comm_split consumes a significant amount of time at larger scales. Further, the table requires vast amounts of memory. Every entry has at minimum three fields: hardware address or rank in the source communicator; color; and key. Using 32-bit integers for each field, this requires 192 Megabytes per process per communicator in a 16-million node system, which is clearly unreasonable. This is especially problematic because per-core performance is growing far more slowly than the number of cores in modern microprocessors [9], and we can only expect memory per-core to grow as quickly as performance per-core for weak scaling.

3 Scalable Communicators

While the memory problem is more concerning, we first present one facet of our solution to the poor performance scaling of MPI_Comm_split.

3.1 Better Performance through Parallel Sorting

As mentioned, much of the time in MPI_Comm_split is spent in sorting the color and key table. In state-of-the-art algorithms, every process sorts the entire table. Our proposal is to sort this table in parallel. The simplest and least-effective way to do this is to have each process sort its table at each step in the recursive doubling operation before exchanging tables with another process. In effect, at each stage, each process merges the sorted table it already has with the sorted table it receives from its partner process for that stage. This turns the $O(n \lg n)$ sort problem into $\lg n$ merge operations of size $2, 4, 8, \ldots, n/4, n/2$, an $O(n)$ problem.

This gives us an $O(\lg n)$ speedup. In the final step, all n processes are merging identical tables. In the second-to-last step, two groups of $n/2$ processes merge identical tables. To do away with this redundant computation, we adapt a parallel sorting algorithm to this problem.

Cheng *et al* present one parallel sorting algorithm [1]. Their algorithm begins with an unsorted n-entry table distributed among p processes. At the conclusion of the algorithm, the table is sorted, and the entries in the the ith process's table have global rank $[(i - 1)\frac{n}{p}, i\frac{n}{p})$.

In brief, their algorithm works as follows:

1. Each process sorts its local table.
2. The processes cooperate to find the exact splitters between each process's portion of the final sorted table using an exact-splitter algorithm [7].

3. The processes forward entries from their input subtables to the correct destination process's subtable using the splitters.
4. Each process merges the p pre-sorted partitions of their subtable.

In step 1, each process takes $O(\frac{n}{p} \lg \frac{n}{p})$ time to sort its subtable. However, we
use recursive doubling and merging to generate the inputs, thus we need not sort
the subtable. Recursive doubling up to the beginning of the parallel sort, where
each process has a table with $\frac{n}{p}$ elements, takes $O(\frac{n}{p})$ time.

In step 2, exact splitters are found after $\lg n$ rounds in which the distance
between the estimate of the splitter and the correct splitter exponentially de-
cays. Each round has an execution-time cost of $O(p + \lg \frac{n}{p})$, for a total time
$O(p \lg n + \lg^2 n)$.

In step 3, each process exchanges at most $\frac{n}{p}$ entries with other processes. (For
random input keys and ranks, the number of entries is expected to be $\frac{n}{p} \times \frac{p-1}{p}$.)
Thus, the time complexity for the exchange is $O(\frac{n}{p})$.

In step 4, each process recursively merges the p partitions of its subtable in
$\lg p$ stages, merging p partitions into $p/2$ partitions in the first stage, then $p/2$
partitions into $p/4$ partitions, and so on, for a cost of $O(\frac{n}{p} \lg p)$.

At the conclusion of the parallel sort, the p processes use recursive doubling
to build full copies of the table.

The total time spent in the parallel sort is $O(\frac{n}{p}) + O(p \lg n + \lg^2 n) + O(\frac{n}{p}) +$
$O(\frac{n}{p} \lg p) = O(p \lg n + \lg^2 n + \frac{n}{p} \lg p)$. There is no closed-form expression for p in
terms of n that minimizes time. If $O(1) < p < O(\frac{n}{\lg n})$, then, asymptotically, the
parallel sort will take less time than the $\lg p$ stages of merging in the conventional
algorithm with merging.

The time spent in the final recursive-doubling stage to build full copies of the
table is $O(n)$.

The amount of communication is more than that in a conventional recursive-
doubling implementation due to the communication incurred in finding the splitters and exchanging table entries between sorting processes. Even so, we observe
performance improvements with a modest number of sorting processes.

3.2   Less Memory Usage with Distributed Tables

For an application to scale to millions of processes, each process must only
exchange point-to-point messages with a small number of unique processes (*e.g.*,
$\lg n$ processes for MPI_Allgather). Thus, most of the entries in a communicator
table are unnecessary.

We propose a distributed table, in which the processes in a communicator
are arranged in a distributed binary tree, sorted by rank. At first, each process
only knows the identities of its parent and two children. When it wishes to
communicate with another process, it asks its children to find the address of the
target process if the target rank is in the process's subtree, otherwise it asks
its parent to find the address. The child or parent process then forwards the
request around the tree until the target process is found. A message containing

the address of the target process is returned along the same path that the target process was found.

To make this efficient, each process maintains a cache of recently accessed processes. The process-finding messages also contain the rank and address of every process along the way, and these are added to the cache. Future lookups first search for the target in the cache, then search for the target's ancestors in the cache to attempt to skip as many hops along the path to the target as possible. This also allows us to take advantage of locality in cache lookups: processes whose ranks are close to each other are close to each other in the tree *most of the time*[1].

Using this strategy, after groups of p processors sort the color and rank tables, they do not have to exchange their subtables so that each has a full copy. Instead, they only have to find the new rank for each of the p processes in the group and the address of their parents: a quick operation. This transforms the $O(n)$ recursive-doubling exchange at the end of the sort into an $O(p \lg \frac{n}{p})$ lookup.

Overall, we reduce the memory usage from an n-entry table per process to a $\frac{n}{p}$-entry table per process during the operation of MPI_Comm_split. The memory needed for each communicator is reduced from a n-entry table for n-member communicators to a small fixed-size cache.

4   Evaluation

There are no systems yet with millions of processors, nor simulators available to simulate millions of processes, we look at a partial simulation. All but the parallel-sort with distributed tables algorithm entail each of n processes receiving an n-entry table to sort. This requires simulating $n^2/flit\text{-}size$ flits.

BigNetSim can simulate up to 2.5 million network events/second using 128 processors [2]. Every hop in the network counts as an event. Thus, one experimental data point would require simulating billions of events and take months.

Instead, we use an approach that is likely very inaccurate, but favors existing algorithms in its inexactitude. We fully implement the parallel-sort phase of the algorithm using 2 through 64 processes. The recursive-doubling exchanges, and the sorting and merging in the conventional algorithm we simulate using only 2 processes. These two processes exchange, merge, or sort the correct amount of sorted or unsorted data with each other at each iteration in each phase of MPI_Comm_split. In the correct algorithm, each process would exchange data with another unique process at each level in the recursive-doubling algorithm.

We do not capture the effect of network contention. Network contention would have the greatest effect during the final stages of the conventional algorithm, when every process exchanges messages on the order of several to tens of Megabytes. Not modeling network contention favors the conventional algorithm in our evaluation.

---

[1] In the worst case, the process at the root of the tree is $\lg n$ hops away from the two processes whose ranks are one away from the root's rank. The average case is much better than this.

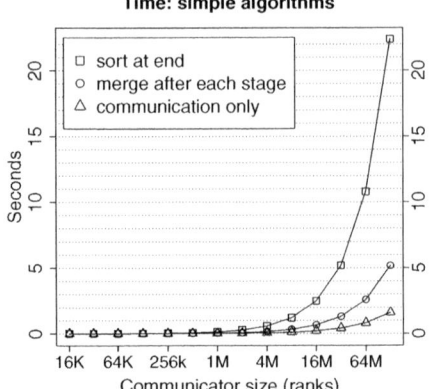

Fig. 1. The total execution time of the conventional algorithm, the conventional algorithm with merging after each step, and the communication time for both

In our evaluation, we generate one new communicator containing all the processes in the input communicator. The keys are randomly-generated. Later, we discuss how the results are likely to change for smaller communicators.

The experiments are run on `blueprint`, a system composed of 120 POWER5+ nodes, each with 16 1.9 GHz Power 5 cores. The nodes are connected with Federation switches. In all of our tests, we use only one core per node to better simulate a million-core system. We use merge and sort functions from the C++ STL, and compile with IBM XLC at optimization level −O4.

Each configuration was run 5 times and the mean is shown. The scale of the standard error was too small to appear on our graphs.

Speedup data for our two parallel-sort `MPI_Comm_split` variations are reported relative to to the conventional `MPI_Comm_split` variation with merging after each step.

In Figure 1, we see the execution time for the conventional `MPI_Comm_split` as implemented in MPICH and Open MPI, along with the time spent in communication for both. The time for one `MPI_Comm_split` call is reduced by a factor of four simply by replacing the $O(n \lg n)$ sort at the end with an $O(n)$ (total) merge after each step.

Figure 2 shows the speedup if we sort the color and key table in parallel and then continue using recursive doubling to create a full copy of the new tree. At best, we get a 2.7x advantage over serial sort.

The speedups presented thus far would be less if network contention in a real system significantly hinders the performance of `MPI_Comm_split`; these optimizations reduce computation costs only.

Figure 3 shows the speedup if we sort in parallel and build the communicator as a distributed tree. We get performance benefits of up to 14x using 64 sorting processors. Compared to the sort-at-the-end MPICH and Open MPI implementation, our algorithm is 60x faster. The parallel sort with distributed

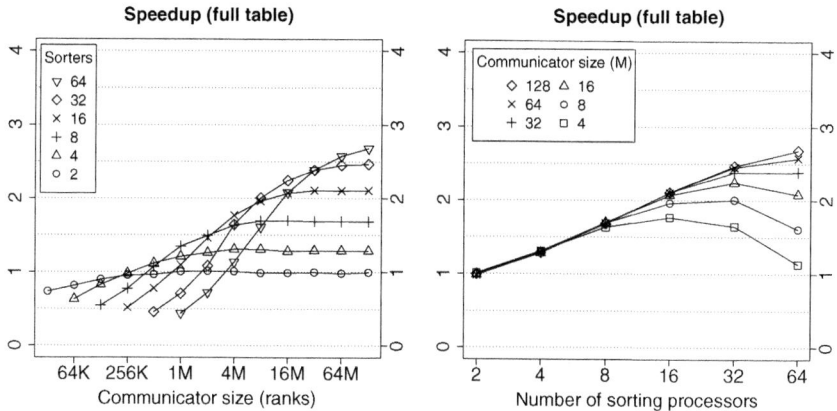

Fig. 2. The speedup of using a parallel-sort with conventional communicator tables

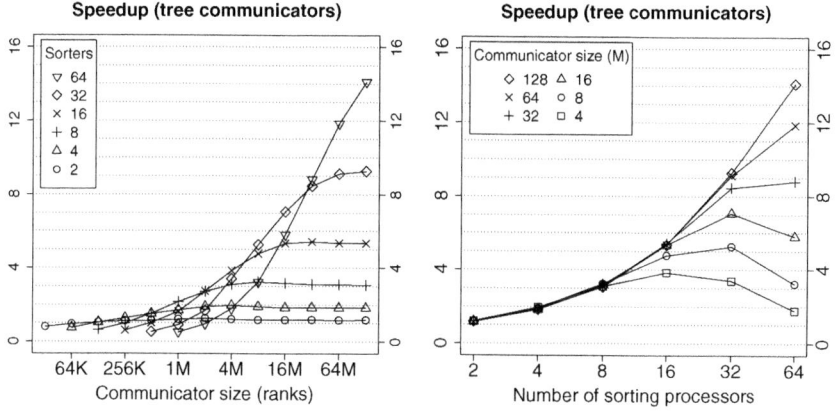

Fig. 3. The speedup of using a parallels-sort with distributed-tree communicators

communicators algorithm significantly reduces the amount of communication and network contention, so we expect the speedups on real systems to be better.

As argued above, our experiments are quite inaccurate since we are not performing a full-system simulation. More important than a single-performance number is the analysis of where the time goes. Table 1 gives a breakdown of the time spent in each operation for the parallel-sort algorithm with distributed tables.

The breakdown matches expectations perfectly. The *collect* rows show the time spent in the recursive-doubling stage before the parallel sort. As expected, it is proportional to the size of each sorting process's subtable: *i.e.*, the number of processors in the MPI_Comm_split call divided by the number of sorting processes.

Table 1. Breakdown of time in `MPI_Comm_split` with parallel sorting and distributed tree communicators

Communicator size	Operation	Sorting processors					
		2	4	8	16	32	64
8M	collect	0.16	0.08	0.03	0.02	0.01	0.00
	splitters	0.00	0.00	0.01	0.01	0.03	0.08
	exchange	0.05	0.03	0.02	0.01	0.01	0.01
	merge	0.06	0.06	0.04	0.03	0.02	0.01
	total	0.27	0.17	0.10	0.07	0.06	0.10
16M	collect	0.32	0.16	0.08	0.03	0.02	0.01
	splitters	0.00	0.00	0.01	0.01	0.03	0.08
	exchange	0.10	0.06	0.03	0.02	0.01	0.01
	merge	0.13	0.13	0.09	0.05	0.03	0.02
	total	0.55	0.35	0.20	0.12	0.09	0.11
32M	collect	0.64	0.32	0.16	0.08	0.03	0.02
	splitters	0.00	0.00	0.01	0.02	0.03	0.08
	exchange	0.20	0.12	0.07	0.04	0.02	0.01
	merge	0.25	0.26	0.19	0.11	0.07	0.04
	total	1.09	0.70	0.42	0.24	0.15	0.15
64M	collect	1.29	0.64	0.32	0.16	0.08	0.03
	splitters	0.00	0.00	0.01	0.02	0.03	0.08
	exchange	0.44	0.24	0.13	0.07	0.04	0.02
	merge	0.51	0.52	0.38	0.24	0.14	0.08
	total	2.24	1.41	0.84	0.48	0.28	0.22
128M	collect	2.58	1.29	0.64	0.32	0.16	0.08
	splitters	0.00	0.00	0.01	0.02	0.03	0.08
	exchange	0.80	0.48	0.27	0.14	0.07	0.04
	merge	1.01	1.05	0.77	0.49	0.29	0.17
	total	4.39	2.82	1.69	0.97	0.56	0.37

The *splitters* rows show the time spent in calculating the median-of-medians. The number of iterations in the exact-splitter algorithm increases as the logarithm of the size of the problem per process. (The distance between the chosen splitter and the exact splitter exponentially decreases as a function of the iteration count.) The time spent per iteration is dominated by a term linear with the number of sorting processes. The time spent in this stage is insignificant until we scale to 32 or more sorting processes.

The *exchange* rows show the time spent exchanging elements between sorting processes once the exact splitters are found. With p sorting processes, on average, we expect $p - 1$ out of every p entries on each process will be exchanged. Thus, this term grows as the size of each sorting process's subtable.

The *merge* rows show the time spent merging the p partitions of each process's subtable together. In the first iteration, $\frac{p}{2}$ pairs of $\frac{n}{p^2}$-entry subtables are merged, in the second iteration, $\frac{p}{4}$ pairs of $\frac{2n}{p^2}$-entry subtables are merged, and so on, for a total cost of $O(\frac{n}{p} \lg p)$ in $\lg p$ iterations. We observe that the times for 2 or 4 sorting processes are the same, since the $\lg p$ term doubles as the $\frac{n}{p}$ term halves.

5   Discussion and Future Work

As we increase the number of sorting processors, the speedup improves for large communicators more than for small communicators (and even degrades with 64 sorting processes and an 8-million process communicator). This is expected due to parallel-sorting overheads, and, in particular, the amount of time spent in the exact splitter algorithm. In the future, we plan to experiment with non-exact splitters. One way is to stop the splitter algorithm when the splitter is good enough. The distance between the test splitter and the exact splitter decreases as the logarithm of the number of iterations of the exact splitter algorithm. We may also choose an inexact splitter based on oversampling, *e.g.*, the algorithm in [8], in which each of the p sorting processes nominates $p-1$ splitters, and the best $p-1$ of all the nominated splitters are used.

We experimented with the case where we take one very large input communicator and create a very large output communicator of the same size but a different rank order. The conventional algorithm collects the entire color and rank table and then only sorts the entries whose color matches that of the desired new communicator. Thus, when the size of the output communicators is much smaller, the performance for the conventional algorithm will not be nearly as bad as that shown here.

However, observe that ignoring the sort cost, the conventional algorithm still spends nearly 2 seconds just in communication time for a 128 million-process input communicator, and that is in our highly-optimistic simulation model. Our parallel algorithm with 128 sorting processes takes 0.37 seconds, including the sort. Moreover, our algorithm does not need 128 million entries worth of temporary storage in each process.

0.37 seconds or less for one MPI_Comm_split call may not sound like much, but since our experimental methodology does not properly model network contention and latency, the true cost on a large system might be much more. Thus, we intend to experiment with optimizing our algorithm for small output communicators. One possibility is to not merge the tables during the parallel reduction and rather than sort the table in parallel, we could have each of the sorting processes scan its subtable for entries that match any of the sorting processes new communicators. Just those entries would be exchanged, and each sorting process would then sort just those entries whose colors match.

6   Conclusion

Existing algorithms for creating MPI communicators do not scale to exascale supercomputers containing millions of cores. They use $O(n)$ memory and take up to $O(n \lg n)$ time in an n-process application, whereas per-core performance and memory are expected to increase sub-linearly in the future.

Our work proposes three techniques that solve the time and space scalability problem: merging after each step in the MPI_Allgather phase, sorting in parallel, and using distributed rather than replicated tables.

These techniques together reduce the time complexity to $O(p \lg n + \lg^2 n + \frac{n}{p} \lg p)$, where we have groups of p sorting processes, and reduce the memory footprint p-fold. In our experiments, this reduces the cost of MPI_Comm_split from over 22 seconds for the largest problem to just 0.37 seconds, a 60x reduction.

Acknowledgments. This work was supported in part by the U.S. Department of Energy under contract DE-FG02-08ER25835 and by the National Science Foundation under grant #0837719.

References

1. Cheng, D.R., Edelman, A., Gilbert, J.R., Shah, V.: A novel parallel sorting algorithm for contemporary architectures. Submitted to ALENEX 2006 (2006)
2. Choudhury, N., Mehta, Y., Wilmarth, T.L., Bohm, E.J., Kalé, L.V.: Scaling an optimistic parallel simulation of large-scale interconnection networks. In: WSC 2005: Proceedings of the 37th conference on Winter simulation, Winter Simulation Conference, pp. 591–600 (2005)
3. Gabriel, E., Fagg, G.E., Bosilca, G., Angskun, T., Dongarra, J.J., Squyres, J.M., Sahay, V., Kambadur, P., Barrett, B., Lumsdaine, A., Castain, R.H., Daniel, D.J., Graham, R.L., Woodall, T.S.: Open MPI: Goals, concept, and design of a next generation MPI implementation. In: Proceedings, 11th European PVM/MPI Users' Group Meeting, Budapest, Hungary, September 2004, pp. 97–104 (2004)
4. Gropp, W., Lusk, E., Doss, N., Skjellum, A.: A high-performance, portable implementation of the MPI message passing interface standard. Parallel Computing 22(6), 789–828 (1996)
5. Gropp, W.D., Lusk, E.: User's Guide for mpich, a Portable Implementation of MPI. Mathematics and Computer Science Division. Argonne National Laboratory (1996), ANL-96/6
6. Meuer, H., Strohmaier, E., Dongarra, J., Simon, H.: TOP500 Supercomputing Sites (2010), http://top500.org (accessed March 24, 2010)
7. Saukas, E.L.G., Song, S.W.: A note on parallel selection on coarse grained multicomputers. Algorithmica 24, 371–380 (1999)
8. Shi, H., Schaeffer, J.: Parallel sorting by regular sampling. J. Parallel Distrib. Comput. 14(4), 361–372 (1992)
9. Sutter, H.: The free lunch is over: A fundamental turn toward concurrency in software. Dr. Dobb's Journal 30(3) (March 2005)
10. Thakur, R., Gropp, W.: Improving the performance of collective operations in mpich. In: Dongarra, J., Laforenza, D., Orlando, S. (eds.) EuroPVM/MPI 2003. LNCS, vol. 2840, pp. 257–267. Springer, Heidelberg (2003)

Enabling Concurrent Multithreaded MPI Communication on Multicore Petascale Systems

Gábor Dózsa[1], Sameer Kumar[1], Pavan Balaji[2], Darius Buntinas[2],
David Goodell[2], William Gropp[3], Joe Ratterman[4], and Rajeev Thakur[2]

[1] IBM T. J. Watson Research Center, Yorktown Heights, NY 10598
[2] Argonne National Laboratory, Argonne, IL 64039
[3] University of Illinois, Urbana, IL 61801
[4] IBM Systems and Technology Group, Rochester, MN 55901

Abstract. With the ever-increasing numbers of cores per node on HPC
systems, applications are increasingly using threads to exploit the shared
memory within a node, combined with MPI across nodes. Achieving
high performance when a large number of concurrent threads make MPI
calls is a challenging task for an MPI implementation. We describe the
design and implementation of our solution in MPICH2 to achieve high-
performance multithreaded communication on the IBM Blue Gene/P.
We use a combination of a multichannel-enabled network interface, fine-
grained locks, lock-free atomic operations, and specially designed queues
to provide a high degree of concurrent access while still maintaining
MPI's message-ordering semantics. We present performance results that
demonstrate that our new design improves the multithreaded message
rate by a factor of 3.6 compared with the existing implementation on the
BG/P. Our solutions are also applicable to other high-end systems that
have parallel network access capabilities.

1   Introduction

Because of power constraints and limitations in instruction-level parallelism,
computer architects are unable to build faster processors by increasing the clock
frequency or by architectural enhancements. Instead, they are building more and
more processing cores on a single chip and leaving it up to the application programmer to exploit the parallelism provided by the increasing number of cores.
MPI is the most widely used programming model on HPC systems, and many
production scientific applications use an MPI-only model. Such a model, however, does not make the most efficient use of the shared resources within the node
of an HPC system. For example, having several MPI processes on a multicore
node forces node resources (such as memory, network FIFOs) to be partitioned
among the processes. To overcome this limitation, application programmers are
increasingly looking at using hybrid programming models comprising a mixture
of processes and threads, which allow resources on a node to be shared among
the different threads of a process.

With hybrid programming models, several threads may concurrently call MPI
functions, requiring the MPI implementation to be thread safe. In order to

R. Keller et al. (Eds.): EuroMPI 2010, LNCS 6305, pp. 11–20, 2010.

achieve thread safety, the implementation must serialize access to some parts of the code by using either locks or advanced lock-free methods. Using such techniques and at the same time achieving high concurrent multithreaded performance is a challenging task [2,3,10].

In this paper, we describe the solutions we have designed and implemented in MPICH2 to achieve high multithreaded communication performance on the IBM Blue Gene/P (BG/P) system [4]. We use a combination of a multichannel-enabled network interface, fine-grained locks, lock-free atomic operations, and message queues specially designed for concurrent multithreaded access. We evaluate the performance of our approach with a slightly modified version of the SQMR message-rate benchmark from the Sequoia benchmark suite [8]. Although implemented on the BG/P, our techniques and optimizations are also applicable to other high-end systems that have parallel network access capabilities.

The rest of this paper is organized as follows. Section 2 provides a brief background of thread safety in MPI and MPICH2 and the architecture of the Blue Gene/P system. Section 3 describes the design and implementation of our solutions in detail. Performance results are presented in Section 4, followed by conclusions in Section 5.

2   Background

We provide a brief overview of the semantics of multithreaded MPI communication, the internal framework for supporting thread safety in MPICH2, and the hardware and software architecture of the Blue Gene/P system.

2.1   MPI Semantics for Multithreading

The MPI standard defines four levels of thread safety: single, funneled, serialized, and multiple [6]. We discuss only the most general level, MPI_THREAD_MULTIPLE, in which multiple threads can concurrently make MPI calls.

MPI specifies that when multiple threads make MPI calls concurrently, the outcome will be as if the calls executed sequentially in some order. Blocking MPI calls will block only the calling thread and will not prevent other threads from running or executing MPI functions. As a result, multiple threads may access and modify internal structures in the MPI implementation simultaneously, thus requiring serialization within the MPI library to avoid race conditions. Logically global resources, such as allocation/deallocation of objects, context ids, communication state, and message queues, must be updated atomically.

Implementing thread safety efficiently in an MPI implementation is a challenging task. The most straightforward approach is to use a single global lock, which is acquired on entry to an MPI function and held until the function returns, unless the function is going to block on a network operation. In that case, the lock is released before blocking and then reacquired after the network operation returns. The main drawback of this approach is that it permits little concurrency in operations.

Optimizations for accessing queues, such as lock-free methods, often require single-reader/single-writer access, which can be a limitation. Since message queues are on the critical path, using simpler (classical) approaches with locks can add significant overhead. Also, locks or lock-free atomic updates themselves are expensive, even in the absence of contention, because of memory-consistency requirements (typically, some data must be flushed to main memory, an action that costs hundreds of cycles in latency).

A further complication is introduced by the feature in MPI that allows "wildcard" (MPI_ANY_SOURCE) receives that can match incoming messages from any sender. For any receive (or for matching any incoming message), this feature requires two logical queues to be searched atomically—receives expecting a specific sender and receives permitting any sender—in a manner that maintains MPI's message-ordering semantics. This requirement makes it difficult to allow for concurrency even in programs written to match receives with specific senders, which in the absence of MPI_ANY_SOURCE could be implemented efficiently with separate queues for separate senders. MPI_ANY_SOURCE implies a shared queue that all threads must check and atomically update, thereby limiting concurrency.

2.2   Framework for Supporting Thread Safety in MPICH2

Thread safety in MPICH2 is implemented by identifying regions of code where concurrent threads may access shared objects and marking them with macros that provide an appropriate thread-safe abstraction, such as a named critical section. For example, updates to message queues are protected by the MSGQUEUE critical section. Most MPI routines, other than a few that are intrinsically thread safe and require no special care, also establish a function-level critical section (ALLFUNC). Different granularities of thread safety (coarse-grained versus fine-grained locks or critical sections) are enabled by simply changing the definitions of these macros in a header file. For example, the simple global lock is implemented by defining the ALLFUNC critical section to acquire and release a global lock and defining the other macros as no-ops. Finer-grained locking is enabled by reversing these definitions, that is, defining the ALLFUNC critical section as a no-op and defining the other named critical sections appropriately.

MPI objects are reference counted internally. This task must be done atomically in a multithreaded environment. The reference-count updates are handled by a macro that can be defined to use a simple update (in the case of the single global lock), processor-specific atomic-update instructions, or a reference-count critical section.

In the few instances where using these macros is not convenient, we use C-preprocessor #ifdefs directly. Since this approach makes the code harder to maintain, however, we try to avoid it and instead rely as much as possible on a careful choice of abstractions with a common set of definitions. Using carefully chosen abstractions makes it easier to switch from a coarse-grained, single-lock approach to a finer-grained approach that permits greater concurrency.

2.3 Blue Gene/P Hardware and Software Overview

The IBM Blue Gene/P is a massively parallel system that can scale up to 3 PF/s of peak performance. Each node of the BG/P has four 850 MHz embedded PowerPC 450 cache-coherent cores on a single ASIC and can achieve a peak floating-point throughput of 13.6 GF/s per node. The nodes are connected with three networks that the application may use: a 3D torus network, which is deadlock free and provides reliable delivery of packets; a collective network, which implements global broadcast and global integer arithmetic operations; and a global interrupt network for fast barrier-synchronization operations. Each node has a direct memory access (DMA) engine to facilitate injecting and receiving packets to and from the torus network. This feature allows the cores to offload packet management and enables better overlap of communication and computation.

The MPI implementation on the BG/P is based on MPICH2 [7] and is layered on top of a lower-level messaging library called the Deep Computing Messaging Framework (DCMF) [5]. DCMF provides basic message-passing services that include point-to-point operations, nonblocking one-sided get and put operations, and an optional set of nonblocking collective calls. MPICH2 is implemented on the BG/P via an implementation of the internal MPID abstract device interface on top of DCMF, called dcmfd. The currently released version supports the MPI_THREAD_MULTIPLE level of thread safety by using the simple, unoptimized approach of a single global lock.

The DMA engine on each node supports 32 injection FIFOs and 8 reception FIFOs per core, an important feature for the work described in this paper. The operating system on the node supports a maximum of four threads, in other words, at most 1 thread per core. Although this limit is much smaller than what is allowed on commodity multicore/SMP platforms, those platforms typically do not offer sufficient parallelism at the network-hardware level that concurrently communicating threads could exploit. On such systems, the MPI implementation would need to serialize accesses to network hardware and operating-system resources and thus would not result in scalable multithreaded communication performance. A commodity-cluster node would need at least four NICs to provide a level of network parallelism comparable to a BG/P node. Therefore, despite the relatively modest number of threads allowed on a BG/P node, the parallelism in the network hardware makes it an interesting platform for studying how to optimize multithreaded MPI communication.

3   Enabling Concurrent Multithreaded MPI Communication on BG/P

To achieve high-performance multithreaded MPI communication on the BG/P, we redesigned multiple layers of the communication-software stack. We enhanced both DCMF and MPICH2 to support multiple communication *channels* between pairs of processes, such that communication from multiple threads on different channels can take place concurrently. We also modified the data structures and algorithms used to implement message queues in MPICH2 in order to enable message-queue manipulations in parallel on a channel basis. In the following subsections, we describe all these optimizations.

3.1   Multichannel Extensions to DCMF

In the existing design of DCMF, only one abstract DMA device is instantiated per MPI process. This single DMA device allocates one injection/reception FIFO group and provides a single access point for the underlying DMA hardware resources. We extended DCMF to have multiple DMA devices that allocate multiple injection/reception groups for each MPI process. For example, in BG/P's SMP mode, where a program runs with one process and up to four threads on each node, four DMA devices are instantiated that allocate four injection and reception groups. In BG/P's dual mode, with two MPI processes with two threads each per node, each process instantiates two DMA devices. Multiple threads of an MPI process can access these DMA software devices independently and in parallel. DCMF encapsulates DMA devices into software abstractions called channels. A channel assigns a mutex to control access to a particular DMA device.

API Changes. To allow threads to lock channels, we added two new calls to the DCMF API: DCMF_Channel_acquire and DCMF_Channel_release. The rest of the API remained unchanged. In particular, we did not add new arguments for DCMF_Send to specify a send channel. Instead, the new DCMF_Channel_acquire call saves the ID of the locked channel in thread-private memory. Subsequent calls to send functions use this thread-private information to post messages to the DMA device currently locked by the executing thread. Send function calls specify the same DMA group ID for both injection and reception of a message; that is, by locking a channel, the sender thread implicitly also defines the reception channel at the destination for outgoing messages.

   This approach to extending the API has the advantage that it requires minimal changes in upper levels of software that call DCMF. It has the drawback of limited flexibility, however; for example, it cannot specify different send and receive channels for a particular message. We plan to explore a more full-featured API that provides greater flexibility.

Progress Engine. The generic DMA progress engine in DCMF ensures that pending outgoing and incoming messages are processed. We extended the progress engine to support multiple channels. Ideally, each channel is advanced by a separate thread, which results in fully parallel progress of the DMA devices. For instance, in the SMP mode, four MPI threads can run on the four cores and make progress on only their corresponding channels. This scheme, however, assumes that all four MPI threads are always active; that is, all of them issue DCMF advance calls eventually, so that pending messages are processed at some point on every channel. If all the threads are not active, this fully parallel progress approach may fail. For instance, a multithreaded MPI application may enter a global barrier by issuing MPI_Barrier calls from threads running on different cores on the different nodes.

Only one thread will call the barrier function on each node, and the other threads may be simply blocked (or not even started yet) until the global barrier completes. This situation can lead to a deadlock if a barrier message arrives at a node on a channel that is not advanced by the thread executing the barrier call.

In order to comply with MPI progress semantics, each thread must eventually make progress on every channel. For thread safety, we also need to prevent multiple threads from accessing the same channel simultaneously. A call to the DMA progress engine causes progress by attempting to lock a channel; if the lock succeeds, the DMA device of the channel is advanced and the channel is unlocked. Making progress on multiple channels instead of just one channel implies higher overhead, which can hurt message latency on the low-frequency BG/P cores. However, it must be done at least occasionally in order to satisfy MPI progress semantics. For this purpose, we use an internal parameter (say, n) to decide how often a thread will attempt to advance other channels. A thread will normally advance only its own channel; but on every nth call to the progress function, it will also try to advance other channels. Thus, all threads can make independent parallel progress most of the time, while still guaranteeing MPI progress semantics.

3.2   Exploiting Multiple Channels in MPICH2

The current MPI 2.2 standard does not directly translate the notion of multiple communication channels into a user-visible concept. For the upcoming MPI-3 standard, Marc Snir has proposed extending MPI to support multiple "endpoints" per process [9], which would map cleanly to our definition of channels. Until such explicit support becomes part of the standard, however, MPI can take advantage of multiple channels only in an application-transparent fashion, that is, by using multiple channels internally without exposing them to the user.

We modified the `dcmfd` device in MPICH2 to select and acquire an appropriate channel by calling `DCMF_Channel_acquire(channel)` immediately before issuing a DCMF send call. After the send call completes, the channel is released. We calculate the channel for a particular message by means of a simple hash function: $channel = (source + dest) \bmod num\_channels$. Selecting the channel based on the source and destination ranks in this manner has the following desirable implications:

- Messages sent on a given communicator from a particular source to the same destination process travel over the same channel. This feature makes it easy for us to support MPI's non-overtaking message-ordering semantics, which require that messages from the same source to the same destination appear in the order in which they were sent.
- Messages sent to a particular destination node from different sources are distributed among the available reception channels. This feature enables incoming messages to be received in parallel.

Truly parallel processing of incoming messages at the MPI level also requires support for parallel message matching, as described below.

3.3   Parallel Receive Queues

MPICH2 has two receive queues implemented via linked lists: a queue of receives posted by the application and a queue of unexpected messages, namely, messages that were received before the application posted the matching receive. When an application posts a receive, the unexpected-message queue is first searched for a matching message. If none is found, the receive is enqueued on the posted-receive queue. Similarly, when a message is received from the network, the posted-receive queue is first searched for a matching receive. If none is found, the message is enqueued on the unexpected queue.

We parallelized the receive queues by providing a separate pair of posted- and unexpected-receive queues for each source rank. In this case, an additional queue is needed to hold posted wildcard receives (source=MPI_ANY_SOURCE). When a message is received from the network and a matching receive is not found in the posted-receive queue for the corresponding channel, the progress engine checks the wildcard queue. If the wildcard queue is not empty, the progress engine acquires the wildcard-queue lock and searches the queue for a match. If a match is not found, the message is enqueued on the channel's unexpected queue.

A complication is introduced when a posted wildcard receive is followed by a non-wildcard receive with a matching tag. Since MPI's message-ordering semantics require that the wildcard receive be matched first and since MPICH2's progress engine first searches the channel-receive queues, we queue the non-wildcard receive in the wildcard queue. When the wildcard receive is matched and removed from the wildcard queue, we move the non-wildcard receive into its channel queue.

4   Performance Results

Currently, there is no canonical benchmark suite or application to measure multithreaded messaging efficiency of MPI. Also, the MPI_THREAD_MULTIPLE mode is often not efficiently supported by existing MPI implementations, which in turn deters applications from using it. The commonly used NAS Parallel Benchmarks (NPB) [1] are not multithreaded. The multi-zone variants of the NAS Parallel Benchmarks (NPB-MZ) [11] do use MPI+threads via OpenMP, but they use only the MPI_THREAD_FUNNELED level of thread safety.

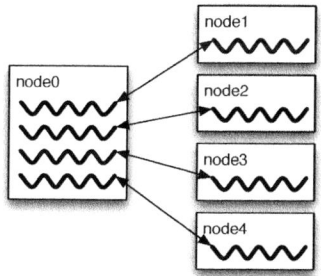

Fig. 1. Communication pattern of the Neighbor Message Rate Benchmark

We chose message rate as a metric to measure messaging performance. We used a slightly modified version of the SQMR Phloem microbenchmark from the Sequoia benchmark suite [8]. Specifically, the original SQMR code runs

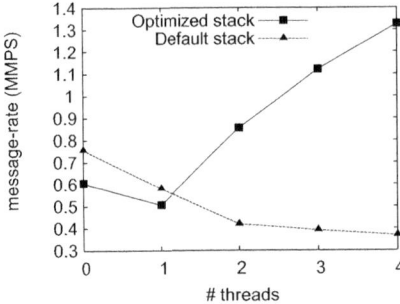
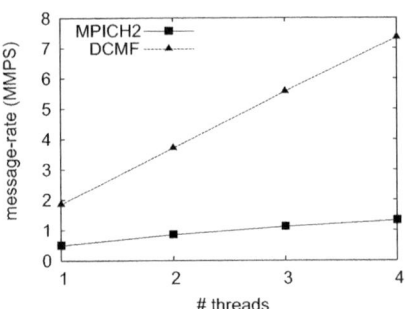

Fig. 2. Message rate performance of default and optimized software stacks on BG/P

Fig. 3. Message rates with the optimized stack when using MPI versus direct DCMF

single-threaded MPI processes; we adapted it for multithreaded processes running in `MPI_THREAD_MULTIPLE` mode.

The modified benchmark, which we call the Neighbor Message Rate Benchmark, measures the aggregate message rate for N threads in a single MPI process, each sending to and receiving from a corresponding peer process on a separate node, as shown in Figure 1. Each iteration of the benchmark involves each thread posting 12 nonblocking receives and 12 nonblocking sends from/to the peer thread, followed by a call to `MPI_Waitall` to complete the requests. Each thread executes 10 warm-up iterations before timing 10,000 more iterations. We used zero-byte messages in order to minimize the impact of data-transfer times on the measurements. The benchmark reports the total number of messages sent (in millions) divided by the elapsed time in seconds. We evaluate our solution based on the overall message rate of the "root" process and the scaling of the message rate with the number of threads.

We ran the benchmark on the BG/P and varied the number of threads from 1 to 4. We also ran it in the `MPI_THREAD_SINGLE` mode to determine the best achievable performance for a single thread without any locking overhead. Figure 2 shows the results with the default production BG/P software stack and with our optimized DCMF and MPICH2. (The case with 0 threads represents the `MPI_THREAD_SINGLE` mode.) The performance with the optimized stack is much better than with the default stack where the message rate actually decreases with the number of threads. With 4 threads, the message rate with optimized stack is 3.6 times higher than with the default stack. Scaling is not perfect though; we observe an average 10% degradation per thread from linear scaling.

To locate the source of this scaling degradation and to measure the MPI overhead in general, we also implemented a DCMF version of the Neighbor Message Rate Benchmark, which directly makes DCMF calls. Figure 3 shows the results of running both the MPI and DCMF versions of the benchmark with our optimized stack. The performance with direct DCMF is much higher than with MPI. We believe this difference is because DCMF is a much simpler, lower-level

API. MPI's message-ordering and matching semantics as well as the notion of communicators, etc., make it more difficult to optimize for multithreading. Nonetheless, the magnitude of the difference suggests room for further optimization, which we plan to investigate. Locking overhead for dynamic channel selection and receive-queue management for message matching are two areas that we specifically plan to optimize further. We also expect that the new proposal for multiple endpoints in MPI-3 [9] will help alleviate some of the bottlenecks at the MPI level.

5   Conclusions

Running MPI applications in fully multithreaded mode is becoming a significant issue as a result of the increasing importance of hybrid programming models for multicore high-end systems. We have presented a solution to achieve high messaging performance in MPICH2 when multiple threads make MPI calls concurrently. We use a combination of a multichannel-enabled network interface, fine-grained locks, lock-free atomic operations, and message queues specifically designed for concurrent multithreaded access. We introduce the "channel" abstraction as the unit of parallelism at the network-interface level and show how MPICH2 can take advantage of channels in a user-transparent way. Applying our optimizations on the Blue Gene/P messaging stack, we demonstrate a factor of 3.6 improvement in multithreaded MPI message rate. Furthermore, the message rate scales reasonably with the number of MPI threads in our optimized stack, as opposed to the default stack where the aggregate message rate decreases with multiple threads. We plan to investigate further optimizations to improve MPI performance compared with native DCMF performance.

The proposed solutions and optimizations for defining and managing multiple network "channels" are also applicable to other high-end systems with parallel network access capabilities. Implementation details will, of course, differ as they depend on the particular messaging software stack, but the techniques we have described for providing access to multiple network channels from concurrent MPI threads and managing progress on multiple channels in parallel should be directly applicable.

Acknowledgments. This work was supported in part by the U.S. Government contract No. B554331; the Office of Advanced Scientific Computing Research, Office of Science, U.S. Department of Energy, under contract DE-AC02-06CH11357 and DE-FG02-08ER25835; and by the National Science Foundation under grant #0702182.

References

1. Bailey, D., Harris, T., Saphir, W., Wijngaart, R.V.D., Woo, A., Yarrow, M.: The NAS parallel benchmarks 2.0. NAS Technical Report NAS-95-020, NASA Ames Research Center, Moffett Field, CA (1995)

2. Balaji, P., Buntinas, D., Goodell, D., Gropp, W., Thakur, R.: Fine-grained multithreading support for hybrid threaded MPI programming. International Journal of High Performance Computing Applications 24(1), 49–57 (2010)
3. Gropp, W., Thakur, R.: Thread safety in an MPI implementation: Requirements and analysis. Parallel Computing 33(9), 595–604 (2007)
4. IBM System Blue Gene solution: Blue Gene/P application development ,
   http://www.redbooks.ibm.com/redbooks/pdfs/sg247287.pdf
5. Kumar, S., Dozsa, G., Almasi, G., Heidelberger, P., Chen, D., Giampapa, M.E., Blocksome, M., Faraj, A., Parker, J., Ratterman, J., Smith, B., Archer, C.J.: The Deep Computing Messaging Framework: Generalized scalable message passing on the Blue Gene/P supercomputer. In: Proceedings of the 22nd International Conference on Supercomputing, pp. 94–103. ACM Press, New York (2008)
6. Message Passing Interface Forum: MPI: A Message-Passing Interface Standard, Version 2.2 (September 2009), http://www.mpi-forum.org
7. MPICH2, http://www.mcs.anl.gov/mpi/mpich2
8. Sequoia benchmark codes, https://asc.llnl.gov/sequoia/benchmarks/
9. Snir, M.: MPI-3 hybrid programming proposal, version 7,
   http://meetings.mpi-forum.org/mpi3.0_hybrid.php
10. Thakur, R., Gropp, W.: Test suite for evaluating performance of multithreaded MPI communication. Parallel Computing 35(12), 608–617 (2009)
11. Wijngaart, R.V.D., Jin, H.: NAS parallel benchmarks, multi-zone versions. NAS Technical Report NAS-03-010, NASA Ames Research Center, Moffett Field, CA (2003)

Toward Performance Models of MPI Implementations for Understanding Application Scaling Issues

Torsten Hoefler[1], William Gropp[1], Rajeev Thakur[2], and Jesper Larsson Träff[3]

[1] University of Illinois at Urbana-Champaign, IL, USA
{htor,wgropp}@illinois.edu
[2] Argonne National Laboratory, Argonne, IL, USA
thakur@mcs.anl.gov
[3] Dept. of Scientific Computing, University of Vienna, Austria
traff@par.univie.ac.at

Abstract. Designing and tuning parallel applications with MPI, particularly at large scale, requires understanding the performance implications of different choices of algorithms and implementation options. Which algorithm is better depends in part on the performance of the different possible communication approaches, which in turn can depend on both the system hardware and the MPI implementation. In the absence of detailed performance models for different MPI implementations, application developers often must select methods and tune codes without the means to realistically estimate the achievable performance and rationally defend their choices. In this paper, we advocate the construction of more useful performance models that take into account limitations on network-injection rates and effective bisection bandwidth. Since collective communication plays a crucial role in enabling scalability, we also provide analytical models for scalability of collective communication algorithms, such as broadcast, allreduce, and all-to-all. We apply these models to an IBM Blue Gene/P system and compare the analytical performance estimates with experimentally measured values.

1 Motivation

Performance modeling of parallel applications leads to an understanding of their running time on parallel systems. To develop a model for an existing application or algorithm, one typically constructs a dependency graph of computations and communications from the start of the algorithm (input) to the end of the algorithm (output). This *application model* can then be matched to a *machine model* in order to estimate the run time of the algorithm on a particular architecture.

Performance models can be used to make important early decisions about algorithmic choices. For example, to compute a three-dimensional Fast Fourier Transformation (3d FFT), one can either use a one-dimensional decomposition where each process computes full planes (2d FFTs) or a two-dimensional decomposition where each process computes sets of pencils (1d FFTs). If we assume

R. Keller et al. (Eds.): EuroMPI 2010, LNCS 6305, pp. 21–30, 2010.

an N^3 box and P processes, the 1d decomposition only requires a single parallel transpose (alltoall) and enables the use of more efficient 2d FFT library calls (for example, FFTW) but is limited in scalability to $P \leq N$. A 2d decomposition scales up to $P \leq N^2$ processes but requires higher implementation effort and two parallel transpose operations going on in parallel on subsets of the processes [6]. Thus, the best choice of distribution mostly depends on the communication parameters and the number of processes.

Application designers typically have a basic understanding and expectation of the performance of some operations on which they base their algorithmic decisions. For example, most application developers would assume that transmitting a message of size S has linear costs in S and that broadcasting a small message to P processes would take logarithmic time in P. Such simple assumptions of *performance models*, or general *folklore*, often guide application models and algorithm development.

One problem with this approach is that inaccuracies in the model often do not influence the medium-scale runs in which they are verified but have significant impact as the parameters (S or P) grow large. One particular example is that several application models assume that broadcast or allreduce scale with $\Theta(Sb\log(P))$ (e.g., [3,17]) while, as demonstrated in Section 4, a good MPI implementation would implement a broadcast or allreduce with $\Theta(S + \log(P))$ [5,13,21]. Generally speaking, performance models for middleware libraries such as MPI depend on the parameters of the network (e.g., bandwidth, latency, topology, routing) **and** the implemented algorithms (e.g., collective algorithms, eager and rendezvous protocols) and are thus hard to generalize.

In this paper, we advocate the importance of communication models for MPI implementations at large scale. We contend that such models should be supplied by each MPI implementation to allow users to reason about the performance. We sketch a hierarchical method to derive communication models of different accuracies so that an implementer can trade off the effort to derive such a model with the accuracy of the model. We provide guidance for MPI modeling by demonstrating that simple asymptotic models can be very helpful in understanding the communication complexity of a parallel application. We also mention some pitfalls in modeling MPI implementations.

2   Previous Work and a General Approach to Modeling

Designing a performance model for MPI communication is a complex task. Numerous works exist that model the performance of a particular MPI implementation on a particular system [20,19,15,1,16]. Other works focus on modeling particular aspects of MPI, such as collective communication [15,18,11].

Those works show that accurate models are only possible in very limited cases and require high effort. Handling the whole spectrum, even for a particular MPI implementation on one particular system, is very challenging. In the extreme case, such a model would require the user to consider all parameters of a parallel application run that are (intentionally) outside the scope of MPI (e.g., process-to-node mappings or contention caused by traffic from other jobs). However, many of those parameters may have little influence on a useful model and might thus be abstracted out. In this work, we build upon common modeling methodologies from previous work and design a hierarchy of models that allows one to trade-off design effort for accuracy. Our work is intended to encourage and guide MPI implementers to specify performance characteristics of each implementation. Below we present different approaches at different levels of detail.

Asymptotic Model. A useful first approximation of communication performance is to give the asymptotic time complexity of the communication operations. We use the standard O, Θ, and o notation for asymptotic models. Asymptotic models can often be deduced with relativey little effort and allow assumptions about the general scalability, but do not allow for absolute statements because the constants remain unspecified. For example, an MPI implementation could state that the implemented broadcast scales with $T_{BC}(S, P) = \Theta(S + \log(P))$.

Dominant term exact model. Sometimes it might be possible (or desirable) to indicate that the significant terms are known, while lower-order terms are either not known or do not play a role asymptotically. An optimal broadcast algorithm that fully exploits the bandwidth of the underlying (strong) interconnect could thus be specified as having running time $T_{BC}(S, P) = \Theta(\log(P)) + \beta S$ [13] in contrast to merely efficient broadcast algorithms with the same asymptotic performance $T_{BC}(S, P) = \Theta(S + \log(P))$, but in reality behaving as $T_{BC}(S, P) = \Theta(\log(P)) + 2\beta S$ [5].

Bounded (Parametrized) Model. It is often possible to specify some of the constants for the asymptotic model and thus allow absolute statements. Such constants often depend on many parameters, and it might be infeasible to specify all of them (e.g., process-to-node mapping or contention). Thus, we propose to specify parameterized upper and lower bounds (e.g., for the worst-possible and best-possible mappings) for those costs. Such models allow the user to make absolute statements about parallel algorithms under worst and best conditions. For example, Equation (2) gives such an estimate for point-to-point messages under congestion.

Exact (Parametrized) Model. It might be possible to define exact models for operations in some specified settings (for example, dedicated network links). This is the most accurate technique in our hierarchy and most convenient for the application developer. For example, the cost of a barrier on a BlueGene/P (BG/P) is $T_{BAR} = 0.95\mu s$ [7], independent of P.

Parameter Ranges. Implementations might adapt the communication algorithms based on the input parameters. For example, point-to-point communications are often implemented with two protocols, eager and rendezvous, depending on the size of the message. Implementers can simply specify different models for different parameter regions as shown in Equation (1).

In the remainder of the paper, we discuss a modeling strategy for point-to-point operations (Section 3) and collective operations (Section 4). We show common pitfalls that are often ignored and demonstrate their influence in practice. We focus on developing guidelines for modeling, similar to the guidelines for performance measurement in [8]. To illustrate the techniques and to demonstrate the importance of certain aspects of modeling, we use a BG/P system as an example where we find it helpful. However, we do not specify a complete communication model for BG/P.

3   The Deficiency of Current Point-to-Point Models

The asymptotic model for point-to-point communication is typically $T(S) = \Theta(S)$. A good parametrized model could be the LogGP model [2]. The simpler latency-bandwidth model $T = \alpha + S\beta$ is covered by setting $g = 0$, $\beta = G$ to the time to transmit a single byte, and $\alpha = L + 2o$ to the start-up overhead. Those models are well understood and thus not further discussed. Figure 1(a) shows an example for an accurate congestion-free point-to-point communication on BG/P. The simple point-to-point model would have three components (we use a piecewise linear latency-bandwidth model):

$$T(S) = \begin{cases} 4.5\mu s + 2.67ns/B \cdot S : & S \leq 256B \\ 5.7\mu s + 2.67ns/B \cdot S : & 256B < S \leq 1024B \\ 9.8\mu s + 2.67ns/B \cdot S : & 1024B < S \end{cases} \qquad (1)$$

However, for modeling real applications such models suffer from the following shortcomings:

Overlap and Progress. A point-to-point model should cover the ability of the MPI library to overlap computation and communication. The LogGP model provides the parameter o to model the per-message overhead; however, this might not be sufficient if the overhead grows with the message size. An additional parameter, for example, O for the overhead per byte, could be introduced to capture those effects [15]. Such a model and its derivation are well understood, and we omit details for brevity.

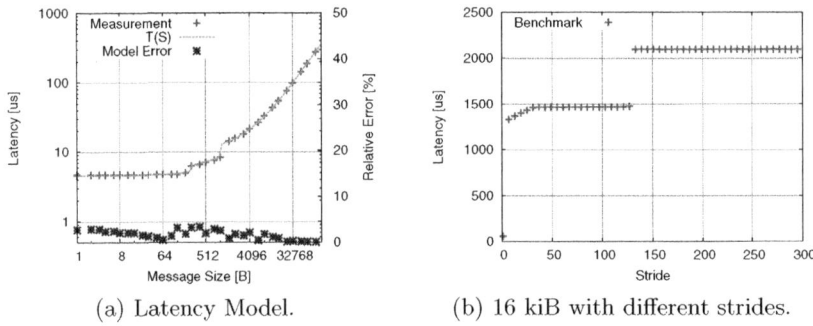

(a) Latency Model.   (b) 16 kiB with different strides.

Fig. 1. Example Models for Latency and Datatypes on BlueGene/P

Synchronization. For some applications, it is important to know about the synchronization properties of messages. MPI implementations typically use receiver buffering for small messages and, for large messages, a rendezvous protocol that delays the sender until the receiver is ready. This effect can be modeled easily and is covered by the parameter S in the LogGPS model [12].

Datatypes. MPI offers a rich set of primitives to define derived datatypes for sending and receiving messages. As these datatypes can reach high complexities, the time to gather all the data from memory can vary significantly for different datatypes. Figure 1(b) shows the influence of the stride in a simple vector datatype when sending 16 kiB MPI_CHAR data. The contiguous case takes $53\mu s$, while a stride of 1 increases the latency to $1292\mu s$ due to the $\approx 2^{14}$ memory accesses. The memory hierarchy in modern computer systems makes modeling datatypes complex; however, one can often provide a worst case that is bound by the slowest memory. In our example, the worst-case ($2090\mu s$) is reached at a stride of 128 when the buffer exceeds the L3 cache on BG/P.

Matching Queue Length. The length of the matching queue can dramatically influence performance as shown in Figure 2(a). We represent the worst-case overhead of a matching queue with R outstanding messages on BG/P as $T_{match}(R) \leq 100ns \cdot R$. A simple 27-point stencil would cause a traversal of an average of 13 requests, adding $T_{match}(13) = 1.3\mu s$ to the latency before the right message is matched.

Topology (Mapping.) The network topology and the process-to-node mapping is also of crucial importance [4]. Information about topology and potentially the mapping is needed for an accurate point-to-point model. For a torus, for example, one could simply multiply the latency with the number of hops. We note that L in the LogP model, by definition, represents the upper bound (i.e., the maximum latency between any two endpoints).

Congestion. Point-to-point models often ignore network congestion, which might be problematic for certain communication patterns. Even networks with full bisection bandwidth are not free of congestion [10]. One way to model congestion would be to define the *effective bisection bandwidth* (formally defined in [10]) as upper bound for $1/G$.

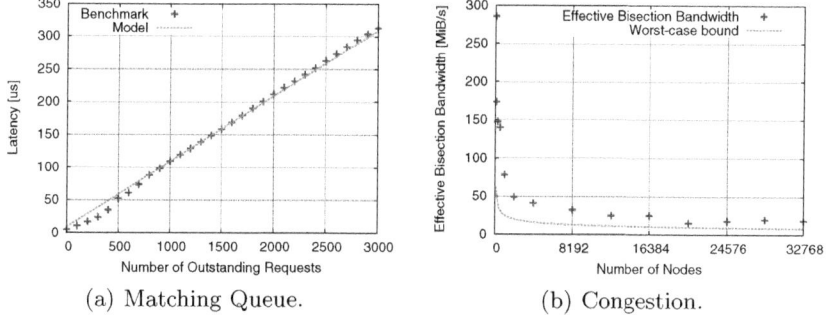

(a) Matching Queue.   (b) Congestion.

Fig. 2. Example Models for Matching and Congestion on BlueGene/P

It is now clear that Equation (1) represents the ideal case: free of congestion, minimal queue lengths, consecutive memory accesses, and an optimal mapping. We will now present an example model for congestion in the general case. First, since we know the upper bound to the bandwidth (Equation (1)), we derive the lower bound, that is, the maximum congestion possible. For this purpose, we assume a cubic allocation on a 3-d torus (k-ary 3-cube) network of size $N_x = N_y = N_z = k$, and (ideal) adaptive routing along shortest paths. Per convention, nodes are identified by three digits ranging from $0 \ldots k-1$. In order to cause the maximum congestion in the network, we choose pairs with maximum distance. In a k-ary 3-cube, the maximum distance between two nodes is $d_3 = 3 \cdot \lfloor k/2 \rfloor$. We assume that each node can inject into all of its six links simultaneously and present the following simple argument: Each node injects six packet streams along the shortest paths to a node at distance d_3 (such a pattern can be generated by connecting each node at coordinate xyz with another node at distance $\lfloor k/2 \rfloor$ in each dimension). Since each stream occupies d_3 links, a total of $6 \cdot k^3 \cdot d_3 = 9k^4$ links would be needed for a congestion-free routing (WLOG, we assume that k is even). Now, we assume that the traffic is spread evenly across the $6 \cdot k^3$ total (unidirectional) links in the steady state of our ideal adaptive routing. This results in a congestion of $3/2k = \mathcal{O}(\sqrt[3]{P})$ per link. Thus, the model presented in Equation (1) must be corrected to reflect congestion on P processes, for example for $S > 1024B$:

$$9.8\mu s + 2.67 ns/B \cdot S \leq T(S, P) < 9.8\mu s + 2.67 ns/B \cdot S \cdot 3/2\sqrt[3]{P} \qquad (2)$$

Upper and lower bounds for the other characteristics can be derived similarly. It is sometimes important to determine the average case congestion/bandwidth, for example, to estimate the scalability of pseudo-random communication patterns as found in many parallel graph computations. The *effective bisection bandwidth* [10] is a good average-case metric and can be measured by benchmarking a huge number of bisection communications between random subsets of processes. Figure 2(b) shows the effective bisection bandwidth as benchmarked on BG/P. The figure also shows the bandwidth bound for the worst-case mapping ($2/3 \sqrt[-3]{P} \cdot 374.5 MiB/s$) assuming ideal routing.

4   Performance Models for Collective Communication

Models for collective communication are of crucial importance for analyzing the scalability of parallel applications. Collective models are often simpler to use than accurate point-to-point models because the algorithms are fixed. Thus, parameters such as topology, synchronization, congestion, and the matching queue can usually be hidden from the user. Asymptotic bounds can often be derived easily from the implementation. A "high quality" MPI implementation should make such statements.

More accurate parametrized models can be specified with a simple extension of point-to-point models as proposed by Xu [22]. Here, one would simply model all parameters, such as L and G in the simplified model, as dependent on the number of processes in the collective and the operation type. An all-to-all communication could be expressed as $T_{a2a} = \alpha(P) + S \cdot \beta(P)$. In a network with full effective bisection bandwidth, one could set $\beta(P) = G$. Startup overheads in α would likely scale linearly in P ($\alpha = \Omega(P)$) for larger messages. However, such models that allow arbitrary functions as parameters in the general case (even tables) are often hard to use to analyze scaling in practice.

Another good method for modeling collective algorithms that build upon point-to-point methods is to construct the model from point-to-point models [11,18,9]. Note that we do not prescribe a specific model rather than a methodology to design such models. Some special networks or topologies might require the addition of terms to describe certain effects (e.g., contention in a torus network). If significant effects are too complex to describe in such a model (e.g., process-to-node mappings) then the upper and lower bounds (best and worst case) should be given.

Process-to-Node Mapping. The process-to-node mapping often plays a role in the performance of collective communication. For instance, the performance of rooted collectives, such as broadcast, can depend on the position of the root and the network topology. This is especially important for multi-core systems. Also, on BG/P the performance varies by the type of allocation. For example, for broadcast, performance was degraded to half for non-cubic allocations.

Datatypes. A similar discussion as for point-to-point models applies. Let us discuss three example models for MPI_Bcast, MPI_Allreduce, and MPI_Alltoall on BG/P. We do this by comparing theoretical bounds based on first principles gathered from the documentation [7] with benchmarked performance on MPI_-COMM_WORLD on full allocations. Models for collective communications on other communicators and non-cubic allocations could be derived with a similar method. All benchmarks used the synchronous BG/P hardware barrier to start the operation on all processes, measured the time for a single execution, and report the average time.

Small Data. First, we look at operations with a single integer (8 bytes), where the specialized collective network is used. The bandwidth of the collective network is 824 MiB/s [7]. IBM, however, did not release latency numbers for this network; thus, we resort to numerical methods for deriving a model.

For the **broadcast** time, we assume $T_{BC}(P, 8) = \alpha_T + \beta_T^{BC} \log_2(P)$ for the collective tree network. α_T models the startup overhead and β_T models the cost (latency) per stage of the tree. With a small broadcast on two processes, we determined $\alpha_T = 13\mu s$ and $\beta_T^{BC} = 0.31\mu s$ from a large run with $P = 32,000$.

For **allreduce**, we used a similar model and determined $\beta_T^{SUM} = 0.37\mu s$ (the difference of $60ns$ is most likely caused by the higher overhead of the Integer sum): $T_{ARE}(P, 8) = \alpha_T^{SUM} + \beta_T^{SUM} \log_2(P)$. We found that for $P \leq 4,096$, the time is a constant $17.77\mu s$, which seems to indicate some other constant overhead in the implementation that probably overlaps with the communication in the tree network.

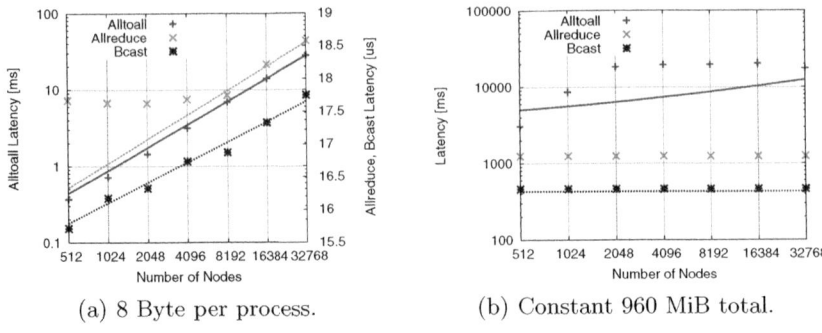

(a) 8 Byte per process.   (b) Constant 960 MiB total.

Fig. 3. Small data (8 Byte process) and large data (960 MiB) scaling for different collective operations

For **alltoall**, the implementation simply sends to all peer processes [7]; thus, we assume $T_{A2A}(P, 8) = \alpha + g(P - 1)$. As noted in Section 3, $\alpha = L + 2o$, and we determined $g = 0.84 \mu s$ for $P = 32,000$. The model functions and the measurement results are shown in Figure 3(a) as points and lines, respectively.

Large Data. For the large-data collectives, we communicated the maximum possible buffer size of $960 MiB$ on BG/P (for alltoall, $960 MiB/P$ per process).

Large-data **broadcasts** use all six links of the Torus network and the deposit-bit feature for communication. The maximum effective bandwidth would be $6 \cdot 374.5 MiB/s = 2247 MiB/s$ and is shown as a line in Figure 3(b). The broadcast would need at least $T_{BC}(P, 8)$ to reach all endpoints. We thus extend our broadcast model with a bandwidth term: $T_{BC}(P, S) = \alpha_T + \beta_T^{BC} \log_2(P) + \frac{2.67}{6} ns/B \cdot S$.

Large **allreduces** use the same message pattern as broadcast, but each stage is slower because of the reduction operation. An experiment showed that the allreduces takes $T_{ARE}(P, 9.6 \cdot 10^8) = 2.68 \cdot T_{BC}(P, 9.6 \cdot 10^8)$.

Let us now recapitulate an argument for the best-case **alltoall** bandwidth assuming ideal adaptive routing. As stated before, alltoall is implemented by simply sending from all processes to all other processes (in some order). We thus assume that all messages hit the network at the same time and are either limited by the injection at the endpoints or by the congestion in the network. Lam et al. showed bounds for one- and two-dimensional tori in [14]. We model the time with LogGP: $T_{A2A} = (P - 1)g + SG \cdot \max\{\frac{P-1}{6}, C(P)\}$ and a congestion factor $C(P)$ (we assume $g > o$ WLOG). For alltoall, we assume that all messages contribute to the worst-case congestion in the network. On a k^3 grid (for odd k, WLOG), those messages occupy different numbers of links, depending on their Euclidean distance, with a maximum of $d_3 = \frac{3(k-1)}{2}$. Let $d = d_1 = \frac{k-1}{2}$, then the total number of occupied links is

$$N(k) \leq k^3 \cdot 2 \sum_{x=0}^{d} 2 \sum_{y=0}^{d} 2 \sum_{z=0}^{d} (x + y + z) = k^3 12d \, (d + 1)^3 = \mathcal{O}(k^7).$$

With a total of $6k^3$ links in the torus, the congestion per link (assuming ideal routing) $C(k) \leq N(k)/6k^3 = 2d(d+1)^3 = \mathcal{O}(k^4)$ and with $k^3 = P$, $C(P) \leq \sqrt[3]{P}(\sqrt[3]{P}/2 + 1)^3 = \mathcal{O}(P\sqrt[3]{P})$. Thus the lower bound for a bandwidth-bound alltoall on a torus would be:

$$T_{A2A} \geq g(P-1) + SG\sqrt[3]{P}(\sqrt[3]{P}/2 + 1)^3.$$

Figure 3(b) shows the bound and benchmark results for alltoall. This analysis indicates that potential for further optimization exists in BG/P's alltoall implementation. Faraj et al. suggest that increasing the number of FIFOs would mitigate the problem [7].

5 Summary and Conclusions

In this work, we describe the importance of analytic performance models for MPI implementations. Such models and their accuracy become more important in the context of algorithm and application design and validation on very large (petascale or exascale) systems. We argue that MPI implementers should supply analytic models with an MPI library in order to allow users to make algorithmic decisions and analyze scalability.

We described a hierarchy of modeling approaches that allow the designer to trade accuracy against effort, and we argue that asymptotic models would already provide important hints to application developers. We demonstrate how simple performance models can be developed, discuss common pitfalls, and show how to address those issues with examples on the BG/P architecture.

Performance is the main motivator for parallelization, and thus, performance modeling is most important in the context of MPI. Our work motivates a discussion of performance models in the MPI community and provides some initial guidance towards more useful modeling for MPI.

Acknowledgments. This work was supported in part by the Office of Advanced Scientific Computing Research, Office of Science, U.S. Department of Energy, under contract DE-AC02-06CH11357 and DE-FG02-08ER25835, and by the Blue Waters sustained-petascale computing project, which is supported by the National Science Foundation (award number OCI 07-25070) and the state of Illinois.

References

1. Al-Tawil, K., Moritz, C.A.: Performance modeling and evaluation of MPI. Journal of Parallel and Distributed Computing 61(2), 202–223 (2001)
2. Alexandrov, A., Ionescu, M.F., Schauser, K.E., Scheiman, C.J.: LogGP: Incorporating long messages into the LogP model for parallel computation. Journal of Parallel and Distributed Computing 44(1), 71–79 (1997)
3. Barker, K.J., Davis, K., Kerbyson, D.J.: Performance modeling in action: Performance prediction of a Cray XT4 system during upgrade. In: Proceedings of the 2009 IEEE Intl. Symp. on Parallel&Distributed Processing, pp. 1–8 (2009)
4. Bhatelé, A., Bohm, E., Kalé, L.V.: Topology aware task mapping techniques: An API and case study. SIGPLAN Not. 44(4), 301–302 (2009)

5. Chan, E., Heimlich, M., Purkayastha, A., van de Geijn, R.A.: Collective communication: theory, practice, and experience. Conc. & Comp. 19(13), 1749–1783 (2007)
6. Eleftheriou, M., Fitch, B.G., Rayshubskiy, A., Ward, T.J.C., Germain, R.S.: Scalable framework for 3D FFTs on the Blue Gene/L supercomputer: implementation and early performance measurements. IBM J. Res. Dev. 49(2), 457–464 (2005)
7. Faraj, A., Kumar, S., Smith, B., Mamidala, A., Gunnels, J.: MPI collective communications on the Blue Gene/P supercomputer: Algorithms and optimizations. In: 17th IEEE Symposium on High-Performance Interconnects, pp. 63–72 (2009)
8. Gropp, W., Lusk, E.L.: Reproducible measurements of mpi performance characteristics. In: Margalef, T., Dongarra, J., Luque, E. (eds.) PVM/MPI 1999. LNCS, vol. 1697, pp. 11–18. Springer, Heidelberg (1999)
9. Hoefler, T., Janisch, R., Rehm, W.: Parallel scaling of Teter's minimization for Ab Initio calculations . In: HPC Nano 2006 in conjunction with the Intl. Conference on High Performance Computing, Networking, Storage and Analysis, SC 2006 (November 2006)
10. Hoefler, T., Schneider, T., Lumsdaine, A.: Multistage Switches are not Crossbars: Effects of Static Routing in High-Performance Networks. In: Proc. of IEEE Intl. Conf. on Cluster Computing, October 2008. IEEE Computer Society Press, Los Alamitos (2008)
11. Hoefler, T., Cerquetti, L., Mehlan, T., Mietke, F., Rehm, W.: A practical approach to the rating of barrier algorithms using the LogP model and Open MPI. In: Proc. of the Intl. Conf. on Parallel Proc. Workshops (ICPP 2005), June 2005, pp. 562–569 (2005)
12. Ino, F., Fujimoto, N., Hagihara, K.: LogGPS: A Parallel Computational Model for Synchronization Analysis. In: PPoPP 2001: Proc. of ACM SIGPLAN symposium on Principles and practices of parallel programming, pp. 133–142 (2001)
13. Jia, B.: Process cooperation in multiple message broadcast. Parallel Computing 35(12), 572–580 (2009)
14. Lam, C.C., Huang, C.H., Sadayappan, P.: Optimal algorithms for all-to-all personalized communication on rings and two dimensional tori. J. Parallel Distrib. Comput. 43(1), 3–13
15. Martínez, D.R., Cabaleiro, J.C., Pena, T.F., Rivera, F.F., Blanco, V.: Accurate analytical performance model of communications in MPI applications. In: 23rd IEEE Intl. Symp. on Parallel and Distributed Processing (IPDPS), pp. 1–8 (2009)
16. Moritz, C.A., Frank, M.: LoGPC: Modeling network contention in message-passing programs. IEEE Trans. on Par. and Distrib. Systems 12(4), 404–415 (2001)
17. Mudalige, G.R., Vernon, M.K., Jarvis, S.A.: A plug-and-play model for evaluating wavefront computations on parallel architectures. In: IEEE International Symposium on Parallel and Distributed Processing, pp. 1–14 (2008)
18. Pjesivac-Grbovic, J., Angskun, T., Bosilca, G., Fagg, G.E., Gabriel, E., Dongarra, J.J.: Performance Analysis of MPI Collective Operations. In: 4th Intl. Workshop on Perf. Modeling, Evaluation, and Optimization of Par. and Distrib. Syst. (2005)
19. Rodríguez, G., Badia, R.M., Labarta, J.: Generation of simple analytical models for message passing applications. In: Danelutto, M., Vanneschi, M., Laforenza, D. (eds.) Euro-Par 2004. LNCS, vol. 3149, pp. 183–188. Springer, Heidelberg (2004)
20. Touriño, J., Doallo, R.: Performance evaluation and modeling of the Fujitsu AP3000 message-passing libraries. In: Amestoy, P.R., Berger, P., Daydé, M., Duff, I.S., Frayssé, V., Giraud, L., Ruiz, D. (eds.) Euro-Par 1999. LNCS, vol. 1685, pp. 183–187. Springer, Heidelberg (1999)
21. Träff, J.L., Ripke, A.: Optimal broadcast for fully connected processor-node networks. Journal of Parallel and Distributed Computing 68(7), 887–901 (2008)
22. Xu, Z., Hwang, K.: Modeling communication overhead: MPI and MPL performance on the IBM SP2. IEEE Parallel Distrib. Technol. 4(1), 9–23 (1996)

PMI: A Scalable Parallel Process-Management Interface for Extreme-Scale Systems*

Pavan Balaji[1], Darius Buntinas[1], David Goodell[1], William Gropp[2], Jayesh Krishna[1], Ewing Lusk[1], and Rajeev Thakur[1]

[1] Argonne National Laboratory, Argonne, IL 60439, USA
[2] University of Illinois, Urbana, IL 61801, USA

Abstract. Parallel programming models on large-scale systems require a scalable system for managing the processes that make up the execution of a parallel program. The process-management system must be able to launch millions of processes quickly when starting a parallel program and must provide mechanisms for the processes to exchange the information needed to enable them communicate with each other. MPICH2 and its derivatives achieve this functionality through a carefully defined interface, called PMI, that allows different process managers to interact with the MPI library in a standardized way. In this paper, we describe the features and capabilities of PMI. We describe both PMI-1, the current generation of PMI used in MPICH2 and all its derivatives, as well as PMI-2, the second-generation of PMI that eliminates various shortcomings in PMI-1. Together with the interface itself, we also describe a reference implementation for both PMI-1 and PMI-2 in a new process-management framework within MPICH2, called Hydra, and compare their performance in running MPI jobs with thousands of processes.

1 Introduction

While process management is an integral part of high-performance computing (HPC) systems, it has historically not received the same level of attention as other aspects of parallel systems software. The scalability of process management is not much of a concern on systems with only a few hundred nodes. As HPC systems get larger, however, systems with thousands of nodes and tens of thousands of processing cores are becoming common; indeed, the largest systems in the world already use hundreds of thousands of processing cores. For such systems, a scalable design of the process-management infrastructure is critical for various aspects such as launching and management of parallel applications, debugging utilities, and management tools. A process-management system must,

---

* This work was supported in part by the Office of Advanced Scientific Computing Research, Office of Science, U.S. Department of Energy, under Contract #DE-AC02-06CH11357; by the DOE grant #DE-FG02-08ER25835; and by the National Science Foundation under grant #0702182.

R. Keller et al. (Eds.): EuroMPI 2010, LNCS 6305, pp. 31–41, 2010.

of course, start and stop processes in a scalable way. In addition, it must provide mechanisms for the processes in a parallel job to exchange the information needed to establish communication among them.

Although the growing scale of HPC systems requires close interaction between the parallel programming library (such as MPI) and the process manager, an appropriate separation between these two components is necessary. This separation not only allows for their independent development and improvement but also keeps the parallel programming library generic enough to be used with any process-management framework. At the same time, these two components must share sufficient information so as to allow the parallel programming library to take advantage of specific characteristics of the system on which it is running.

With these requirements in mind, we initially designed PMI, a generic process-management interface for parallel applications. In this paper, we start by describing the first generation of PMI (PMI-1). PMI-1 is widely used in MPICH2 [1] and other MPI implementations derived from it, such as MVAPICH2 [4], Intel MPI [6], SiCortex MPI [12], and Microsoft MPI [7] (for the programming library side) as well as in many process-management frameworks including MPICH2's internal process managers (Hydra, MPD, SMPD, Gforker, Remshell), and other external process managers such as SLURM [15], OSC mpiexec [9], and OSU mpirun [13] (for the process-manager side).

While extremely successful, PMI-1 has several limitations, particularly when applied to modern HPC systems. These limitations include issues related to scalability for large numbers of cores on a single node and efficient interaction with hybrid programming models that combine MPI and threads, amongst others. Building on our experiences with PMI-1, we recently designed a second-generation interface (PMI-2) that overcomes the shortcomings of PMI-1. The second part of the paper describes this new interface and a reference implementation of both PMI-1 and PMI-2 in a new process-management framework within MPICH2, called Hydra [5]. We also present performance results comparing PMI-2's capabilities to that of PMI-1 and other process-management interfaces on system scales of nearly 6,000 processes.

2   Requirements of a Process-Management Interface

In this section we provide a brief overview of what is required of a process-management interface for scalable parallel process management on large systems.

2.1   Decoupling the Process Manager and the Process-Management Interface

In our model, process management comprises three primary components: *(1)* the parallel programming library (such as MPI), *(2)* the PMI library, and *(3)* the process manager. These components are illustrated in Figure 1 with examples of different MPI libraries, PMI libraries, and process managers.

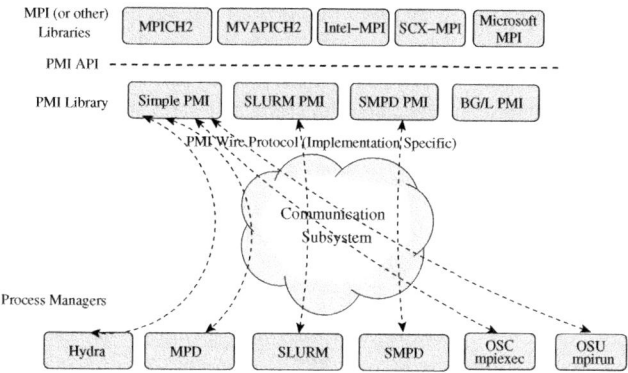

Fig. 1. Interaction of MPI and the process manager through PMI

The process manager is a *logically* centralized process (but often a distributed set of processes in practice) that manages *(1)* process launching (including starting/stopping processes, providing the environment information to each process, stdin/out/err forwarding, propagating signals) and *(2)* information exchange between processes in a parallel application (e.g., to set up communication channels). Several process managers are available (e.g., PBS [8], SUN Grid Engine [14], and SSH), that already provide such capabilities.

The PMI library provides the PMI API. The implementation of the PMI library, however, might depend on the system itself. In some cases, such as for IBM Blue Gene/L (BG/L) [3], the library may use system-specific features to provide PMI services. In other cases, such as for processes on a typical commodity cluster, the PMI library can communicate with the process manager over a communication path (e.g. TCP). While the PMI library can be implemented in any way that the particular implementation prefers, in both PMI-1 and PMI-2 there is a predefined "wire protocol" where data is exchanged through the sockets interface. The advantage of using this protocol is that any application that uses the PMI API with the predefined PMI wire protocol is compatible with *any* PMI process manager implementation that accepts the wire protocol.

We note that the PMI API and the PMI wire protocol are separate entities. An implementation may choose to implement both, or just one of them. For example, the PMI library on BG/L provides the PMI API but does not use the sockets-based wire protocol. Thus, the library is compatible with any programming model using the PMI API, but it is not compatible with process managers that accept the sockets-based PMI wire protocol.

2.2   Overview of the First-Generation PMI (PMI-1)

Processes of a parallel application need to communicate with each other. Establishing this communication typically requires publishing a contact address, which may be an IP address, a remotely accessible memory segment, or any other interconnect-specific identifier. Since the process manager knows where all the processes are, and because it is (probably) managing some communication with the processes to handle standard I/O (stdin, stdout, and stderr), it is natural to have the process-management system also provide the basic facilities for information interchange. This is the key feature of our process-management interface, PMI—a recognition that these two features are closely related and can be effectively provided through a single service.

While PMI itself is generic for any parallel programming model and not just MPI, for ease of discussion we consider only the MPI programming model here. In the case of MPI, PMI deals with aspects such as providing each MPI process with information about itself (such as its rank) as well as about the other processes in the application (such as the size of MPI_COMM_WORLD). Furthermore, each PMI process manager that launches parallel applications is expected to maintain a database of all such information. PMI defines a portable interface that allows the MPI process to interact with the process manager by adding information to the database ("put" operations) and querying information added by other processes in the application ("get" operations). The PMI functions are translated into the appropriate wire protocol by the PMI provider library and exchanged with the process manager. Most of the database is exchanged by using "key-value" pairs. Together with "put" and "get" operations, PMI also provides collective "fence" operations that allow efficient, collective data-exchange capabilities (the use of fence is described in more detail in Section 2.3).

As an example interaction between the MPI library, the PMI library, and the process manager, consider a parallel application with two processes, P_0 and P_1, where P_0 wants to send data to P_1. In this example, during MPI initialization, each MPI process adds to the PMI database information about itself that other processes can use to connect to it. When P_0 calls an MPI_Send to P_1, the MPI library can look up information about P_1 from the PMI database, connect to P_1 (if needed) by using this information, and send the data.

2.3   PMI Requirements for the Process Manager

In designing the process-management interface, there are two primary requirements for the process manager. First, a careful separation of features is needed to enable layering on a "native" process manager with the lowest possible overhead. This requirement arises because many systems already have some form of a process manager (often integrated with a resource manager) that is tightly tied to the system. A portable PMI must make effective use of these existing systems without requiring extra overhead (e.g., requiring no additional processes beyond what the native system uses). For example, an interface that requires asynchronous processing of data or interrupts to manage data might cause additional overhead for applications even when they are not interacting with the PMI services; this can be a major issue on large-scale systems. Second, a scalable data-interchange approach for the key-value system is needed.

This second requirement has a number of aspects. Consider a system in which each process in a parallel job starts, creates a "contact id," and wishes to make it available to the other processes in the parallel job. A simple way to do this is for the process to provide the data to central server, for example, by adding the data expressed as a (key,value) pair into a simple database. If all p processes do this with a central server, the time complexity is $\mathcal{O}(p)$; the time for all processes to extract just a single value is also $\mathcal{O}(p)$. This approach is clearly not scalable. Using multiple servers instead of a single one helps, but it introduces other problems.

Our solution in PMI is to provide a collective abstraction, permitting the use of efficient collective algorithms to provide more scalable behavior. In this model, processes put data into a key-value space (KVS). They then *collectively* perform a fence operation. Following completion of the fence, all processes can perform a get operation against the KVS. Such a design permits many implementations. Most important, the fence step, which is collective over all processes, provides an excellent opportunity for the implementation to distribute the data supplied by the put operations in a scalable manner. For example, a distributed process manager implementation with multiple processes can use this opportunity to allow these processes to share their local information with each other.

3   Second-Generation PMI (PMI-2)

While the basic design of PMI-1 was widely adopted by a large number of PMI libraries and process managers, as we move to more advanced functionality of MPI as well as to larger systems, several limitations of PMI-1 have become clear. The second-generation PMI (PMI-2) addresses these limitations.

The complete details of the PMI-2 interface (including function names), and wire protocol are available online [10,11]. To avoid dilution, we do not explicitly mention them in this paper. Instead, we describe the major areas in which PMI-2 improves on PMI-1.

Lack of Query Functionality: PMI-1 provides a simple key-value database that processes can put values into and get values from. While the process manager is best equipped to understand various system-specific details, PMI-1 does not allow it to share this information with the MPI processes. In other words, the process manager itself cannot add system-specific information to the key-value database; thus MPI processes cannot query such information from the process manager through PMI-1.

An example issue created by this limitation occurs on multicore and multiprocessor systems, where the MPI implementation must determine which processes reside on the same SMP node (e.g., to create shared-memory segments or for hierarchical collectives). Each process gathers this information by fetching the contact information for all other MPI processes and determining which contact addresses are local to itself. While this approach is functional, it is extremely inefficient because it results in $\mathcal{O}(p)$ PMI get operations for each MPI process ($\mathcal{O}(p^2)$ total operations).

PMI-2 introduces the concept of *job attributes*, which are predefined keys provided by the process manager. Using such keys, the process manager can pass system-specific information to the MPI processes; that is, these keys are added into the key-value database directly by the process manager with system layout information, allowing each MPI process to get information about the layout of all MPI processes in a single operation. Further, since the process manager knows that such attributes are read-only, it can optimize their storage by caching copies on local agents, thus allowing the number of PMI requests to be reduced from $\mathcal{O}(p^2)$ (in the case of PMI-1) to nearly zero (in the case of PMI-2)[1].

Database Information Scope: PMI-1 uses a flat key-value database. That is, an MPI process cannot restrict the scope of a key that it puts into the database; all information is global. Thus, if some information needs to be local only to a subset of processes, PMI-1 provides no mechanism for the MPI processes to inform the process manager about it. For example, information about shared-memory keys is relevant only to processes on the same node; but the process manager cannot optimize where such information is stored or replicated.

To handle this issue, PMI-2 introduces "scoping" of keys as node-level and global (further restriction of the scope is not supported in PMI-2, in favor of simplicity as opposed to generality). For example, keys corresponding to shared memory segments on a node can be restricted to a node-level scope, thus allowing the process manager to optimize retrieval.

Hybrid MPI+Threads Programs: PMI-1 is not thread safe. Therefore, in the case of multithreaded MPI programs, the MPI implementation must protect calls to PMI-1 by using appropriate locking mechanisms. Such locking is often coarse-grained and serializes communication between the PMI library and the process manager. That is, until the PMI library sends a query to the process manager and gets a response for it (a round-trip communication), no other thread can communicate over the same socket. PMI-2 functions are thread-safe. Thus, multiple threads can communicate with the server in a more fine-grained manner, thereby pipelining requests better and improving performance.

Dynamic Processes: Each process group in PMI-1 maintains a separate database, and processes are not allowed to query for information across databases. For dynamically spawned processes, this is a severe limitation because it requires such processes to manually exchange their database information and load them into their individual databases. This procedure is cumbersome and expensive (with respect to both performance and memory usage). PMI-2 recognizes the concept of a "job" that can contain multiple applications connected to each other or where one is spawned from another. This allows such jobs to share database information without the need to explicitly replicate it.

Fault Tolerance: PMI-1 does not specify any mechanism for respawning processes when a fault occurs. Note that this is different from spawning an MPI-2 dynamic process, since such a process would form its own process group (MPI_COMM_WORLD) and not just replace a process in the existing process group. PMI-2 provides

---

[1] The read-only attributes still need to be fetched, which causes the number of requests with PMI-2 to be non-zero.

a concept of respawning processes, where a new process essentially replaces the original process within the same process group.

PMI-2 has been implemented as part of a new process-management framework in MPICH2, called Hydra [5].

4   Experimental Evaluation and Analysis

In this section, we present the results of several experiments that compare the performance of PMI-2 with that of PMI-1.

4.1   System Information Query Functionality

As described in Section 3, PMI-1 provides only a simple key-value database that processes can put values into and get values from, so a process manager cannot provide system specific information to the MPI processes. Thus, in order to determine which processes reside on the same SMP node, $\mathcal{O}(p^2)$ PMI operations are required. With PMI-2's job attributes, this reduces to nearly zero.

This behavior is reflected in the launch time of MPI applications. Figure 2(left) shows this behavior for a simple MPI application (that just calls MPI_Init and MPI_Finalize) with PMI-1 and PMI-2 on a 5760-core SiCortex system. As shown in the figure, the overall launch time increases rapidly with system size for PMI-1. With PMI-2, on the other hand, the time taken is significantly less. Figure 2(right) shows further analysis of the two PMI implementations with respect to the number of PMI requests observed by the process manager. This figure illustrates the reason for the performance difference between the implementations: PMI-1 has several orders of magnitude more PMI requests than does PMI-2.

4.2   Impact of Added PMI Functionality over the Native Process Manager

As described in Section 2.3, some systems already have a process manager (often integrated with a resource manager) that is tightly tied to the system. While

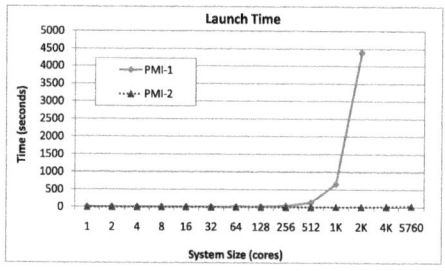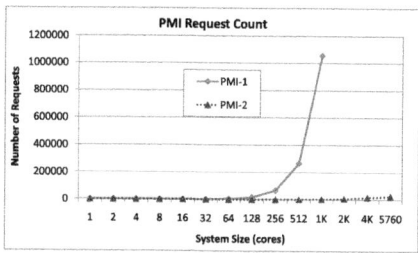

Fig. 2. Process launching on a 5760-core SiCortex system: (left) launch time and (right) number of PMI requests

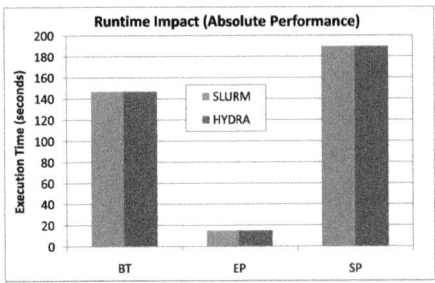 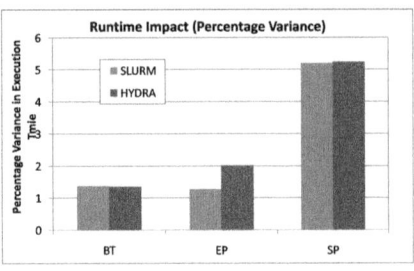

Fig. 3. Runtime impact of separate PMI server daemons: (left) absolute runtime; and (right) percentage variance in runtime

some process managers might natively provide PMI functionality, others do not. An efficient implementation of the PMI interface must make effective use of such "native process managers" without requiring extra overhead (for example, by requiring heartbeat operations that wake up additional processes, thus disturbing the core computation). In this section, we evaluate this "noise impact" on 1600-cores of the SiCortex system using Class C NAS parallel benchmarks in two modes: *(1)* using the native process manager on the system, SLURM, that already provides PMI-1 services, and *(2)* using the Hydra process manager that internally uses SLURM for process launching and management, but separately provides PMI services on top of it using an extra process daemon.

As shown in Figure 3, the impact of having additional PMI services (legend "Hydra") on top of the native process manager (legend "SLURM") on the system does not add any significant overhead. Figure 3(left) shows the impact on runtime, where there is no perceivable overhead. Figure 3(right) shows the percentage difference between the highest and lowest execution times noticed on a large number of runs of the application. Again, in most cases this difference is close to 0%, with a maximum of 0.5% for the EP application.

The primary reason for such lack of overhead is that the PMI design completely relies on synchronous activity, and thus there is no asynchronous waking of PMI service daemons. That is, once the initialization is complete, unless the MPI process sends a PMI request, there is no additional overhead.

4.3   Performance of Multithreaded MPI Applications

With an increasing number of cores on each node, researchers are studying approaches for using MPI in conjunction with threads, in which case MPI functions might be called from multiple threads of a process. Since PMI-1 is not thread safe, all PMI calls must be protected by coarse-grained external locks; thus, only one thread can communicate with a process manager at a given time. PMI-2, on the other hand, is thread-safe, allowing for multiple threads to communicate with the process manager in a fine-grained manner.

In this experiment, we measure the concurrency of PMI operations by using a benchmark that continuously publishes and unpublishes services to the process manager[2]. With PMI-1, each thread obtains a lock, sends a publish request and waits for a response from the process manager before releasing the lock. With PMI-2, each PMI request contains a thread ID; so the PMI library can send one request and release the lock (allowing

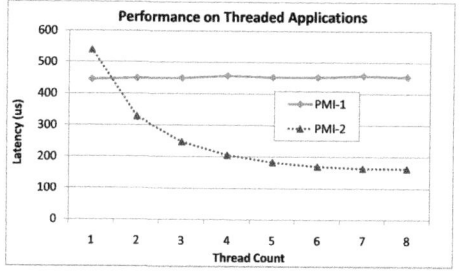

Fig. 4. Multithreading Performance

other threads to send requests) even before it gets its response. When the process manager responds to the publish requests, it sends back the original thread ID with the response, allowing it to be forwarded to the appropriate thread.

The impact of such threading capability is illustrated in Figure 4. As shown in the figure, the average time taken by each PMI request for PMI-1 does not reduce with increasing number of threads since all requests are serialized (the total amount of work is fixed, but shared between all the threads). With PMI-2, however, when multiple threads make concurrent PMI requests, all requests are pipelined in a fine-grained manner, allowing for better concurrency. Thus the average PMI latency perceived by each thread would be lesser. We notice that for the single-threaded case, PMI-2 has additional overhead compared with PMI-1. This result is unexpected and is being investigated.

4.4   Comparison with Alternative Process Management Frameworks

In this section, we compare an implementation of the PMI-2 interface with an alternate process-management framework, Open-RTE. OpenRTE [2] (ORTE) is the process-management system used in Open MPI. It is designed to provide robust and transparent support for parallel process management. Like PMI, it includes a system of (key,value) pairs that are exchanged between the MPI processes and the process-management system.

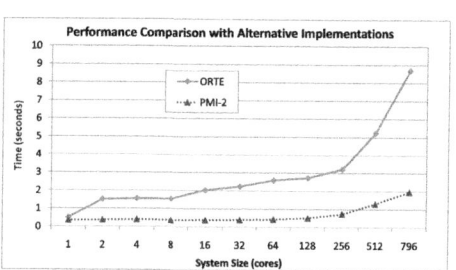

Fig. 5. Job launch time comparison with alternative process managers

Figure 5 compares launch time of MPI applications for the Hydra implementation of PMI-2 and OpenRTE. As shown in the figure, PMI-2 performs slightly better than ORTE[3] on this 796-core commodity cluster.

---

[2] These operations are used in `MPI_Publish_name` and `MPI_Unpublish_name`.

[3] Open MPI v1.4.1 was used for the measurements presented here.

5   Related Work

Improvements to the process-management framework for parallel programming models is not a new research topic. However, most efforts have focused on improving the process manager itself with respect to how it launches and manages processes. The OSC mpiexec [9], OSU mpirun (also known as SceLA) [13], and SLURM [15] are examples of such work. OSC mpiexec is a process manager for MPI applications that internally uses PBS [8] for launching and managing jobs. It is a centralized process manager that communicates using multiple process-management wire protocols, including PMI-1. OSU mpirun is based on SSH; it uses a hierarchical approach to launch processes and interacts with PMI-1. SLURM [15] differs from other process managers primarily in that it provides an entire infrastructure that launches and manages processes; it also provides its own PMI-1 implementation to interact with the processes. While all these implementations seek to improve process management on large-scale systems, our work differs in that none of these implementations study the requirements and limitations of the interface between the MPI library and the process manager, which is the PMI API and the wire protocol.

ORTE provides a mechanism for MPI processes to interact with the integrated process manager. However, it does not explicitly decouple these functionalities, as do PMI and its associated wire protocol.

In summary, our work differs from other process-management systems with respect to its capabilities and underlying architecture. At the same time, PMI-2 provides a complementary contribution to those systems in that it can be used with them simultaneously.

6   Concluding Remarks

We presented a generic process-management interface, PMI, that allows different process-management frameworks to interact with parallel libraries such as MPI. We first described PMI-1, which is currently used in MPICH2 and all its derivatives. We then described PMI-2, the second generation of PMI that eliminates various shortcomings in PMI-1 on modern HPC systems, including scalability issues for large multi-core systems and interaction with hybrid MPI-and-threads models. Our performance results demonstrate significant advantages of PMI-2 compared with PMI-1.

References

1. Argonne National Laboratory: MPICH2, http://www.mcs.anl.gov/research/projects/mpich2
2. Castain, R., Woodall, T., Daniel, D., Squyres, J., Barrett, B., Fagg, G.: The Open Run-Time Environment (OpenRTE): A transparent multi-cluster environment for high-performance computing. In: Di Martino, B., Kranzlmüller, D., Dongarra, J. (eds.) EuroPVM/MPI 2005. LNCS, vol. 3666, pp. 225–232. Springer, Heidelberg (2005)

3. Gara, A., Blumrich, M., Chen, D., Chiu, G., Coteus, P., Giampapa, M., Haring, R., Heidelberger, P., Hoenicke, D., Kopcsay, G., Liebsch, T., Ohmacht, M., Stein-macherBurow, B., Takken, T., Vranas, P.: Overview of the Blue Gene/L system architecture. IBM Journal of Research and Development 49(2/3) (2005)
4. Huang, W., Santhanaraman, G., Jin, H., Gao, Q., Panda, D.: Design of high per-formance MVAPICH2: MPI2 over InfiniBand. In: Proceedings of the sixth IEEE International Symposium on Cluster Computing and the Grid, Singapore Manage-ment University, Singapore, May 16–19 (2006)
5. Hydra process management framework, http://wiki.mcs.anl.gov/mpich2/index.php/Hydra_Process_Management_Framework
6. Intel MPI, http://software.intel.com/en-us/intel-mpi-library/
7. Microsoft MPI: http://msdn.microsoft.com/en-us/library/bb524831VS.85.aspx
8. PBS: Portable batch system , http://www.openpbs.org
9. OSC Mpiexec, http://www.osc.edu/~djohnson/mpiexec
10. PMI-2 API ,http://wiki.mcs.anl.gov/mpich2/index.php/PMI_v2_API
11. PMI-2 Wire Protocol , http://wiki.mcs.anl.gov/mpich2/index.php/PMI_v2_Wire_Protocol
12. SiCortex Inc., http://www.sicortex.com
13. Sridhar, J., Koop, M., Perkins, J., Panda, D.K.: ScELA: Scalable and Extensible Launching Architecture for Clusters. In: Sadayappan, P., Parashar, M., Badrinath, R., Prasanna, V.K. (eds.) HiPC 2008. LNCS, vol. 5374, pp. 323–335. Springer, Heidelberg (2008)
14. Sun Grid Engine , http://www.sun.com/software/sge/
15. Yoo, A.B., Jette, M.A., Grondona, M.: SLURM: Simple Linux utility for resource management. In: Feitelson, D.G., Rudolph, L., Schwiegelshohn, U. (eds.) JSSPP 2003. LNCS, vol. 2862, pp. 44–60. Springer, Heidelberg (2003)

Run-Time Analysis and Instrumentation for Communication Overlap Potential

Thorvald Natvig and Anne C. Elster

Norwegian University of Science and Technology(NTNU)
Sem Sælands vei 9, NO-7491 Trondheim, Norway
{thorvan,elster}@idi.ntnu.no

Abstract. Blocking communication can be runtime optimized into non-blocking communication using memory protection and replacement of MPI functions. All such optimizations come with overhead, meaning no automatic optimization can reach the performance level of hand-optimized code.In this paper, we present a method for using previously published runtime optimizers to instrument a program, including measured speedup gains and overhead.The results are connected with the program symbol table and presented to the user as a series of source code transformations. Each series indicates which optimizations were performed and what the expected saving in wallclock time is if the optimization is done by hand.

Keywords: MPI Overlap Communication Instrumentation Analysis.

1 Introduction

For point-to-point communication, MPI [1] offers two methods of communication; blocking and non-blocking. Blocking communication is the easiest to use, as communication is complete by the time the function call returns. When using blocking communication, all communication time is overhead, and this limits effective speedup.

Non-blocking communication offers an alternative, where control immediately returns to the program, while the communication operation is performed asynchronously. The program must not reference the data area until communication is complete, which makes this method of communication harder to use correctly. If the programmer forgets to explicitly wait for communication to finish before accessing data, the program may read unreceived data or write to unsent data, causing data corruption. Often, these problems do not appear in small test runs, but only appear when the application is scaled to larger problem sizes on a larger number of nodes. However, non-blocking communication allows overlap both between inidividual communication requests and with computation, greatly reducing the effective overhead of parallelization.

1.1 Previous Work

We have previously shown that it is possible to run-time optimize applications by turning blocking communication into non-blocking communication, using

R. Keller et al. (Eds.): EuroMPI 2010, LNCS 6305, pp. 42–49, 2010.

memory protection to ensure data integrity [2]. We have also demonstrated how real-time network performance can be modeled and how the tradeoff between optimization savings and optimization overhead can be quickly decided at run-time [3]. The work presented here builds directly on these, extendig the run-time analysis of function calls and presenting the result to the user as potential source code transformations.

Itzkovitz and Schuster [4] introduce the idea of mapping the same physical memory multiple times into the virtual address space of a process in order to reduce the number of page protection changes.

Keller *et.al* [5] have integrated Valgrind memory checking with Open MPI. This does datatype analysis and parameter validation, but has higher overhead than our method presented here. It also does no direct overlap analysis.

Danalis *et.al* have developed ASPhALT [6], a tool that does automatic transformation at the compiler level, but only for specific recognized problems.

Breakpad [7] is a crash reporter utility, which includes the tools we use for extracting uniform symbol tables from many different platforms.

1.2 Outline

Section 2 details our run-time method for communication analysis and performance measurements. Section 3 shows how this information is post-processed to predict wallclock savings from manual optimization. Section 4 shows an example of performance optmization results. Sections 5 and 6 present our future work and conclusions.

2 Instrumentation Method

We inject a small library into the application using *LD_PRELOAD*, which intercepts all MPI and memory allocation function calls the application performs. When the application allocates memory, it is allocated from a shared pool which is physically mapped twice in memory. This allows us to manipulate the memory protection seen by the application, while simultaneously having a private, non-protected view of the same physical memory.

For all function calls, we record a *context*. The context is the calling address, the parameters for the function and the three previous stack return addresses. This ensures that the *context* is fairly unique to a single execution point in the application, even if the MPI functions are wrapped inside other functions in the application. A series of related calls (such as multiple send/recieve in a border exchange phase) are called a *chain*. Our injected library keeps track of *contexts* and *chains* in memory.

2.1 Startup

On application startup, we perform a number of quick benchmarks. We time memory protection overhead for various memory sizes. On some architectures this is very cheap, while on others it can be prohibitively expensive. However, we need to know the magnitude of this overhead to give accurate speedup predictions.

We also benchmark the point-to-point bandwidth of all the nodes for transfers up to $512kB$ to have a baseline prediction of request transfer time between any two nodes. The transfer time is linear with transfer size beyond $512kB$ for most systems.

2.2   Sends and Receives

When an MPI send request is issued by the application, our library will intercept the call and protect the application-visible memory of the request with *read-only* protection. We then start the request as a non-blocking request, operating on our private, non-protected view of the memory. The memory protection will ensure that the application does not alter the request until the transfer is finished. We note the *context* of the call and append it to the current *chain*.

When an MPI receive request is issued by the application, the call is likewise intercepted, and the application-visible memory is protected *no-access*. The request is then started as a non-blocking request on our private view of the memory and, exactly as for read requests, we note the *context* and append it to the current *chain*.

Any number of send requests are allowed to operate in parallel. When a receive request is issued on pages which already have active requests, the request buffer address and data type are analyzed. If the requests only have overlapping pages, but no actual overlapping cells, the call is started non-blocking as above. This is especially common for non-contigous datatypes, which can have interleaving cells without actually sharing any. When requests do overlap, we mark this as the end of the current *chain*. This means that we wait for all non-blocking requests to finish, unprotect the associated memory and start this receive request as a non-blocking request. The request will begin a new *chain*.

Note that we also intercept collective communication functions. As long as their buffer area does not overlap a non-blocking request, they are performed but otherwise ignored. Non-blocking collective communication will unfortunately not be available until MPI 3.0.

2.3   Page Faults

If the application accesses memory a page fault will occur. This typically occurs at the end of a communication phase when the application accesses data to compute. We handle this page fault, checking if the faulting address is one we have protected. If it is not, we restore the original page fault handler and allow the crash mechanism of the application to handle it. If it is, we mark the end of the current *chain*, wait for all requests to finish and unprotect the memory before allowing the application to finish.

2.4   Other Function Calls

We also intercept OS function calls, such as file reads, socket sends etc. While no non-blocking improvement is made, we need to ensure that their buffer area does not overlap any request we have made non-blocking.

Similarily, we intercept functions such as *MPI_Pack()*, which operate on buffers with datatypes. This allows us to analyze if the actual memory cells overlap or just the pages. As in the above, overlapping cells indicate the end of the *chain*, whereas overlapping pages without overlapping cells are ignored.

2.5 Performance Measurements

If the same *chain* of *contexts* is observed multiple times, we start performing speedup measurements. This is done by alternating function calls between two modes.

For both modes, the memory for the entire *chain* is protected when the first request in the chain is seen, with the most restrictive access of any request in the *chain*. This ensures that there will, ideally, be only two page protection calls; at the start of the chain and at the end. If we have mis-detected the *chain* and the application does not follow the pattern we expect, our library will identify this on the next function call or page fault. It will then unprotect the memory and mark the *chain* as broken.

In the first mode, MPI function calls are performed exactly as they are originally written. This gives us the original communication time for a chain of requests. In the second mode, all functions are optimized to their non-blocking versions. This gives a real-world measurement of the effect of overlapping multiple communication requests. Both values are recorded as part of the *chain*.

We also measure the time between the end of a chain and the start of the next request to have an estimate of the compute time between each chain of requests.

2.6 Non-blocking Verification

If the application already uses non-blocking MPI function calls, the memory areas for these functions are protected similarly to the blocking calls. The memory is not unprotected until all the non-blocking operations have been *MPI_Wait()*ed for.

This allows our method to verify race conditions from non-blocking communication, which is important if our recommendations are implemented in the original application.

3 Post-processing and Presentation

Once the application calls MPI_Finalize, we start our analysis pass. The list of *contexts* and *chains* are stored in shared memory, and an external application is started to analyze them. The use of an external application allows the injected library to stay lightweight.

3.1 Chain Merges

We analyze *chains* which follow each other with little or no appliction computation between them. If the first chain ended because of a receive request that overlaps memory areas, we scan the second chain for any requests that do not overlap the terminating receive request's memory. If found, these are noted as code that could be reordered and moved into the first chain.

3.2   Expected Savings

All *chains* are analyzed for their expected wallclock time savings, which our measurements have hopefully revealed. For *chains* which have not been seen sufficiently many times, we have no relevant performance measurements, so we use the network measurements to estimate the non-blocking performance.

Each *chain*'s savings is then multiplied by the number of times it is called, and results are presented in the order of most savings first. By default, transformations need to yield at least a 5% overall speedup to be reported.

3.3   Code Parser

Our analysis tool uses the symbol files for the application to find the correlation between calling address and program source code line. A source code parser will then open the relvant source code file and read the corresponding lines.

With the exception of the request pointer parameter, most blocking to non-blocking transformations have identical parameters. Our analysis tool will declare an array of requests at the start, rewrite the blocking communication functions to non-blocking, and use *MPI_Waitall()* at the end of the *chain*.

It should be noted that our code parser is not a full syntax tree parser, and fails if it encounters macros that expand to commas or other unusual code constructs. In this case, it simply reports the speedup it achieved.

The "before" and "after" results are presented to the user along with an explanation of what the transform does.

3.4   Scalability Analysis

Our analysis tool will dump the *chains* to disc when it is done. It can optionally be informed about the base problem size, which will be noted in the file. A number of such analysis passes, with varying problem sizes, constitute an analysis set.

Once a sufficiently large set has been obtained, the analysis can be switched to projection mode. In this mode, the analyzer will work on theoretical problem sizes. This is done by curve fitting the data size of the various requests to the problem size, and similarily curve fitting the execution time of each compute phase to the problem size.

We can extrapolate the expected wallclock time for any problem size, and our analysis tool supports prioritizing improvement areas based on the extrapolated results rather than actual measurements. This allows quick benchmarking on small problem sizes in a few minutes, with the goal to improve wallclock time for large problem sizes which take hours or days to compute.

4   Results

The majority of development of the instrumentation was done on 2D and 3D code, but we've chosen to include a 1D border exchange here so the output will fit in the paper. Here is an example of output when applied to such an application:

Example of Analysis Output

```
/* Chain #2, seen 11713 times: 60.3us per chain, 0.7 sec total savings.
 * Please change line 72-78 from
 */

MPI_Sendrecv(& local[1 * g], g, MPI_FLOAT, prev, 0,
             & local[(1+1) * g], g, MPI_FLOAT, next, 0,
             MPI_COMM_WORLD, &a);

MPI_Sendrecv(& local[1 * g], g, MPI_FLOAT, next, 0,
             & local[0 * g], g, MPI_FLOAT, prev, 0,
             MPI_COMM_WORLD, &b);

/* to */

MPI_Request req_72[4];
MPI_Status status_72[4];
MPI_Isend(& local[1 * g], g, MPI_FLOAT, prev, 0, MPI_COMM_WORLD, &req_72[0]);
MPI_Irecv(& local[(1+1) * g], g, MPI_FLOAT, next, 0, MPI_COMM_WORLD, &req_72[1]);
MPI_Isend(& local[1 * g], g, MPI_FLOAT, next, 0, MPI_COMM_WORLD, &req_72[2]);
MPI_Irecv(& local[0 * g], g, MPI_FLOAT, prev, 0, MPI_COMM_WORLD, &req_72[3]);
MPI_Waitall(4, req_72, status_72);
*(&a) = status_72[1];
*(&b) = status_72[3];
```

We've trimmed the output slightly, removing the per-transfer scalability analysis as well as the marker for the code-line which triggered the end of chain. The

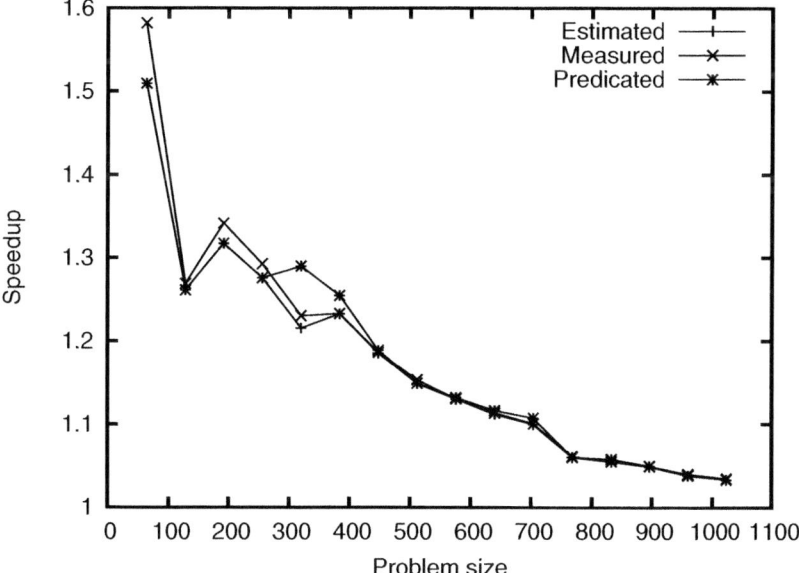

Fig. 1. Speedup predictions for the optimized code, measured optimized code and scalability-extrapolated results

transformed code aims to be functionally identical to the original code, including the assignment of MPI status variables.

Figure 1 shows the speedup of the original test program on various problem sizes, comparing the estimated and real speedup of the transformations. Also shown is the speedup for large problem sizes, estimated using only problem sizes $n \leq 256$. What is not clear from the graph is that the wallclock time predictor mispredicts the actual completion time for sizes of $n > 768$. These sizes are large enough that the computation no longer fits in L2 cache, which makes the computation phase more expensive. As this affects all implementations equally, the effect is not clear when looking only at speedup.

Please note that speedup of 2D and 3D cases are better, as there are more simultaneous requests which potentially reduce the effect of latency even more. Please see our previously published papers for optimization results of the underlying technique.

5   Current and Future Work

Our overlap analysis currently only does analysis for the problem size actually tried. There might be rounding or interpolation errors which cause overlap states to change as the problem size changes. Hence, while we know that the transformations presented to the user are safe for the specific problem size, we cannot guarantee they are safe for *any* problem size.

It would be interesting to do the entire analysis as a Valgrind module. This should enable more suggestions for code reordering. It would also allow us to switch from *chain* termination to waiting for only the request that is needed. This should allow larger overlap of computation and communication.

As it is, our code parser only works with C code. It would be interesting to extend this to use Open64 and hence work with a lot more languages.

The scalability predictor currently ignores the number of nodes, and assumes it will remain constant with the problem size as the only variable. This is naturally not the case for most applications, and it should be extended to cover runs both with varying number of nodes and varying problem sizes.

6   Conclusion

We have developed a method for run-time analysis of potential communication overlap improvements, with presentation of these to the user as transformations that are easily applied to their application source code. Each transformation includes the potential speedup of the application, an analysis of scalability, and also includes suggestions to replace the original code.

Acknowledgments. Thanks to NTNU and NOTUR for access to the computational clusters we have performed this work on. Thanks to Jan Christian Meyer and Dr. John Ryan for constructive feedback.

References

1. Message Passing Interface Forum: MPI: A Message-Passing Interface Standard, UT-CS-94-230 (1994)
2. Natvig, T., Elster, A.C.: Automatic and transparent optimizations of an application's MPI communication. In: Kågström, B., Elmroth, E., Dongarra, J., Waśniewski, J. (eds.) PARA 2006. LNCS, vol. 4699, pp. 208–217. Springer, Heidelberg (2007)
3. Natvig, T., Elster, A.C.: Using context-sensitive transmission statistics to predict communication time. In: PARA (2008)
4. Itzkovitz, A., Schuster, A.: MultiView and MilliPage – fine-grain sharing in page-based DSMs. In: Proceedings of the third USENIX symposium on operating system design and implementation (1999)
5. Keller, R., Fan, S., Resch, M.: Memory debugging of MPI-parallel Applications in Open MPI. In: Proceedings of ParCo 2007 (2007)
6. Danalis, A., Pollock, L., Swany, M.: Automatic MPI application transformation with ASPhALT. In: Parallel and Distributed Processing Symposium (2007)
7. Google: Breakpad - An open-source multi-platform crash reporting system, http://code.google.com/p/google-breakpad/

Efficient MPI Support for Advanced Hybrid Programming Models

Torsten Hoefler[1,*], Greg Bronevetsky[2], Brian Barrett[3],
Bronis R. de Supinski[2], and Andrew Lumsdaine[4]

[1] University of Illinois at Urbana-Champaign, Urbana, IL, USA
htor@illinois.edu
[2] Lawrence Livermore National Laboratory, Center for Applied Scientific Computing,
Livermore, CA, USA
{bronevetsky1,bronis}@llnl.gov
[3] Sandia National Laboratories, Albuquerque, NM, USA
bwbarre@sandia.gov
[4] Indiana University, Open Systems Lab, Bloomington, IN, USA
lums@cs.indiana.edu

Abstract. The number of multithreaded Message Passing Interface
(MPI) implementations and applications is increasing rapidly. We discuss how multithreaded applications can receive messages of unknown
size. As is well known, combining MPI_Probe/MPI_Recv is not thread-
safe, but many assume that trivial workarounds exist. We discuss those
workarounds and show how they fail in practice by either limiting the
available parallelism unnecessarily, consuming resources in a non-scalable
way, or promoting global deadlocks. In this light, we propose two funda-
mentally different efficient approaches to enable thread-safe messaging
in MPI-2.2: fine-grained locking and matching outside of MPI. Our approaches provide thread-safe probe and receive functionality, but both
have deficiencies, including performance limitations and programming
complexity, that could be avoided if MPI would offer a thread-safe
(stateless) interface to MPI_Probe. We propose such an extension for the
upcoming MPI-3 standard, provide a reference implementation, and
demonstrate significant performance benefits.

1 Introduction

Current processor trends are leading to an abundance of clusters composed of
multi-core nodes. While the Message Passing Interface (MPI [1]) remains a viable
programming model to use all processors in these systems, multi-core systems
naturally lead to increased use of shared memory programming models based on
threading. Hybrid MPI/threaded programs can decrease the surface to volume
ratio between MPI processes, which can result in more efficient use of the interconnection network [2]. Thus, these hybrid programs are becoming increasingly
common [3]. As a result, it is critical for MPI to support the model well.

---

* The first author performed most of this work at Indiana University.

R. Keller et al. (Eds.): EuroMPI 2010, LNCS 6305, pp. 50–61, 2010.
© Springer-Verlag Berlin Heidelberg 2010

MPI-2 includes a mechanism to request a level of thread support. Previously, most hybrid programs could conform to the MPI_THREAD_FUNNELED level. With the increase in hybrid programs, applications that use shared memory task parallelism and, thus, require MPI_THREAD_MULTIPLE support, are more likely. This trend not only motivates the implementation of that support [4] but also an examination of how well the MPI standard supports those programs. We find that the support is generally sufficient [5] although one glaring weakness exists: The semantics of probing for messages (e.g., in order to receive messages of unknown size) does not interact properly with realistic uses in threaded programs.

In this work, we discuss the issue of receiving messages of unknown size in multithreaded MPI programs. We explain the problem and show why obvious approaches to its solution are not feasible. We then discuss two elaborate techniques that would work with MPI-2.2. Despite the complex implementation of such techniques, which could be done in a library, we show that all proposed solutions limit performance significantly. Finally, we discuss an addition to the MPI standard that would enable the desired functionality. We describe a reference implementation, discuss issues in the context of hardware-optimized implementations, and present benchmark results which show the benefits of this approach.

2 Multithreaded MPI Messaging

We discuss several options for MPI version 2.2 to receive messages of unknown size in multithreaded environments. Unknown size messages in MPI are received with the sequence of *probe* (determine the size), *malloc* (reserve buffer), and *receive* (receive message). We investigate the issue of *false matching*, in which two threads perform a probe, malloc and a subsequent receive *concurrently*. Two actions happen concurrently if they happen completely independently (e.g., without synchronization or code flow dependencies) so that they could interleave in any way. Assume that two threads, A and B, perform a probe, malloc, and receive, denoted by A_p, A_m, A_r and B_p, B_m, B_r respectively. If those calls happen concurrently, then they could interleave as the series: $A_p, B_p, B_m, B_r, A_m, A_r$ that leads to incorrectly matching a message in thread B that was probed in thread A. We show that simple workarounds either limit parallelism unnecessarily or require structural changes to the application. Therefore, we propose two more sophisticated approaches and advocate for extensions or changes to the MPI standard to improve support for probing in threaded environments.

Separating Threads with Tags or Communicators. False matching could be avoided by using different virtual channels to address each thread in each process. A virtual channel in MPI is uniquely identified by the tuple (c, s, τ) (communicator, source, tag) on the receiver side and (c, r, τ) (communicator, receiver, tag) on the sender side. False matching can be avoided by using different tags (or communicators) for each thread. However, one would need $t \cdot p$ communicators (or tags) in order to address all threads in an MPI universe with p processes, each with t threads. This mechanism is not scalable (binds

$\Omega(p)$ resources) and not flexible enough for many applications. For example, a multithreaded master in a master/worker implementation can no longer use automatic load-balancing in which any idle thread probes and receives the next message to arrive. Similarly, a reentrant library that calls MPI with a variable (not predetermined) number of threads cannot use tag-based thread-addressing. Thus, such thread-addressing schemes seem unsuitable for most applications.

2.1   A Fine-Grained Locking Mechanism

Clearly, with MPI's matching semantics, coarse-grained locking (e.g., protecting the access to probe/malloc/recv at the communicator) overly limits parallelism. For example, a probe/receive pair with tag=4 and src=5 does not conflict with a probe/receive pair with tag=5 and src=5. However, another probe/receive pair with tag=4, src=5 would conflict with the first pair. Thus, we could lock each possible (communicator, source, tag) tuple separately. In the following, we assume that each lock is associated with a specific communicator and we limit the discussion to (source, tag) pairs.

One could arrange locks for (source, tag) pairs in a two-dimensional matrix. However, storing a $max(source) \cdot max(tag)$ matrix in main memory is infeasible. A sparse matrix representation with a hash table or map [(source, tag) $\rightarrow$ lock] seems much more efficient.

We show a simple locking strategy that minimizes the critical region with a nonblocking receive in Listing 1. However, this strategy does not cover wildcard receives.

```
lock map(src,tag)
probe(src, tag, comm, stat)
buf = malloc(get_count(stat)*sizeof(datatype))
irecv(buf, get_count(stat), datatype, src, tag, comm, req)
unlock map(src,tag)
wait(req)
```

Listing 1. Simple (limited) receive locking protocol

Probe/receive pairs with wildcards must be performed mutually exclusively within a set of channels. Thus, if a wildcard is used, we must lock a full row or column of the matrix. If both fields are wildcards, we must lock the whole matrix. As a result, we consider four (source, tag) cases in order to implement a fine-grain locking strategy: (1) (int,int), (2) (any_src,int), (3) (int,any_tag), and (4) (any_source,any_tag). We denote any_src or any_tag with an asterisk (*) in the following. In order to support each case fully, we need a sparse two-dimensional (src, tag) and thread-safe data structure with the following operations:

(un)lock(x,y) acquires/releases (x,y)
(un)lock(x,*) acquires/releases all entries on src x
(un)lock(*,y) acquires/releases all entries on tag y
(un)lock(*,*) acquires/releases the whole matrix

Our sparse two-dimensional locking protocol differentiates among these four cases, using three levels of locks: A two-dimensional map of locks for all points (source, tag), two one-dimensional maps of locks for each source and tag line, and one lock for the whole matrix. It uses lists of held locks per (source, tag) pair, for each source and each tag and for the whole matrix. Listing 2 shows a possible algorithm that implements a sparse two-dimensional locking structure. The code shown in Listing 2 is a critical region that is protected with locks itself!

```
if (source != MPI_ANY_SOURCE and tag != MPI_ANY_TAG)
  check if either whole matrix, source, tag, (source, tag) is locked
  if (nothing is locked)
    lock (source, tag) and increase usage count of source, tag, matrix
if (source != MPI_ANY_SOURCE and tag == MPI_ANY_TAG)
  check if either whole matrix or source is locked
  check if any_source or some tag for source is in use
  if (nothing is locked/used)
    lock source and increase usage count of source and matrix
if (source == MPI_ANY_SOURCE and tag != MPI_ANY_TAG)
  check if either whole matrix or tag is locked
  check if any_tag or if some source for tag is in use
  if (nothing is locked/used)
    lock tag and increase usage count of tag and matrix
if (source == MPI_ANY_SOURCE and tag == MPI_ANY_TAG)
  check if whole matrix is locked or in use
  if (nothing is locked/used)
    lock matrix
```

Listing 2. Function to lock the 2d_sparse_map. Unlock is equivalent

However, while this local locking scheme ensures correct and parallel message reception, it can unexpectedly influence global synchronization. For example, rank 0 sends two messages to rank 1 in which sending of the second message depends on a reply to the first message. The first message has tag 1, and the second message has tag 2. The receiver, rank 1, has two threads A and B. Thread A receives from channel (0, 2) and thread B from channel (0, any_tag). Thread A sends the needed reply after the message is received. We show pseudo-code for rank 0 in Listing 3 and for rank 1 in Listing 4.

```
A:                                A:
  send(..., 1, 1, comm)             probe/recv(0, 2, comm)
  recv(..., 1, 1, comm)           B:
  send(..., 1, 2, comm)             probe/recv(0, ANY_TAG, comm)
  ...                               send(..., 0, 1, comm)
```

Listing 3. Rank 0   **Listing 4.** Rank 1

This program must terminate in a correct MPI implementation that supports MPI_THREAD_MULTIPLE. However, if A locks (0, 2) first and enters MPI_Probe then B cannot lock (0, any_tag). Thus, ranks 0 and 1 cannot proceed and the presented algorithm can cause spurious deadlocks.

In general, a receive with an explicit (integer) source and tag can block ones with wildcards, for example, receiving on channel (0, 1) blocks receives on (any_src, 1), (0, any_tag), and (any_src, any_tag). Thus, wildcard probes and receives must dominate more specific ones, which requires that MPI_Probe has not yet been called for the more specific one. Since MPI calls cannot be aborted, we must poll with multiple probes/receives. Only the most general probe/receive (any_src, any_tag) is allowed to block. We can implement the required polling with the same two-dimensional locking scheme to enable maximum concurrency. Listing 5 shows the polling (nonblocking) algorithm.

```
while(!stat)
  lock 2d_sparse_map(src,tag) /* see previous listing */
  iprobe(src, tag, comm, stat)
  if(stat)
    buf = malloc(get_count(stat)*sizeof(datatype))
    irecv(buf, get_count(stat), dtatype, src, tag, comm, req)
  unlock 2d_sparse_map(src,tag)
  if(stat) wait(req)
```

Listing 5. Polling receive locking protocol

We note that requiring polling is a fundamental problem that prevents an efficient implementation of many multithreaded implementations.

Most parallel MPI applications only use a subset of the possible parameter combinations during a program run. For example, an application might not use any_src or any_tag at all, which enables the use of the simple locking scheme described in Listing 1. Other applications might use any_tag in all probes and receives, and enable a much simpler, one-dimensional locking of source (even though this limits possible parallelism).

Table 1 lists all combinations and possible optimizations. An x in the column any_src or any_tag means that any_src or any_tag is used during the program run. An x in "direct" indicates that at least one call does not use any_src and any_tag.

Table 1. Possible parameter combinations

Scenario	any_src	any_tag	Specific	Strategy
1	-	-	x	simple 2d, blocking
2	-	x	-	simple 1d, blocking
3	-	x	x	2d lock, polling
4	x	-	-	simple 1d, blocking
5	x	-	x	2d lock, polling
6	x	x	-	2d lock, polling
7	x	x	x	2d lock, polling

For example, under scenario 4, all calls use any_src as an argument and thus a simple one-dimensional locking scheme can be used. Scenario 7, the most general one under which a program run could use all combinations, requires the polling scheme (Listing 5). Different scenarios can be defined for each communicator. Thus, performing all calls with various wildcards on distinct communicators simplifies locking requirements but might lead to other problems as discussed in the introduction.

Further, although most applications only use a subset of the possible parameter combinations, which allows for a specialized implementation, a library-based solution must provide the general implementation. Similarly, an implementation of language bindings such as MPI.NET [6] must assume the general case (scenario 7) so using the fine grained locking approach likely entails a high cost.

2.2   Matching Outside of MPI

If polling is infeasible, we can instead perform MPI source/tag matching outside of the MPI library in order to provide correct threaded semantics for MPI_Probe. This solution uses a helper thread that repeatedly calls MPI_Probe with any_src and any_tag. When MPI_Probe returns, the thread allocates a message buffer, into which it then receives the probed message with MPI_Irecv. The associated MPI_Request is stored in a data structure for use when an application thread issues a matching receive operation. This data structure is similar to the two-dimensional locking structure from Listing 2. For each (source, tag) pair (including wildcards) it maintains the count of threads that are waiting to receive a messages with that pair as well as two lists of messages. The first list tracks "expected" messages – newly arrived messages that match this (source, tag) pair and have been matched to waiting threads but not yet been picked up by those threads. The second list tracks "unexpected" messages – newly arrived messages for which a receiver thread has not yet been identified. Since all such messages match four different (source, tag) pairs (including wildcards), each is placed into four such lists, one for every pair. The data structure maintains a lock and a condition variable for each pair to synchronize access to the count and the message lists.

This method can also take advantages of the previously described matching lock mechanism. However, it requires implementation of the complete matching semantics in a thread-safe way (including thread synchronization) on top of MPI and introduces additional buffering, which is clearly suboptimal. The implementation would also require eager and rendezvous protocols for performance reasons and would also lose potential optimizations such as matching in hardware. Thus, such an implementation is highly undesirable from a user's perspective.

3   Extending the MPI Standard: Matched Probe

We have discussed issues with tread-safe matching in MPI and pointed at a problem in the specification. We have shown that all simple workarounds are either infeasible or incorrect (deadlock). Although our two mechanisms support correct semantics of MPI_Probe in threaded environments, they are nontrivial and limit either performance or concurrency significantly in the general case (any_src and any_tag possible). Thus, a general library implementation, as is required for new language bindings, cannot limit those scenarios and must pay the cost of general support. Further, both mechanisms duplicate work that an MPI implementation performs internally and limit hardware offload capabilities.

For these reasons, we must modify the MPI standard to eliminate the need to use these mechanisms, which would entail deprecating the existing probe operations. One possible solution would replace those operations with thread-safe versions that return a request that the application can later complete (in the original thread or not, but under application control) [7]. While in MPI-2.2, matching is done in probe and then again in receive, we decouple matching and receiving. We propose to add two new calls, mprobe and mrecv (and their non-blocking versions) to the MPI standard. We sketch the proposal here; a detailed version is available elsewhere [7]. The proposed *mprobe* returns a **message handle** that identifies a message (which is then unavailable in any other matching context). The proposed *mrecv* can then receive such a matched message. Listing 6 shows an example for thread-safe matching with a matched probe.

```
MPI_Message msg; MPI_Status status;
/* Match a message */
MPI_Mprobe(MPI_ANY_SOURCE, MPI_ANY_TAG, MPI_COMM_WORLD, &msg, &status);

/* Allocate memory to receive the message */
int count; MPI_get_count(&status, MPI_BYTE, &count);
char* buffer = malloc(count);

/* Receive this message. */
MPI_Mrecv(buffer, count, MPI_BYTE, &msg, MPI_STATUS_IGNORE);
```

Listing 6. Matched probe example

This mechanism reduces the user burden and minimizes the total number of locks required. We also enable efficient hardware matching and eager protocols. We discuss an implementation and possible issues in the following.

3.1 A Reference Implementation of Matched Probe

The matched probe proposal has been implemented as a proof of concept using Open MPI. Open MPI provides two mechanisms for message matching: One in which matching occurs inside the MPI library (used with network APIs such as Open Fabrics, TCP, and shared memory) and one in which matching occurs either in hardware or in a lower-level library (used with network APIs such as Myrinet/MX and Portals). The implementation of matched probe presented in this paper is based on MPI-level message matching. Issues with hardware level matching are discussed in Section 3.2.

The matched probe implementation does not significantly change the message matching and progression state machine of Open MPI. It adds an exit state from

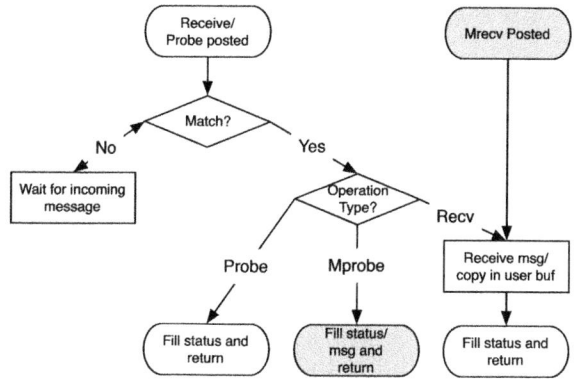

Fig. 1. High-level state diagram of MPI receive matching

message matching (MPROBE in addition to PROBE and RECV), and adds an entry point back into the state machine. Open MPI tracks all unexpected messages (those that the matched probe operation can impact) as a linked list of message fragment structures, which includes source and tag. Communicators are separate channels and use separate lists. The list of unexpected messages is walked in an identical fashion for a probe and a receive. However, the fragment is removed from the unexpected message list and processed in the receive case.

In the case of matched probe, the message fragment is removed from the unexpected message list (similar to a receive). It is then stored in the MPI_Message structure returned to the user. When the user calls MPI_Mrecv or MPI_Imrecv, the message fragment is retrieved from the MPI_Message structure that the user provided and the normal receive state machine is started from the point right after message matching.

3.2 Low-Level Message Matching

The previously described implementation of Matched Probe for MPI-level matching, while straightforward, will not work if the lower level communication API provides message matching (such as Portals on the Cray XT line, Myrinet/MX, TPorts on Quadrics, and PSM on Qlogic). In these cases, the message matching state engine is not exposed to the MPI implementation, and may be executed on NIC hardware. In these cases, we must extend the interface of the lower-level API to support Matched Probe, likely with an implementation of similar complexity to the Open MPI implementation. Likewise, firmware based hardware matching (TPorts and Accelerated Portals), adding entry points out-of and back in-to the firmware state machine should be straightforward.

Hardware assisted matching presents a more complicated situation. Hardware designs would require modifications to support a matched probe. In addition, carrying the extra state to restart the state machine for a partially matched message could be cumbersome in hardware. However, since these designs are not in use, such designs have no bearing on the practical cost of this MPI extension.

Thus, adoption of our extension requires a trade off between the benefits of making future designs of this type compatible our extension.

4   Performance Evaluation

We use two benchmarks that assess the performance and concurrency of the different mechanisms for thread-safe message reception. Both benchmarks and the two-dimensional locking (Section 2.1) are integrated in the publicly available Netgauge tool [8]. The benchmarks were run on Sif at Indiana University. Sif consists of Xeon L5320 1.86 GHz CPUs with a total of 8 processing cores per node running Linux 2.6.18 connected with Myrinet 10G. We used Open MPI revision 22973[1] using the TCP transport layer, configured with `--enable-mpi-thread-multiple`.

4.1   Receive Message Rate

Our first benchmark compares the message receive rate at a multithreaded receiving process with two-dimensional locking (2D, cf. Section 2.1) and matching outside MPI (OUT, cf. Section 2.2) for MPI-2.2 and the new matched probe (MPROBE, cf. Section 3) mechanism. In this test, 8 processes send to process 0, which uses 8 threads to receive the messages. Each process i sends its messages with tag i and each thread j either receives messages from process $j + 1$ or any_src, with tag $j + 1$ or any_tag.

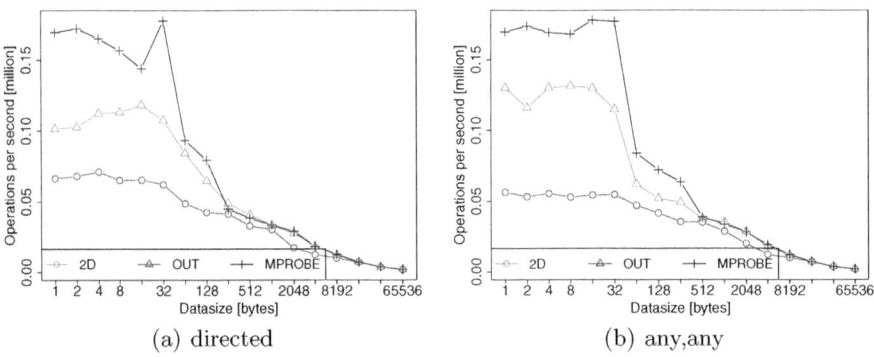

(a) directed   (b) any,any

Fig. 2. Message rate of different options for Open MPI on Sif

Figure 2 shows the different message rates achieved by the two locking schemes and wrong matching with 8 processes sending to 8 threads on process 0. Figure 2(a) shows results for *directed* (i.e., neither any_src nor any_tag) and Figure 2(b) shows *any* (any_src and any_tag). The OUT and 2D implementations exploit knowledge of which wildcard pattern (any,any or directed) to expect (cf. Table 1). Both figures show significant performance differences between the

---

[1] Available at: http://svn.open-mpi.org/svn/ompi/tmp-public/bwb-mprobe

approaches for small message-sizes. The rate of larger messages is bandwidth-bound and thus similar for all approaches. The two-dimensional locking scheme is faster than the matching outside of MPI, which must copy each message. However, our matched probe implementation outperforms both approaches and achieves the highest message rates.

4.2   Threaded Roundtrip Time

Our threaded roundtrip time (RTT) benchmark measures the time to transmit n messages between two processes with t threads each. It is thus somewhat similar to the overhead benchmark proposed by Thakur et al. [9]. Process 0 synchronizes its t threads with `pthread_barrier_wait` before each thread $j \in \{0..t-1\}$ sends n messages with tag j to process 1. The t threads at process 1 receive and send n messages from/to process 0 and each thread in process 0 receives n messages. The receives either use a specific tag $j \in \{0..t-1\}$ or any_tag and a specific source $s \in \{0,1\}$ or any_src.

Figure 3 shows the latency overhead of the different locking schemes. For the any,any case in Figure 3(a), the current implementation of MProbe results in higher latency than both the 2D locking and matching outside MPI schemes. Latency increases mainly due to 2d-locking and outside MPI locking only using a single lock (cf. Table 1) based on the knowledge that only any,any receives are used while the matched probe implementation in MPI must handle the general case. As an aside, this example demonstrates the potential of additional info objects in MPI in which users could specify such constraints.

Figure 3(b) shows the latencies for the directed case (using integer tag and src values). For small messages, Mprobe is faster than 2-d locking due to the explicit removal from the queue (it only needs to be locked once). The outside MPI version is even faster for small messages because it receives the messages immediately and the copy overhead is low. However, for large messages, the copy overheads are dominating.

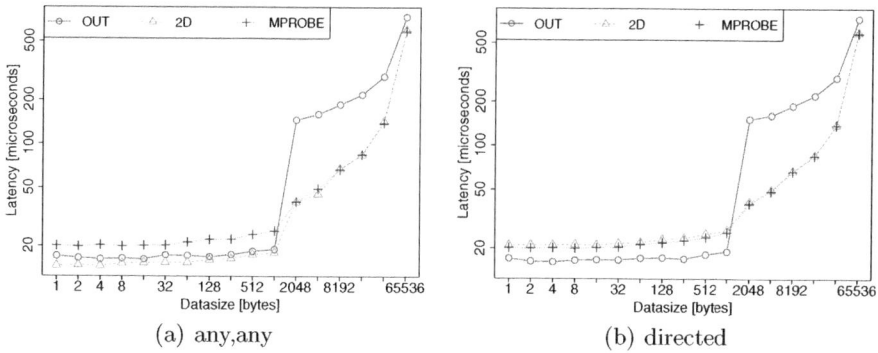

(a) any,any   (b) directed

Fig. 3. Latency of different options for Open MPI on Sif

In Open MPI itself, there are two sources of unnecessary latency in the current MProbe implementation that we could remove with further development: creation of an additional request structure and an additional call to the progress engine. The extra request structure results in a small overhead, approximately 10 ns. The significant latency hit is from the additional progress engine calls, which could probably be mitigated through additional optimization. The issue is exacerbated by the high cost of entering Open MPI's progress engine when multithreaded support is enabled.

5   Summary and Conclusions

In this paper we describe the problem of receiving messages of unknown size in threaded environments. We show that often assumed simple solutions to the problem either introduce significant overheads or may lead to spurious deadlocks. We propose two advanced protocols to solve the problem in MPI-2.2. However, both protocols add various overheads to the critical paths. We then propose an extension of the MPI-3 standard that solves the matching problems. We show a reference implementation in Open MPI and discuss issues that might arise in hardware implementations.

Our performance analysis shows the benefits with regard to the message rates of the matched probe approach over the other protocols. We also analyze latencies for a multithreaded ping-pong benchmark. This analysis demonstrates that protocols on top of MPI can take advantage of special domain knowledge (only any,any calls), which serves as another good example for adding user assertions to the MPI standard.

Acknowledgments. The authors thank all members of the MPI Forum that were involved in the discussions about matched probes, Douglas Gregor (Apple), and the anonymous reviewers. Sandia National Laboratories is a multiprogram laboratory operated by Sandia Corporation, a Lockheed Martin Company, for the United States Department of Energy's National Nuclear Security Administration under contract DE-AC04-94AL85000. This work was partially performed under the auspices of the U.S. Department of Energy by Lawrence Livermore National Laboratory under Contract DE-AC52-07NA27344. (LLNL-CONF-434306).

References

1. MPI Forum: MPI: A Message-Passing Interface Standard. Version 2.2 (2009)
   http://www.mpi-forum.org/docs/mpi-2.2/mpi22-report.pdf
2. Itakura, K., Uno, A., Yokokawa, M., Ishihara, T., Kaneda, Y.: Scalability of Hybrid Programming for a CFD Code on the Earth Simulator. Parallel Comput. 30, 1329–1343 (2004)
3. Rabenseifner, R.: Hybrid Parallel Programming on HPC Platforms. In: Proc. of the Fifth European Workshop on OpenMP, EWOMP 2003, Aachen, Germany (2003)

4. Gropp, W.D., Thakur, R.: Issues in Developing a Thread-Safe MPI Implementation. In: Mohr, B., Träff, J.L., Worringen, J., Dongarra, J. (eds.) PVM/MPI 2006. LNCS, vol. 4192, pp. 12–21. Springer, Heidelberg (2006)
5. Balaji, P., Buntinas, D., Goodell, D., Gropp, W., Thakur, R.: Toward Efficient Support for Multithreaded MPI Communication. In: European PVM/MPI Users' Group Meeting, pp. 120–129 (2008)
6. Gregor, D., Lumsdaine, A.: Design and Implementation of a High-Performance MPI for C# and the Common Language Infrastructure. In: Proceedings of PPoPP 2008, New York, NY, USA, pp. 133–142 (2008)
7. Gregor, D., Hoefler, T., Barrett, B., Lumsdaine, A.: Fixing Probe for Multi-Threaded MPI Applications (Revision 4). Technical report, Indiana University (2009)
8. Hoefler, T., Mehlan, T., Lumsdaine, A., Rehm, W.: Netgauge: A Network Performance Measurement Framework. In: Perrott, R., Chapman, B.M., Subhlok, J., de Mello, R.F., Yang, L.T. (eds.) HPCC 2007. LNCS, vol. 4782, pp. 659–671. Springer, Heidelberg (2007)
9. Thakur, R., Gropp, W.: Test Suite for Evaluating Performance of MPI Implementations That Support MPI_THREAD_MULTIPLE. In: Cappello, F., Herault, T., Dongarra, J. (eds.) PVM/MPI 2007. LNCS, vol. 4757, pp. 46–55. Springer, Heidelberg (2007)

An HDF5 MPI Virtual File Driver for Parallel In-situ Post-processing

Jerome Soumagne[1,3], John Biddiscombe[1], and Jerry Clarke[2]

[1] Swiss National Supercomputing Centre (CSCS), SCR Department,
Galleria 2, Via Cantonale, 6928 Manno, Switzerland
[2] US Army Research Laboratory (ARL), CIS Directorate,
Aberdeen Proving Ground, MD, USA
[3] INRIA Bordeaux Sud-Ouest, HiePACS Research Team,
351 cours de la Liberation, 33405 Talence, France

Abstract. With simulation codes becoming more powerful, using more and more resources, and producing larger and larger data, monitoring or post-processing simulation data *in-situ* has obvious advantages over the conventional approach of saving to – and reloading data from – the file system. The time it takes to write and then read the data from disk is a significant bottleneck for both the simulation and subsequent post-processing. In order to be able to post-process data as efficiently as possible with minimal disruption to the simulation itself, we have developed a parallel virtual file driver for the HDF5 library which acts as an MPI-IO virtual file layer, allowing the simulation to write in parallel to remotely located distributed shared memory instead of writing to disk.

Keywords: Distributed Shared Memory, Parallel Systems, Large Scale Post-processing, Virtual File Layer.

1 Introduction

The HDF5 library [1] already provides the user with several different file drivers, which are the core pieces of code responsible for the transfer of user controlled memory onto disk. They act as an abstraction layer between the high level HDF5 API and the low level file system API. The drivers provided by the HDF5 package include the `core` (memory based), `sec2` (posix compliant serial IO), `mpio` (parallel file IO) and `stream` drivers. The `stream` driver [3] has been created for the purpose of providing live access to simulation data by transfer to remote grid servers or via sockets to a waiting application. The `mpio` driver uses an HDF5 layer on top of MPI-IO to write data in parallel to the file system – our driver emulates this behaviour but instead routes the data in parallel to a Distributed Shared Memory (DSM) buffer over multiple TCP connections using either an MPI or a socket based protocol. Compared to other systems such as ADIOS [11] which defines a common API interface to a number of IO libraries (including HDF5), we instead use the existing HDF5 interface, making it possible for a large number of existing applications to switch to our framework with a simple (one line) code change and link to our `H5FDdsm` library.

R. Keller et al. (Eds.): EuroMPI 2010, LNCS 6305, pp. 62–71, 2010.
© Springer-Verlag Berlin Heidelberg 2010

The original DSM implementation (upon which this work is based), referred to as the Network Distributed Global Memory (NDGM), was created by Clarke [6], it was used for CFD code coupling between applications modelling fluid-structure interactions [8] using very different models (and hence partitioning schemes) to represent that domain. Since the DSM can be considered as a flat memory space, one of the principle advantages that it provides is that coupled simulations do not need to be aware of the parallel domain decomposition used by the other partners in the simulation/analysis. Separate codes may write their data using any HDF5 structures suitable for the representation, providing the other coupled processes are able to understand the data and read it with their own partitioning scheme. This effectively abstracts the data model used by either partner away and leaves the HDF5 API as the mediator between the coupled applications. The original NDGM implementation supported the transfer of data between processes using only a single channel of serial MPI based traffic and therefore had a limited capacity. Our new DSM based virtual file driver (VFD) allows very high speed *parallel* transfer of data directly between coupled simulations, or a simulation and a post-processing application such as ParaView [10] – for which we have created a custom plugin which allows full control of the visualization of live data. Our design is intended to address three principal objectives: require minimal modification to existing HDF enabled simulation codes; be portable enough to allow use on the widest range of systems; provide excellent performance.

2 Architecture

The distributed shared memory (DSM) model operates such that by using the common HDF5 API, data is transparently sent (across the network) to the DSM which is distributed among several nodes and is seen by the simulation writing data as a uniform memory space (figure 1). The current implementation allocates an identical memory buffer on each participating DSM host/server node, however, this is simply a reflection of the fact that the systems used for hosting consist of homogeneous arrays of nodes and is not a fundamental restriction.

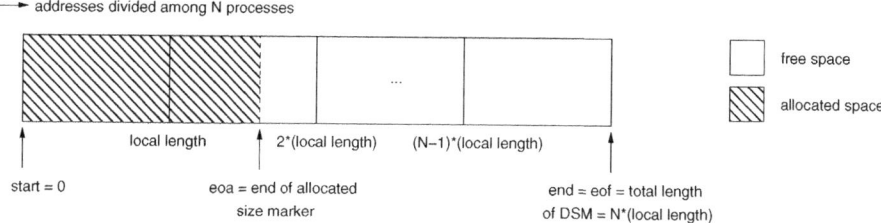

Fig. 1. DSM overview : The DSM memory spans the hosting nodes and is divided equally among them, as the simulation requests space to write datasets into, the allocated space grows to fill the available memory. Once the host process has read the contents, the space may be freed by wiping the objects from the memory.

2.1   A Streaming MPI Virtual File Layer

The DSM driver itself is derived from the `core` and the `mpio` drivers. The `core` driver as defined in the HDF5 library allows files to be written in memory instead of disk, whilst the `mpio` driver uses MPI-IO to write files in parallel and gives support to the user for collective/independent IO operations – as defined in the MPI specification. Since HDF5 metadata must be synchronized across processes, it is necessary to have collective calls for all operations which modify metadata. These include calls to file, group and dataset creation, during which, processes share information about the position, size and layout of datasets within the file. The actual writing of data may be done independently from any process since (providing processes are not reading back data written by other nodes) the sending of information into the file has no effect upon the other nodes.

2.2   Operating Modes

The DSM uses a client/server model where the simulation writing the data is the client and the nodes receiving the data are the server. However, the DSM VFD itself is linked directly to the HDF5 library in the form of an object which is capable of operating either as client or server depending solely on a single configuration flag. This means that there is considerable flexibility in how the DSM is managed – depending on how the DSM is configured, the storage and transfer of memory may take several forms:

1. The DSM memory is allocated on the remote machine which is receiving the simulation data; the receiving machine may be a visualization cluster with *fat* memory nodes intended for exactly this kind of purpose. When data is written, it is transferred across the network and the read process on the remote machine operates directly on the local DSM network. This *is* the default operating mode.
2. The DSM memory buffer resides on the same machine as the simulation (thereby increasing memory usage on that machine); data is written immediately into memory and transfers across the network are only initiated when the remote machine makes read requests.
3. Both simulation and DSM may reside on the same machine; this means that only a single network is traversed and the DSM is used as a convenient means of interfacing the partner codes. In this configuration, the DSM is symmetric with data transfer into the DSM initiated from the simulation and read by the coupled process.

It is important to note that whether the DSM operates as a client or server in either of the above operating modes, is transparent and (generally) unknown to the actual applications running. They only see an HDF5 file and perform write/read operations exactly as if they were using a local disk system and have no knowledge of the internal/external transfers which are taking place underneath.

2.3   Communication

We define two communicators, one called `intracommunicator` which handles data locally between processes defining the DSM, and the other called `intercommunicator` which handles the transmission of data over the network between the coupled applications. The `intracommunicator` will usually be the same as the `MPI_COMM_WORLD` for whichever application is hosting the DSM, though if the application uses only a subset of nodes for IO, it will be the communicator defined by those nodes. The `intercommunicator` however, can be created either by joining the (IO involved) communicators of the connected applications, or by the use of another protocol such as a socket based connection.

2.4   Synchronization

As with any distributed shared memory system, processes involved in the distributed memory interface need to share information and therefore, all write and read operations need a synchronization mechanism in order to keep coherency of the memory. When the simulation has finished writing data, it may send a signal, via the DSM VFD driver to the remote application to say that transfer is complete – however, though this is permitted in our implementation, it breaks compatibility with the simple HDF5 based API within which we wish to operate. We have therefore added hooks into certain HDF5 API calls which can be used to signal events to the DSM. Specifically, we note that the majority of applications with which we are working, compute time steps of data in discrete steps and after writing the results to disk, close the file before beginning the next operation. We therefore intercept calls to `H5Fclose` and the driver sets a flag inside the DSM (using the `intercommunicator`) that can be read to indicate that the data within the DSM is ready and valid. Once the host DSM begins read operations, further writes can be blocked until the flag is cleared and the DSM is available again. In this way, the simulation writes data, resumes computation and providing the coupled application does not block further writes, overlapped processing may take place within each application.

2.5   Distribution Strategies

It is evident from figure 1, that a simple allocation of space using a flat/linear memory model requires a careful matching of the size of the DSM buffer created to the expected amount of data sent into it. A very large DSM buffer combined with small data transfers will result in the majority of traffic being routed to a single low rank node on the DSM server. One can see from figure 2 that as the amount of data sent to the DSM increases and the number of nodes being utilized rises, the bandwidth increases accordingly, this is discussed further in section 4. The current implementation makes use of fixed size buffer allocations on each DSM node, however we intend to implement a more sophisticated allocation making use of virtual addresses to data offsets, which will allow datasets to be placed on different nodes even if they would normally fit on the first node.

In fact if the HDF5 chunking interface is used (whereby datasets are broken into pieces prior to their allocation within the file), a mechanism already exists to achieve exactly this goal, and we shall implement different methods for allocation of chunks between host nodes by manipulating the chunk start addresses individually.

3   Implementation

The implementation of the driver is designed to place the majority of the burden of operations on the DSM host. The default mode of operation is that the simulation is client and has no additional computational or other burden placed upon it, other than the send of data to the remote DSM which takes the place of the send to disk. The DSM host however must perform servicing operations and handle data requests for write and read operations.

3.1   DSM Service Thread

The request mechanism is implemented using posix threads, all the nodes hosting the DSM have two threads. The first thread is the normal application thread which will perform whatever computation of post-processing is required. The second "service thread" receives IO requests and responds to them within the VFD/DSM buffer code itself. On first initialization, the DSM service thread enters a wait state where it remains until the simulation VFD is first initialized, at this point a connection is established between the two codes and an inter-communicator is created to handle the parallel transfer of data between nodes. Upon reception of a (remotely issued) data write, the DSM service thread places the data into memory. Once the simulation has finished writing, either by issuing an H5Fclose or a specific IO completed call (via the DSM API), the DSM sets its internal state flag to indicate *data ready* and allows IO requests from the local application. During local reads, remote access is blocked and any incoming data writes will be delayed until reading has completed. The DSM therefore performs a ping-pong type operation between remote and local machines as it handles requests. Asynchronous parallel read/write modifications are not permitted, though this would be possible if both local and remote applications were sharing an MPI communicator and the file was opened and modified collectively by both applications (this would be infeasible for the majority of cases since collective operations require knowledge about the specific data which one can assume would not be worthwhile to exchange).

MPI intercommunicator The MPI inter-communicator is intended to be used when the simulation code and the DSM are on the same machine, or on machines that have compatible MPI process managers. To establish the connection between the applications we make use of the MPI_Comm_connect, MPI_Comm_accept set of functions. Unfortunately, some large machines such as the Cray XT^{TM} series or IBM Blue Gene® are unable to use this communicator as they do not support the *spawn* set of functions defined in the MPI specification. For the same reason, the MPI_Comm_join function which also allows applications to share communicators cannot be used on these machines.

Sockets intercommunicator To allow our driver to be used on machines without full MPI support, we have introduced the socket intercommunicator which uses a single socket to initialize the connection between both applications, then creates additional sockets to link every node of one application to every node of the other. Many OS implementations currently limit the number of open socket connections to around 1024, placing a (configurable) limit on the number that can be maintained at any given time. A future solution will be to manage connections created on the fly. The main advantage currently given by this intercommunicator is that it does not depend on the MPI implementation used within the connected applications, therefore it is possible to create connections between any combination of cluster or machine.

Additional intercommunicator modules may be developed and added to the driver, which will allow future versions to take advantage of other low level connection libraries - such as those used in recent MPI distributions which are optimized for very large numbers of processes and hybrid shared/distributed memory architectures. Portability to any machine requires only the inclusion of a communication module suitable for the target platform.

3.2   Configuration of the DSM

When writing in parallel using the HDF5 API, it is necessary to select a parallel file driver from within the application code, the only one currently available is the mpio driver, which would normally be selected by setting a file access property list using the function:

```
herr_t H5Pset_fapl_mpio(hid_t fapl_id, MPI_Comm comm,
    MPI_Info info)
```

To use the dsm driver from C (1) or Fortran90 (2):

```
(1) herr_t H5Pset_fapl_dsm(hid_t fapl_id, MPI_Comm comm,
        void *dsmBuffer)
```

```
(2) h5pset_fapl_dsm_f(prp_id, comm, hdferr)
        INTEGER(HID_T) prp_id
        INTEGER comm, hdferr
```

Setting the dsmBuffer variable to NULL in (1), or using (2) instruct the H5FDdsm library to auto-allocate and manage a singleton DSM for the user. The communicator comm is used as the intercommunicator. If the DSM is configured by the user application, then it may be supplied as the dsmBuffer parameter (currently C interface only). If the connected processes share a file system, the DSM server can write all configuration details to a location specified in a H5FD_DSM_CONFIG_PATH environment variable; settings are then picked up automatically by the client process on initialization. Parameters which can be configured (either via the configuration file or in code) are:

1. If host, how much DSM memory should be allocated per node (or in total);
2. If using MPI intercommunicator, the port name given by MPI_Comm_accept;
3. If using socket intercommunicator, the host and port number used in the socket binding operation.

4   Performance

To measure write bandwidth and to emphasize the current distribution strategy, the DSM is fixed at $5GB$ each on 8 post-processing host nodes – giving a total memory size of $40GB$. The number of processes writing to the DSM is varied (from 1 to 32), and a variable size of dataset is written using simple lists of double precision particle positions $N \times \{x, y, z\}$, ranging from $1 \cdot 10^3$ up to $5 \cdot 10^7$ particles per node. The maximum data size is given by $32 \times 3 \times 8 \times 5 \cdot 10^7 = 36GB$. Results are shown in figures 2 using MPICH2 between two clusters on an SDR Infiniband link. Also shown is the result of 64 nodes writing $2.5 \cdot 10^7$ particles ($36GB$).

As mentioned in section 2.5, the current performance is representative of the linear distribution strategy used. With the DSM size fixed, writing a small amount of data sends everything to the first process (low address range). Matching the DSM size to the amount of data to be written makes use of the network to all DSM nodes. The right side of figure 2 shows that when the full $36GB$ of data is written the nodes show a relatively even distribution of data – and this is when the maximum bandwidth is achieved – corresponding to the highest points of figure 2 (left). In this test the 8 nodes used are connected to the host application via an SDR Infiniband switch with individual links rated at $10Gb/s$, and we obtain a peak bandwidth close to $3GB/s$.

For a second test, matching the DSM size to the data written, the DSM performs much better than a parallel file system (GPFS or Lustre). Figure 3 shows the performance achieved on two different types of machine with different communication systems. Generally speaking, the more DSM nodes used (and hence network links), the higher the overall bandwidth. Using MVAPICH2 on a QDR Infiniband interface (therefore using RDMA) for the inter-communication gives better results than the socket inter-communicator. On the other hand, as shown in figure 3 (right), using a socket inter-communicator on a Cray XT5 machine scales well as more nodes are used to saturate the network. In fact both machines used for the test are multi-core machines, which complicates the relationship between node counts writing and receiving data. In this test, the receiving nodes were allocated exclusively (8 receive links active), but sending processes used 12 cores per node, so maximum traffic would not be expected until much more than $8 \times 12 = 96$ nodes are sending. The increasing performance mirrors existing results from HDF5 studies [5, 12], where bandwidth continues to increase until the network is saturated – which depends, in the case of DSM traffic, on the number of links to host nodes available. A parallel file system on

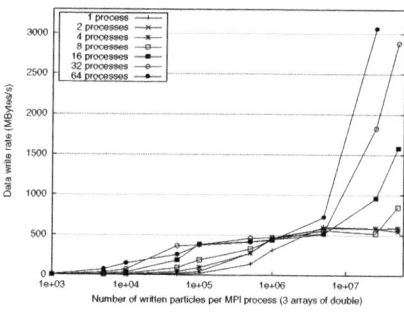 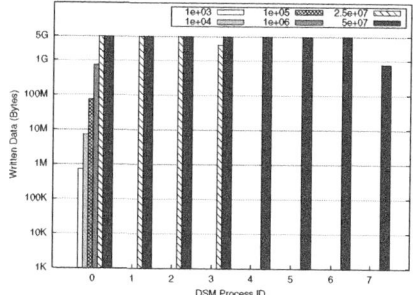

Fig. 2. (Left) Write speed test from a regular cluster to DSM distributed among 8 post-processing nodes with a fixed DSM buffer size of $40GB$. (Right) The amount of data received on each of the 8 DSM host nodes is shown, when 32 processes write N particles – the high rank nodes are only active with the largest data writes.

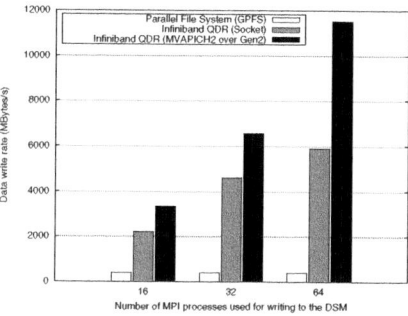 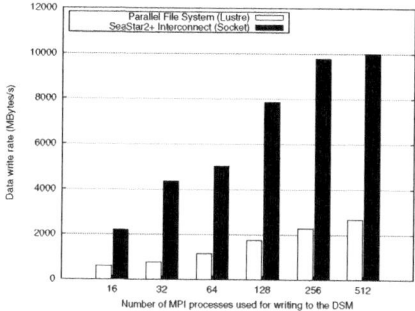

Fig. 3. (Left) Write test with a 20GB DSM distributed among 8 post-processing nodes on a regular cluster. (Right) Write test with a 20GB DSM distributed among 8 post-processing nodes on a Cray XT5 machine (in production mode).

the other hand will have a fixed number of service nodes which once saturated, cannot be exceeded. The Lustre filesystem (20 OSS nodes) used here was in regular production mode and the results therefore reflect typical usage.

5   Application: Integration of the DSM Interface within ParaView

Whilst the original design of the DSM based VFD was intended for code coupling, it works equally well for post-processing and visualization. We have therefore integrated the DSM host into a plugin for the widely used ParaView package which enables automatic display and analysis of data from any simulation making use of the HDF5 driver. To better support visualization of diverse simulation data, we have integrated into the plugin several readers which accept HDF5 data, including a reader based on Xdmf, the eXtensible Data Model and Format [7].

We have also enabled the generation of XML descriptions of the data (required by Xdmf to create topological connectivity from the raw HDF5 arrays) from a simple template which can be supplied by the user and which is filled in by helper code inside the DSM after data has arrived. The plugin checks for data ready synchronization messages and automatically updates the active pipelines in the ParaView GUI so that the visualization and any user generated analyses automatically track the ongoing simulation.

6   Related Work and Discussion

Asynchronous IO has been shown to offer significant improvements in performance [4] on large simulations by reducing the need for processes to wait until the filesystem is ready. ADIOS is able to issue non blocking writes, leaving the application free to perform overlapped IO and computation (at the expense of double-buffered memory). ADIOS goes further, by allowing the use of server side services such as DataStager [2], and DART [9] which schedule and *prioritize* RDMA requests using asynchronous IO so that network traffic is optimized and data may be directed either to the file system, or to an awaiting application. Whilst our library does not offer the range of features/flexibility of these systems, it does achieve high throughput by removing the filesystem bottleneck and allow the simulation to overlap computation with post-processing in the DSM server application. Providing the server does not lock the DSM by reading whilst the simulation is waiting to write, the performance is limited only by the underlying network and number of links between the two. DataStager offers a solution to the need for collective dataset creation by the scheduling of transfers ordered by rank so that file offsets are known when the domain decomposition is not straightforward. Such a facility is not available within HDF5 and is a potential bottleneck for very large process counts.

7   Conclusion and Future Work

We have developed a parallel MPI streaming driver based on a distributed memory buffer which provides an easy way to couple applications already using the HDF5 API. It performs much better than file systems when used with dedicated links and is ideally suited for in-situ visualization and post-processing of data, rather than saving to disk and post-processing off-line. To improve performance of writes when the data size is significantly smaller than the DSM buffer, we shall take advantage of the HDF5 chunking interface and virtual addressing to distribute packets of data more evenly between host nodes, improving network efficiency. An additional improvement will be to allow dynamic resizing of the DSM buffer on individual nodes and enabling pointer sharing between the DSM storage and the post-processing code so that duplication of memory is reduced when reading. For computational steering, a second channel/buffer will be added so that it becomes possible to send commands back to the simulation code, making use of the same HDF5 read/write API.

Acknowledgments. This work is supported by the "NextMuSE" project receiving funding from the European Community's Seventh Framework Programme (FP7/2007-2013) under grant agreement 225967.

References

1. Hierarchical Data Format (HDF5), http://hdf.ncsa.uiuc.edu/
2. Abbasi, H., Wolf, M., Eisenhauer, G., Klasky, S., Schwan, K., Zheng, F.: Datastager: scalable data staging services for petascale applications. In: HPDC 2009: Proceedings of the 18th ACM international symposium on High performance distributed computing, pp. 39–48. ACM, New York (2009)
3. Allen, G., Benger, W., Dramlitsch, T., Goodale, T., Christian Hege, H., Lanfermann, G., Merzky, A., Radke, T., Seidel, E.: Cactus Grid Computing: Review of Current Development (2001)
4. Borrill, J., Oliker, L., Shalf, J., Shan, H.: Investigation of leading HPC I/O performance using a scientific-application derived benchmark. In: Lumpe, M., Vanderperren, W. (eds.) SC 2007, pp. 1–12. ACM, New York (2007), doi:10.1145/1362622.1362636
5. Chilan, C.M., Yang, M., Cheng, A., Arber, L.: Parallel I/O Performance Study with HDF5, a Scientific Data Package. Tech. rep., National Center for Supercomputing Applications, University of Illinois, Urbana-Champaign (2006)
6. Clarke, J.A.: Emulating Shared Memory to Simplify Distributed-Memory Programming. IEEE Comput. Sci. Eng. 4(1), 55–62 (1997)
7. Clarke, J.A., Mark, E.R.: Enhancements to the eXtensible Data Model and Format (XDMF). In: HPCMP-UGC 2007: Proceedings of the 2007 DoD High Performance Computing Modernization Program Users Group Conference, pp. 322–327. IEEE Computer Society Press, Washington (2007)
8. Clarke, J.A., Namburu, R.R.: A Generalized Method for One-Way Coupling of CTH and Lagrangian Finite-Element Codes With Complex Structures Using the Interdisciplinary Computing Environment. Tech. rep., US Army Research Laboratory, Aberdeen Proving Ground, Md. (November 2004), ARL-TN-230
9. Docan, C., Parashar, M., Klasky, S.: DART: a substrate for high speed asynchronous data IO. In: HPDC 2008: Proceedings of the 17th international symposium on High performance distributed computing., pp. 219–220. ACM, New York (2008), doi:10.1145/1383422.1383454
10. Henderson, A.: ParaView Guide, A Parallel Visualization Application. Kitware Inc. (2005), http://www.paraview.org
11. Lofstead, J.F., Klasky, S., Schwan, K., Podhorszki, N., Jin, C.: Flexible IO and integration for scientific codes through the adaptable IO system (ADIOS). In: CLADE 2008: Proceedings of the 6th international workshop on Challenges of large applications in distributed environments, pp. 15–24. ACM, New York (2008)
12. Yang, M., Koziol, Q.: Using collective IO inside a high performance IO software package - HDF5. Tech. rep., National Center for Supercomputing Applications, University of Illinois, Urbana-Champaign (2006)

Automated Tracing of I/O Stack*

Seong Jo Kim[1], Yuanrui Zhang[1], Seung Woo Son[2], Ramya Prabhakar[1],
Mahmut Kandemir[1], Christina Patrick[1], Wei-keng Liao[3], and Alok Choudhary[3]

[1] Department of Computer Science and Engineering
Pennsylvania State University, University Park, PA, 16802, USA
[2] Mathematics and Computer Science Division
Argonne National Laboratory, Argonne, IL, 60439, USA
[3] Department of Electrical Engineering and Computer Science
Northwestern University, Evanston, IL 60208, USA

Abstract. Efficient execution of parallel scientific applications requires
high-performance storage systems designed to meet their I/O requirements. Most high-performance I/O intensive applications access multiple
layers of the storage stack during their disk operations. A typical I/O request from these applications may include accesses to high-level libraries
such as MPI I/O, executing on clustered parallel file systems like PVFS2,
which are in turn supported by native file systems like Linux. In order to
design and implement parallel applications that exercise this I/O stack,
it is important to understand the flow of I/O calls through the entire
storage system. Such understanding helps in identifying the potential
performance and power bottlenecks in different layers of the storage hierarchy. To trace the execution of the I/O calls and to understand the
complex interactions of multiple user-libraries and file systems, we propose an automatic code instrumentation technique, which enables us to
collect detailed statistics of the I/O stack. Our proposed I/O tracing tool
traces the flow of I/O calls across different layers of an I/O stack, and
can be configured to work with different file systems and user-libraries.
It also analyzes the collected information to generate output in terms of
different user-specified metrics of interest.

Keywords: Automated code instrumentation, Parallel I/O, MPI-IO,
MPICH2, PVFS2.

1 Introduction

Emerging data-intensive applications make significant demands on storage system performance and, therefore, face what can be termed as *I/O Wall*, that
is, I/O behavior is the primary factor that determines application performance.
Clearly, unless the I/O wall is properly addressed, scientists and engineers will
not be able to exploit the full potential of emerging parallel machines when

* This work is supported in part by NSF grants 0937949, 0621402, 0724599, 0821527,
0833126, 0720749, 0621443, 0724599, and 0833131 and DOE grants DEAC02-
06CH11357, DE-FG02-08ER25848, DE-SC0002156, and DESC0001283.

R. Keller et al. (Eds.): EuroMPI 2010, LNCS 6305, pp. 72–81, 2010.

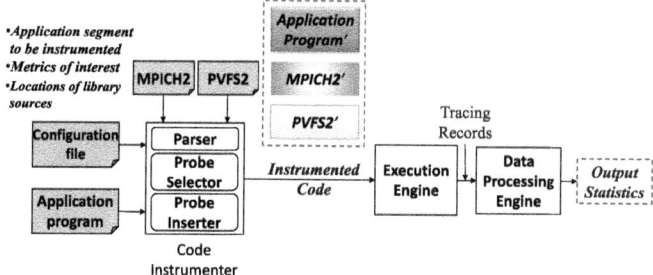

Fig. 1. Our automated I/O tracing tool takes as input the application program, I/O stack information and a configuration file which captures the metrics of interest, locations of target sources, and a description of the region of interest in the code. It automatically generates and runs instrumented code, and finally collects and analyzes all the statistics on the fly.

running large-scale parallel applications from bioinformatics, climate prediction, computational chemistry, and brain imaging domains.

The first step in addressing the I/O wall is to *understand* it. Unfortunately, this is not trivial as I/O behavior today is a result of complex interactions that take place among multiple software components, which can be referred to, collectively, as *I/O Stack*. For example, an I/O stack may contain an application program, a high-level library such as MPI-IO [8], a parallel file system such as PVFS [3], and a native file system such as Linux. A high-level I/O call in an application program flows through these layers in the I/O stack and, during this flow, it can be fragmented into multiple smaller calls (sub-calls) and the sub-calls originating from different high-level calls can contend for the same set of I/O resources such as storage caches, I/O network bandwidth, disk space, etc. Therefore, understanding the I/O wall means understanding the flow of I/O calls over the I/O stack.

To understand the behavior of an I/O stack, one option is to let the application programmer/user to instrument the I/O stack manually. Unfortunately, this approach (manual instrumentation) is very difficult in practice and extremely error prone. In fact, tracking even a single I/O call may necessitate modifications to numerous files and passing information between them.

Motivated by this observation, in this work, we explore automated instrumentation of the I/O stack. In this approach, as shown in Figure 1, instead of instrumenting the source code of applications and other components of the I/O stack manually, an application programmer specifies what portion of the application code is to be instrumented and what statistics are to be collected. The proposed tool takes this information as input along with the description of the target I/O stack and the source codes of the application program and other I/O stack software, and generates, as output, an instrumented version of the application code as well as instrumented versions of the other components (software layers) in the I/O stack. All necessary instrumentation of the components in the I/O stack (application, libraries, file systems) are carried out automatically.

A unique aspect of our approach is that it can work with different I/O stacks and with different metrics of interest (e.g., I/O latency, I/O throughput, I/O power). Our experience with this tool is very encouraging so far. Specifically, using this tool, we automatically instrumented an I/O-intensive application and collected detailed performance and power statistics on the I/O stack.

Section 2 discusses the related work on code instrumentation and profiling. Section 3 explains the details of our proposed automated I/O tracing tool. An experimental evaluation of the tool is presented in Section 4. Finally, Section 5 concludes the paper with a summary of the planned future work.

2   Related Work

Over the past decade many code instrumentation tools that target different machines and applications have been developed and tested. ATOM [24] inserts probe code into the program at compile time. Dynamic code instrumentation [1,2,17], on the other hand, intercepts the execution of an executable at runtime to insert user-defined codes at different points of interest. HP's Dynamo [1] monitors an executable's behavior through interpretation and dynamically selects hot instruction traces from the running program.

Several techniques have been proposed in the literature to reduce instrumentation overheads. Dyninst and Paradyn use fast breakpoints to reduce the overheads incurred during instrumentation. They both are designed for dynamic instrumentation [12]. In comparison, FIT [5] is a static system that aims retargetability rather than instrumentation optimization. INS-OP [15] is also a dynamic instrumentation tool that applies transformations to reduce the overheads in the instrumentation code. In [27], Vijayakumar et al. propose an I/O tracing approach that combines aggressive trace compression. However, their strategy does not provide flexibility in terms of target metric specification. Tools such as CHARISMA [20], Pablo [23], and TAU (Tuning and Analysis Utilities) [19] are designed to collect and analyze file system traces [18]. For the MPI-based parallel applications, several tools, such as MPE (MPI Parallel Environment) [4] and mpiP [26], exist. mpiP is a lightweight profiling tool for identifying communication operations that do not scale well in the MPI-based applications. It reduces the amount of profile data and overheads by collecting only statistical information on MPI functions. Typically, the trace data generated by these profiling tools are visualized using tools such as Jumpshot [28], Nupshot [14], Upshot [11], and PerfExplorer [13].

Our work is different from these prior efforts as we use source code analysis to instrument the I/O stack automatically. Also, unlike some of the existing profiling and instrumentation tools, our approach is not specific to MPI or to a pre-determined metric; instead, it can target an entire I/O stack and work with different performance and power related metrics.

3   Our Approach

3.1   High-Level View of Automated Instrumentation

Our goal is to provide an automated I/O tracing functionality for parallel applications that exercise multiple layers of an I/O stack, with minimal impact to the performance. To this end, we have implemented in this work an automated I/O tracing tool that, as illustrated in Figure 1, comprises three major components: code instrumenter, execution engine, and data processing engine.

As shown in Figure 1, the code instrumenter consists of the parser, the probe selector, and the probe inserter. In this context, a *probe* is a piece of code being inserted into the application code and I/O stack software (e.g., in the source codes of MPI-I/O and PVFS2), which helps us collect the required statistics. The code instrumenter takes as input the application program, high level I/O metrics of interest written in our specification language, and the target I/O stack (which consists of the MPI library and PVFS2 in our current testbed).

The parser parses I/O metrics of interest from the configuration file, extracts all necessary information to instrument the I/O stack in a hierarchial fashion from top to bottom, and stores it to be used later by other components of the tool. After that, the probe selector chooses the most appropriate probes for the high-level metrics specified by the user. Finally, the probe inserter automatically inserts the necessary probes into the proper places in the I/O stack. Note that, depending on the target I/O metrics of interest, our tool may insert multiple probes in the code. Table 1 lists a representative set of high-level metrics that can be traced using our tool.

Table 1. Sample high-level metrics that can be traced and collected using the tool

I/O latency experienced by each I/O call in each layer (client, server, or disk) in the stack
Throughput achieved by a given I/O read and write call
Average I/O access latency in a given segment of the program
Number of I/O nodes participating in each collective I/O
Amount of time spent during inter-processor communication in executing a collective I/O call
Disk power consumption incurred by each I/O call
Number of disk accesses made by each I/O call

The execution engine compiles and runs the *instrumented* I/O stack, and generates the required trace. Finally, the data processing engine analyzes the trace log files and returns statistics based on the user's queries. The collected statistics can be viewed from different perspectives. For example, the user can look at the I/O latency/power breakdown at each server or at each client. The amount of time spent by an I/O call at each layer of the target I/O stack can also be visualized.

3.2   Technical Details of Automated Instrumentation

In this section, we discuss details of the code instrumenter component of our tool. Let us assume, for the purpose of illustration, that the user is interested

```
-A [application.c, application]
-L [$MPICH2, $PVFS2, ClientMachineInfo, $Log]
-O [w]
-C [100-300]
-S [4, <3 max>, <3 max>, <3 max>, <3 max>, <3 max>]
-T [4, <3, mpi.0.log , mpi.1.log, mpi.2.log>, <3, client.0.log, client.1.log, client.2.log>,
   <3, server.0.log, server.1.log,server.2.log>, <3, disk.0.log, disk.1.log,disk.2.log>]
-Q [latency, inclusive, all, list:, list:*, list:*, list:*]
-P [App;common;App-probe1;-l main;before]
-P [App;common;App-probe2;-l MPI_Comm_rank;after]
-P [App;common;App-Start-probe;-l MPI_File_read;before]
-P [App;common;App-Start-probe4;-l MPI_File_write;before]
-P [MPI;latency;MPI-Start-probe;-n 74;$MPICH2/mpi/romio/mpi-io/read.c]
-P [MPI;latency;MPI-End-probe;-n 165;$MPICH2/mpi/romio/mpi-io/read.c]
-P [MPI;latency;MPI-Start-probe;-n 76;$MPICH2/mpi/romio/mpi-io/read_all.c]
-P [MPI;latency;MPI-End-probe;-n 118;$MPICH2/mpi/romio/mpi-io/read_all.c]
-P [MPI;latency;MPI-End-probe;-n 73;$MPICH2/mpi/romio/mpi-io/write.c]
-P [MPI;latency;MPI-End-probe;-n 168;$MPICH2/mpi/romio/mpi-io/write.c]
-P [MPI;latency;MPI-Start-probe;-n 75;$MPICH2/mpi/romio/mpi-io/write_all.c]
-P [MPI;latency;MPI-End-probe;-n 117;$MPICH2/mpi/romio/mpi-io/write_all.c]
-P [MPI;latency;MPI-probe;-n 62;$MPICH2/mpi/romio/mpi-io/adio/ad_pvfs2_read.c]
-P [MPI;latency;MPI-probe;-n 295;$MPICH2/mpi/romio/mpi-io/adio/ad_pvfs2_write.c]
-P [PVFSClient;latency;Client-Start-probe;-n 372;$PVFS2/client/sysint/sys-io.sm]
-P [PVFSClient;latency;Client-End-probe;-n 397;$PVFS2/client/sysint/sys-io.sm]
-P [PVFSClient;latency;Client-probe;-n 670;$PVFS2/client/sysint/sys-io.sm]
-P [PVFSServer;latency;.Server-Start-probe;-n 153;$PVFS2/server/io.sm]
-P [PVFSServer;latency;.Server-End-probe;-n 5448;$PVFS2/io/job/job.c]
-P [PVFSServer;latency;.Disk-start-probe;-n 1342;$PVFS2/io/flow/flowproto-bmi-trove/flowproto
   -multiqueue.c]
-P [PVFSServer;latency;.Disk-end-probe1;-n 1513;$PVFS2/io/flow/flowproto-bmi-trove/flowproto-
   multiqueue.c]
-P [PVFSServer;latency;.Disk-end-probe2;-n 1513;$PVFS2/io/flow/flowproto-bmi-trove/flowproto-
   multiqueue.c]
```

Fig. 2. An example configuration file

in collecting statistics about the execution *latency* of each I/O call in each layer of the I/O stack, that is, the amount of time spent by an I/O call in MPI-I/O, PVFS2 client, PVFS2 server, and disk layers. A sample configuration file that captures this request is given in Figure 2. This file is written in our specification language, and Table 2 describes the details of each parameter in the configuration file. Let us now explain the contents of this sample configuration file.

Table 2. Flags used in a configuration file

Parameter	Description
-A	Application file name or path
-L	Path for I/O libraries
-O	Operation of interest
-C	Code segment of interest to trace
-S	I/O stack specification
-T	Tracing file location generated by our tool
-Q	Metric of interest
-P	Probe name and inserting location

In this example, the user wants to collect the execution *latency* of MPI-IO write operations (indicated using -O[w]) that occur between lines 100 to 300 of an application program called, *application.c*. Also, the user specifies three I/O stack layers, which are MPI-IO, PVFS2 client, and PVFS2 server (below the application program). Finally, the user describes the trace log file names and their locations for the data processing engine. Based on the target metric of interest, that is *latency*, the most appropriate latency probes can be automatically inserted into the designated places in the probe specification.

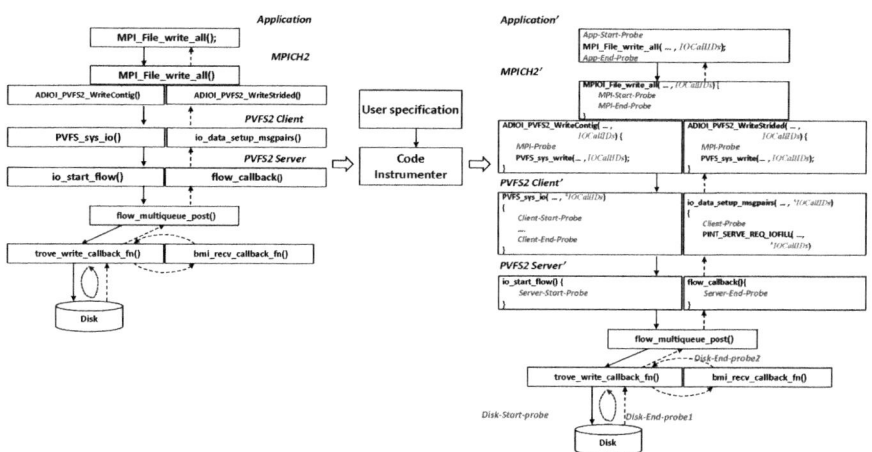

Fig. 3. Illustration of inserting probes into the application program and I/O stack components by our code instrumenter. Application′, MPICH2′, PVFS2 Client′, and PVFS2 Server′ represent the instrumented I/O stack.

Figure 3 illustrates how the code instrumenter works. It takes as an input the user configuration file along with MPI-IO and PVFS2. The parser parses this configuration file and extracts all the information required by the other components such as the probe inserter and the data processing engine. Based on the specified target metric, i.e., the execution *latency* of MPI *write* operations in all the layers including MPI-IO, PVFS2 client, PVFS2 server, and disk, the probe selector employs only the write-related latency probes, which helps to minimize the overheads associated with the instrumentation. Then, following the call sequence of MPI write function, from the MPI-IO library though the PVFS2 client to the PVFS2 server in Figure 3, the probe inserter selectively inserts the necessary probes into the start point and the end point of each layer described in the configuration file.

After the instrumentation, the probe inserter compiles the instrumented code. During the compilation, it also patches a small array structure, called *IOCallID*, to the MPI-IO and PVFS2 functions to be matched for tracing. IOCallIDs contain information about each layer such as the layer ID and the I/O type. When IOCallIDs are passed from the MPI-IO layer to the PVFS2 client layer, the inserted probe extracts the information from them and generates the log files with the latency statistics at the boundary of each layer.

Note that a high-level MPI-IO call can be fragmented into multiple small sub-calls. For example, in two-phase I/O [6], which consists of an I/O phase and a communication phase, tracing an I/O call across the layer boundaries in the I/O stack is not trivial. In our implementation, each call has a unique ID in the current layer and passes it to the layer below. This help us to connect the high-level call to its sub-calls in a hierarchical fashion. It also helps the data processing engine (see Figure 1) to combine the statistics coming from different layers in a

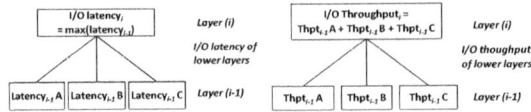

Fig. 4. Computation of I/O latency and I/O throughput metrics

systematic way (for example, all the variables that hold latency information at different layers are associated with each other using these IDs).

In the PVFS2 server layer, our tool uses a unique structure, called *flow_desciptor*, to perform the requested I/O operations from the PVFS2 client. The probe inserted into the start point in the server layer extracts the information in the IOCallIDs passed from the PVFS2 client and packs it into the flow_descriptor. Since the flow_descriptor is passed to the entire PVFS2 server, the probe in the server extracts the necessary information from it to collect the latency related statistics without much complexity.

The execution engine runs the instrumented code and generates the trace log files in each layer. Finally, the data processing engine analyzes all the trace log files and collects the execution I/O latency induced by each MPI operation in each layer. The I/O latency value computed at each layer is equal to the maximum value of the I/O latencies obtained from different layers below it. However, the computation of I/O throughput value is additive, i.e., the I/O throughput computed at any layer is the sum of I/O throughputs from different sub-layers below it. Figure 4 illustrates the computation of these metrics. To compute the I/O power, we use the power model described in [9].

4   Evaluation

To demonstrate the operation of our tracing tool, we ran a benchmark program using three PVFS2 servers and three PVFS2 clients on a Linux cluster that consists of 6 dual-core processor nodes, AMD Athlon MP2000+, connected through Ethernet and Myrinet. Each node of this system runs a copy of PVFS2 and MPICH2. To measure disk power consumption per I/O call, we used the disk energy model [9] based on the data sheets of the IBM Ultrastar 36Z15 disk [25]. Table 3 gives the important metrics used to calculate power consumption.

Table 3. Important disk parameters for power calculation

Parameter	Default Value
Disk drive module	IBM36Z15
Storage capacity (GB)	36.7
Maximum disk speed (RPM)	15000
Active power consumption (Watt)	13.5
Idle power consumption (Watt)	10.2

In our evaluation, we used the FLASH I/O benchmark [7] that simulates the I/O pattern of FLASH [29]. It creates the primary data structures in the FLASH code and generates three files: a checkpoint file, a plot file for center data, and a plot file for corner data, using two high-level I/O libraries: PnetCDF [16] and HDF5 [10].

The in-memory data structures are 3D sub-blocks of size 8x8x8 or 16x16x16. In the simulation, 80 of these blocks are held by each processor and are written to

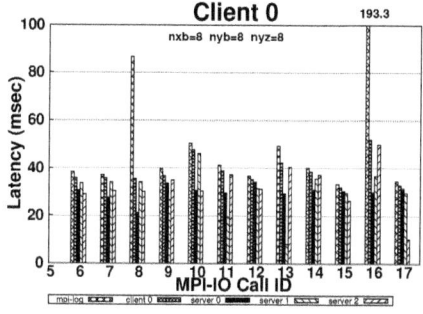
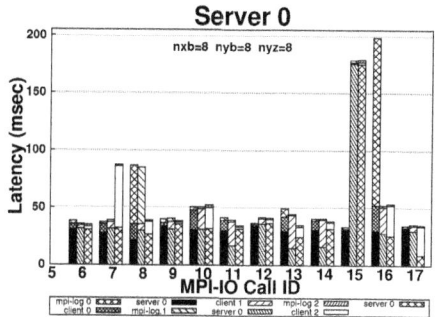

Fig. 5. Latency of Client 0 using PnetCDF **Fig. 6.** Latency of Server 0 using PnetCDF

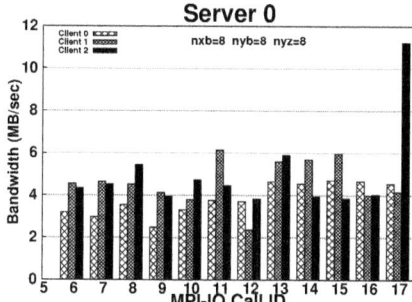
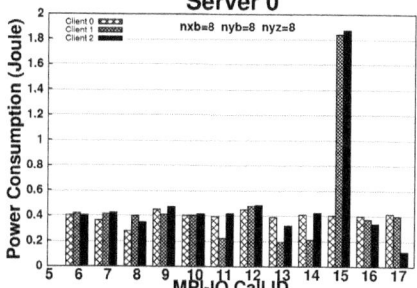

Fig. 7. Disk throughput of Server 0 using PnetCDF

Fig. 8. Power consumption of Server 0 using PnetCDF

three files with 50 MPI_File_write_all function calls. We used the sub-blocks of size 8x8x8. nxb, nyb, and nyz in Figures 5 through 9 represent these sub-block sizes.

First, Figure 5 shows the I/O latencies experienced by each MPI-IO call from Client 0's perspective when PnetCDF is used. MPI-IO calls 6-17 are shown with latencies (in milliseconds) taken in mpi-log, Client 0, Server 0, Server 1, and Server 2, from left to right. We see that MPI-IO call 16 takes 193.3 milliseconds in mpi-log, but only 32 milliseconds in Server 1. In this experiment, some of the MPI-IO calls (0-5, 18-45) calls are directed to the metadata server to write header information of each file. These calls were not recorded in the I/O server log file. Figure 6, on the other hand, plots the latencies observed from Server 0's perspective. The three bars for every call ID represent cumulative latencies from each client (Client 0, Client 1 and Client 2 from left to right). Further, each bar also gives a breakdown of I/O latency (in milliseconds) taken for the request to be processed in the MPI-IO, PVFS client, and PVFS server layers, respectively. From this result, one can see, for example, that Client 1 and Client 2 spend less time than Client 0 in the Server 0 as far as call ID 16 is concerned. These two plots in Figure 5 and Figure 6 clearly demonstrate that our tool can be used to study the I/O latency breakdown, from both clients' and server's perspectives.

Figure 7 illustrates the I/O through-put in Server 0. One can observe from this plot the detailed I/O throughput patterns of different clients regarding this server. Comparing Figure 6 with Figure 7, one can also see that the bottleneck I/O call in the application code depends on whether I/O latency or I/O throughput is targeted. Figure 8, on the other hand, presents the power consumption results for Server 0. We see that most of the power is consumed by I/O call 15, and except for this call, power consumptions of Client 0 and Client 1 are similar on this server.

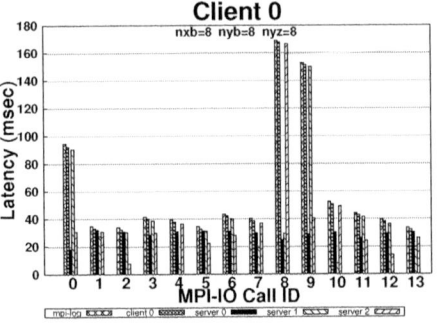

Fig. 9. Latency of Client 0 using HDF5

Our final set of results are given in Figure 9 and depict the I/O latency values observed from Client 0's viewpoint when using HDF5, instead of PnetCDF. Overall, these sample set of results clearly show that our tool can be used to collect and analyze detailed latency, throughput, and power statistics regarding the I/O stack.

5   Concluding Remarks and Future Work

Performing code instrumentation manually is often difficult and could be error-prone. Hence, we propose an automatic instrumentation technique that can be used to trace and analyze scientific applications using high level I/O libraries like PnetCDF, HDF5, or MPI I/O over file systems like PVFS2 and Linux. The tracing utility uses existing MPI I/O function calls and therefore adds minimum overhead to the application execution. It takes target high level metrics like I/O latency, I/O throughput and I/O power as well as a description of the target I/O stack as input and analyzes the collected information to generate output in terms of different user-specified metrics. As our future work, we plan to extend our analysis to other available I/O benchmarks, such as S3D-IO [22] and GCRM [21], to characterize their I/O behavior. We also plan to investigate techniques for dynamic code instrumentation that makes use of information available at run-time to generate/restructure code for data optimization.

References

1. Bala, V., et al.: Dynamo: A transparent dynamic optimization system. In: PLDI (2000)
2. Bruening, D.L.: Efficient, transparent, and comprehensive runtime code manipulation. PhD thesis, MIT, Cambridge, MA, USA (2004)
3. Carns, P.H., et al.: PVFS: A parallel file system for linux clusters. In: Proceedings of the Annual Linux Showcase and Conference (2000)
4. Chan, A., et al.: User's guide for MPE: Extensions for MPI programs (1998)

5. De Bus, B., et al.: The design and implementation of FIT: A flexible instrumentation toolkit. In: Proceedings of PASTE (2004)
6. del Rosario, J.M., et al.: Improved parallel I/O via a two-phase run-time access strategy. SIGARCH Comput. Archit. News 21(5), 31–38 (1993)
7. Fisher, R.T., et al.: Terascale turbulence computation using the flash3 application framework on the IBM Blue Gene/L system. IBM J. Res. Dev. 52(1/2) (2008)
8. Gropp, W., et al.: MPI — The Complete Reference: the MPI-2 Extensions, vol. 2. MIT Press, Cambridge (1998)
9. Gurumurthi, S., et al.: DRPM: Dynamic speed control for power management in server class disks. In: ISCA (2003)
10. HDF (Hierarchical Data Format) , http://www.hdfgroup.org
11. Herrarte, V., Lusk, E.: Studying parallel program behavior with upshot. Technical Report ANL–91/15, Argonne National Laboratory (1991)
12. Hollingsworth, J.K., et al.: MDL: A language and compiler for dynamic program instrumentation. In: Malyshkin, V.E. (ed.) PaCT 1997. LNCS, vol. 1277, Springer, Heidelberg (1997)
13. Huck, K.A., Malony, A.D.: PerfExplorer: A Performance Data Mining Framework For Large-Scale Parallel Computing. In: SC (2005)
14. Karrels, E., Lusk, E.: Performance analysis of MPI programs. In: Workshop on Environments and Tools For Parallel Scientific Computing (1994)
15. Kumar, N., et al.: Low overhead program monitoring and profiling. In: Proceedings of PASTE (2005)
16. Li, J., et al.: Parallel netCDF: A high-performance scientific I/O interface. In: SC (2003)
17. Luk, C.-K., et al.: Pin: Building customized program analysis tools with dynamic instrumentation. In: PLDI (2005)
18. Moore, S., et al.: Review of performance analysis tools for MPI parallel programs. In: Cotronis, Y., Dongarra, J. (eds.) PVM/MPI 2001. LNCS, vol. 2131, p. 241. Springer, Heidelberg (2001)
19. Moore, S., et al.: A scalable approach to MPI application performance analysis. In: Di Martino, B., Kranzlmüller, D., Dongarra, J. (eds.) EuroPVM/MPI 2005. LNCS, vol. 3666, pp. 309–316. Springer, Heidelberg (2005)
20. Nieuwejaar, N., et al.: File-access characteristics of parallel scientific workloads. IEEE Transactions on Parallel and Distributed Systems 7, 1075–1089 (1996)
21. Randall, D.A.: Design and testing of a global cloud-resolving model (2009)
22. Sankaran, R., et al.: Direct numerical simulations of turbulent lean premixed combustion. Journal of Physics: Conference Series 46(1), 38 (2006)
23. Simitci, H.: Pablo MPI Instrumentation User's Guide. University of Illinois. Tech. Report (1996)
24. Srivastava, A., Eustace, A.: ATOM: A system for building customized program analysis tools. In: PLDI (1994)
25. Ultrastar, I.: 36Z15 Data Sheet (2010), http://www.hitachigst.com/hdd/ultra/ul36z15.htm
26. Vetter, J., Chambreau, C.: mpiP: Lightweight, scalable MPI profiling (2010), http://mpip.sourceforge.net/
27. Vijayakumar, K., et al.: Scalable I/O tracing and analysis. In: Supercomputing PDSW (2009)
28. Zaki, O., et al.: Toward scalable performance visualization with jumpshot. Int. J. High Perform. Comput. Appl. 13(3), 277–288 (1999)
29. Fryxell, B., et al.: FLASH: Adaptive Mesh Hydrodynamics Code. The Astrophysical Journal Supplement Series 131 (2000)

MPI Datatype Marshalling: A Case Study in Datatype Equivalence

Dries Kimpe[1,2], David Goodell[1], and Robert Ross[1]

[1] Argonne National Laboratory, Argonne, IL 60439
[2] University of Chicago, Chicago, IL 60637
{dkimpe,goodell,rross}@mcs.anl.gov

Abstract. MPI datatypes are a convenient abstraction for manipulating complex data structures and are useful in a number of contexts. In some cases, these descriptions need to be preserved on disk or communicated between processes, such as when defining RMA windows. We propose an extension to MPI that enables marshalling and unmarshalling MPI datatypes in the spirit of `MPI_Pack`/`MPI_Unpack`. Issues in MPI datatype equivalence are discussed in detail and an implementation of the new interface outside of MPI is presented. The new marshalling interface provides a mechanism for serializing all aspects of an MPI datatype: the typemap, upper/lower bounds, name, contents/envelope information, and attributes.

1 Introduction

Since its inception, MPI has provided *datatypes* to describe the location of data in memory and files for communication and I/O. These datatypes are a flexible and powerful abstraction, capable of efficiently expressing extremely sophisticated data layouts. While MPI offers facilities to simply and efficiently transmit, store, and retrieve data described by these datatypes, however, it does not provide any direct mechanism to transmit the datatype description itself.

We originally set out to develop a library capable of *marshalling* MPI datatypes. We define marshalling to be the act of generating a representation of an MPI datatype that can be used to recreate an "equivalent" datatype later, possibly in another software context (such as another MPI process or a postprocessing tool). Such functionality is useful in many cases, such as the following:

- Message logging for fault-tolerance support
- Self-describing archival storage
- Type visualization tools
- Argument checking for collective function invocations
- Implementing "one-sided" communication, where the target process does not necessarily know the datatype that will be used
- Message or I/O tracing for replay in a tool or simulator.

When viewed in the abstract or from the perspective of a particular use case, datatype marshalling appears to be a well-defined problem with a number of

R. Keller et al. (Eds.): EuroMPI 2010, LNCS 6305, pp. 82–91, 2010.

direct solutions. As we considered the problem from several different angles, however, we consistently came up with different, sometimes incompatible, requirements. These requirements stem from the lack of a clear definition for "equivalent" MPI datatypes.

The rest of this paper is organized as follows. In Section 2 we discuss the thorny issue of MPI datatype equivalence. In Section 3 we present the design and implementation of our datatype marshalling library. In Section 4 we briefly evaluate the time and space performance of our implementation. In Section 5 we discuss related work, and our conclusions in Section 6.

2   MPI Datatype Equivalence

Pragmatically, two MPI datatypes might generally be considered equivalent when one can be substituted for another in MPI operations. However, datatypes are characterized by several independent dimensions that may constitute a concrete definition of equivalence.

The MPI standard [5] provides one definition for datatype equivalence (MPI-2.2 §2.4):

> Two datatypes are equivalent if they appear to have been created with the same sequence of calls (and arguments) and thus have the same typemap. Two equivalent datatypes do not necessarily have the same cached attributes or the same names.

Capturing the extent of the type is critical in cases where a count ≥ 1 is used. Section 4.1.7 states that `MPI_Type_create_resized` does the following:

> Returns in newtype a handle to a new datatype that is identical to old-type, except that the lower bound of this new datatype is set to be lb, and its upper bound is set to be lb + extent. Any previous **lb** and **ub** markers are erased, and a new pair of lower bound and upper bound markers are put in the positions indicated by the lb and extent arguments.

If one sensibly interprets this as stating that `MPI_Type_create_resized`'s effect is to insert `MPI_LB` and `MPI_UB` markers into the typemap,[1] then the MPI standard definition provides an adequate definition for point-to-point, collective, one-sided (RMA), and I/O operations. If the typemaps match, then the MPI operations will access the same data items.[2]

In explanation, consider two types, A and B, with identical typemaps but differing extents, E_A and E_B. The code to create these types is shown in Listing 1. If `MPI_Send` is invoked with count $= 2$ and alternately with A and B, different data will be sent (Figure 1). The typemap for B must incorporate the `MPI_UB` defined by the resize.

---

[1] It is interesting that `MPI_LB` and `MPI_UB`, while deprecated for being error-prone to use, are extremely helpful in understanding the equivalence of datatypes.

[2] The implicit pad (ϵ) used in an MPI executable is intended to mimic the alignment behavior of the compiler used. This can vary based on architecture, compiler, and compiler flags, and it is not explicitly captured in the typemap.

```
1  MPI_Aint lb, extent;
2  MPI_Datatype A, B;
3  MPI_Type_vector(2, 1, 2, MPI_BYTE, &A);
4  MPI_Type_get_extent(A, &lb, &extent);
5  MPI_Type_create_resized(A, 0, extent+1, &B);
```

Listing 1. MPI Code to Create MPI Datatypes A and B

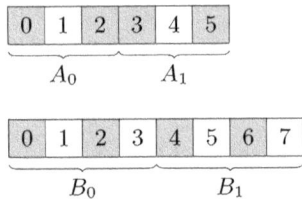

Fig. 1. A and B used in MPI_Send with count $= 2$. (shaded boxes indicate transmitted bytes)

We note that no facility for comparing datatypes is provided in the MPI standard, as it is for comparing communicators (i.e., MPI_Comm_compare). This omission complicates the construction of external libraries that would marshal datatypes, as it is impossible to detect equivalent datatypes without dissecting the datatype via the envelope and contents calls.

In many contexts, several additional characteristics besides the typemap may determine the semantic equivalence of two types. These include the type names, construction sequence, and attribute values.

2.1   Type Name Equivalence

The name associated with an MPI datatype may be changed by the MPI_Type_set_name routine. This information is not considered in the MPI standard's definition of type equivalence. For example, a type used to represent the layout of a dataset in a file may be named by that dataset's name. In some cases, an application or library may not consider two types to be equivalent unless the types' names are also equivalent.

2.2   Constructor Equivalence

The MPI-2 standard introduced two functions and a handful of constants that together provide a form of type introspection. The MPI_Type_get_envelope function returns the "combiner" used to create the type. Combiner examples include MPI_COMBINER_VECTOR, MPI_COMBINER_RESIZED, and MPI_COMBINER_NAMED, corresponding to MPI_Type_vector, MPI_Type_create_resized, and a named predefined datatype such as MPI_INT. The complementary function, MPI_Type_get_contents, returns information sufficient to recreate the call to the combiner routine, such as the input datatypes, counts, and indices. The MPI standard requires that "the actual arguments used in the creation call for a datatype can be obtained," including zero-count arguments. This requirement goes beyond the MPI standard's definition of equivalence, as elements with a zero count do not appear in the typemap of the constructed datatype.

Other than a heavyweight, noncomposable scheme involving the MPI profiling interface (PMPI_* functions), the envelope and contents routines are the only available mechanisms for determining the make-up of an MPI datatype. Thus, any external scheme for marshalling datatypes will use this interface.

2.3   Attribute Equivalence

MPI provides attributes to allow applications and libraries to attach process-local information to communicator, window, and datatype objects. We limit our discussion of attributes to datatypes. An attribute attached to a datatype object is a (key,value) pair, with exactly one value for a given key. Keys are integer values, allocated in a process-local context via MPI_Type_keyval_create and deallocated by MPI_Type_keyval_free. Attribute values are void pointers[3] and can be queried/set/deleted with the MPI_Type_{get,set,delete}_attr functions.

Creating a keyval both reserves a key for later use and associates a set of function pointers with that key. The corresponding function pointer is invoked by the MPI implementation when types are copied (via MPI_Type_dup) or deleted (via MPI_Type_free) as well as when attributes themselves are explicitly replaced or deleted. These function pointers are responsible for copying underlying attribute values and cleaning up associated storage according to the semantics of that attribute's usage.

For example, the MPITypes library [8] uses attributes to cache high-performance dataloop [9] representations of MPI datatypes on the datatypes themselves. This strategy allows the MPITypes library to avoid recomputing the dataloop representation on every use. The dataloop information could be stored externally, without the use of the attribute code, but the attribute system provides two advantages. First, if a type is duplicated via MPI_Type_dup, the dataloop representation can be trivially copied, or shared and reference counted. Second, the dataloop can be easily freed when the type is freed. Otherwise the MPITypes library has no easy means to identify when a type is no longer in use; hence, it must use an external caching scheme with bounded size and an eviction policy, or memory usage will grow without bound.

Two MPI datatypes that are equivalent modulo their attributes may or may not be semantically equivalent, depending on the particular usages of those attributes. Consider the case of marshalling a type, t_1 followed by unmarshalling the obtained representation into a second type, t_2. If the marshalling system naïvely fails to preserve attributes during this round trip, any attributes, such as the dataloop from the MPITypes example, must be recalculated for t_2 when accessed later. If the attribute value is essential for correct operation and cannot be recalculated, erroneous program behavior may occur.

---

[3] Attribute values are address-sized integers (KIND=MPI_ADDRESS_KIND) in Fortran.

```
1   int MPIX_Type_marshal(const char *typerep, MPI_Datatype type,
2          void *outbuf, MPI_Aint outsize, MPI_Aint *num_written);
3   int MPIX_Type_marshal_size(const char *typerep, MPI_Datatype type,
4          MPI_Aint *size);
5   int MPIX_Type_unmarshal(const char *typerep, void *inbuf,
6          MPI_Aint insize, MPI_Datatype *type);
```

Listing 2. Marshalling and Unmarshalling Function Prototypes

Therefore, any complete MPI datatype marshalling solution should provide the capability to also marshal a datatype's attributes. Section 3 details one approach to maintaining attributes despite serialization.

3   MPI Datatype Marshalling

Listing 2 shows the function prototypes of the marshalling and unmarshalling functions. We modeled our prototypes on those of the packing and unpacking functions (MPI_Type_{un}pack) defined by the MPI standard. MPIX_Type_marshal_size returns an upper bound for the space required to marshal the given type. MPIX_Type_unmarshal reconstructs the datatype. If the type passed to MPIX_Type_marshal was a named type, such as MPI_INT, the same named type will be returned when unmarshalling. The committed state of the returned datatype is undefined, and the user is responsible for freeing the type if it is not a built-in type.

The representation parameter allows the user to choose which encoding will be used to marshal the type definition. Our library currently defines three type representations: **internal**, **external**, and **compressed**.

A datatype marshalled by using "internal" representation can be unmarshalled only by a process of the same parallel program as the marshalling process. As such, datatypes using "internal" encoding cannot be stored on disk to be retrieved later. The main advantage of using "internal" encoding is that it enables MPI library specific optimizations. For example, an MPI implementation could use its internal type description as the "internal" encoding, avoiding repeated calls to the MPI type construction and introspection functions to marshal and unmarshal a datatype. In addition, any optimizations performed by MPI_Type_commit could be captured and stored as well, making sure these optimizations don't need to be repeated for the unmarshalled type.

Similar to the "external32" data representation in MPI-IO, the "external" type representation has a well-defined layout ensuring the marshalled type can be unmarshalled by an MPI program using another MPI implementation. The "external" format is described in Section 3.1. The "compressed" format reduces the space consumed by a marshalled type at the expense of additional computation to marshal and unmarshal the type. The "compressed" type representation is described in Section 3.4.

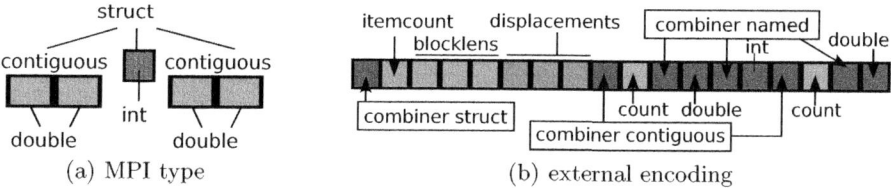

(a) MPI type   (b) external encoding

Fig. 2. External datatype encoding

3.1 External Type Representation

We chose to use the eXternal Data Representaion standard (XDR) [10] to portably store a datatype description. The "external" type format follows a top-down model: the MPI datatype is first broken down into its combiner and associated data. For example, the type shown in Figure 2 has a top-level combiner of MPI_COMBINER_CONTIGUOUS. The combiner is converted into an integer by using a lookup table.[4] The integer is stored in XDR encoding. MPI_Type_contiguous has two more parameters: a base type and a count describing how many times the base type is to be repeated. The count is stored using XDR. Next, the process repeats but this time for the base type, essentially descending into the datatype. Since no cycles are possible in MPI datatypes, this process will eventually end at a leaf node of the datatype. As all derived datatypes are ultimately built from predefined types, this leaf node must be a named predefined type or un-named Fortran predefined type. Named types are marshalled simply by storing the code for a MPI_COMBINER_NAMED, followed by an integer identifying the named type, ending the recursion.

In general, each MPI datatype constructor will be marshalled to a combiner and zero or more integers, addresses, and MPI datatypes. By defining a portable way to store the combiner, integers, and addresses (which is provided by XDR), and the set of named datatypes, using recursion, any MPI datatype can be portably marshalled and unmarshalled.

3.2 Marshalling Type Names

Marshalling type names is relatively straightforward. The name can easily be obtained using MPI_Type_get_name, and stored using the XDR representation for character strings. When unmarshalling, any name found in the data stream is reattached to the type handle using MPI_Type_set_name.

3.3 Marshalling Attributes

As discussed in Section 2.3, supporting marshalling and unmarshalling of at-tributes is desirable and can often simplify libraries or optimize type handling.

---

[4] This conversion is done because the actual value of MPI_COMBINER_CONTIGUOUS is not specified by the MPI standard and thus might differ between MPI implementations. The same is true for the value of the named datatypes. Such link-time constants may not be used as labels in a C-language switch statement.

Marshalling and unmarshalling attributes is not straightforward, however, as there is no easy way to obtain the attributes associated with a datatype. The MPI standard does not provide a function capable of retrieving the set of attributes associated with a particular handle.

In addition, attributes have a user-defined meaning and cannot be portably interpreted by a library. `MPI_Type_dup`, which must copy attributes to the new handle, faces similar issues. The solution adopted by the standard requires that when a new keyval is registered, two function pointers are provided. One is called when an attribute needs to be copied, and one is called when an attribute needs to be destructed (for example, when the object it is associated with is freed).

For each keyval that needs to be marshalled, a call to `MPI_Type_register_marshalled_keyval` must be made. Since `keyvals` have limited process local scope, their actual value cannot be marshalled. Instead, the library associates each `keyval` with a *canonical name*. When unmarshalling the attribute, this mapping is used to retrieve the correct local keyval for the attribute.

Listing 3 presents the C binding function prototypes for our proposed attribute marshalling interface. It closely mirrors the datatype marshalling interface; the corresponding function will be called by the datatype marshalling system when a type is marshalled or unmarshalled.

In order to store attributes, the "external" representation described in Section 3.1 is extended by following the type description with an integer indicating the number of attributes that follow in the data stream. Each attribute is stored by storing its canonical name, followed by the data provided by its `MPIX_Type_marshal_function`. Note that this data is not modified in any way by

```
1   /* provides upper bound on buffer size */
2   typedef int MPIX_Type_marshal_attr_size_function(int keyval,
3       const char *canonical_name, const char *typerep,
4       MPI_Datatype type, MPI_Aint *size);
5   /* Marshals attribute value specified by keyval/canonical_name into
6       outbuf. Sets ((*num_written)=0) if outsize isn't big enough. */
7   typedef int MPIX_Type_marshal_attr_function(int keyval,
8       const char *canonical_name, const char *typerep,
9       MPI_Datatype type, void *outbuf, MPI_Aint outsize,
10      MPI_Aint num_written);
11  /* responsible for setting attribute on type */
12  typedef int MPIX_Type_unmarshal_attr_function(
13      const char *canonical_name, const char *typerep,
14      MPI_Datatype type, void *inbuf, MPI_Aint insize);
15  int MPIX_Type_register_marshalled_keyval(int keyval,
16      const char *canonical_name,
17      MPIX_Type_marshal_attr_size_function *marshal_size_fn,
18      MPIX_Type_marshal_attr_function *marshal_fn,
19      MPIX_Type_unmarshal_attr_function *unmarshal_fn);
```

Listing 3. Attribute Marshalling Function Prototypes

the library and is stored as opaque XDR data. Therefore, the user is responsible for serializing the attribute in a portable fashion.

3.4 Compression

A given base type might be used multiple times in constructing a derived datatype. For example, the type shown in Figure 2 contains multiple copies of a type composed out of two doubles. The marshalled representation of the complete type, C, should ideally contain only one copy. Unfortunately, there is no easy way to detect reuse of types. According to the MPI standard, `MPI_Type_get_contents` returns handles to datatypes that are *equivalent* to the datatypes used to construct the type. There is no guarantee that the returned handle will be the same as the handle used in constructing the datatype.

Therefore, our library relies on a non-type-specific compression function (zlib) to remove duplicate datatypes from the marshalled representation. Compression can be requested by passing "compressed" as the requested encoding to the marshalling functions.

4 Evaluation

We evaluated the marshalling and unmarshalling functions for a number of MPI datatypes, using both "external" and "compressed" representations. We timed 10,000 iterations for each operation and reported the mean time per iteration. Table 1 shows the results.

The first type evaluated ("named") refers to any predefined MPI datatype. As each predefined type is treated equally by the library, the numbers listed are valid for any predefined type. The second type tested ("indexed") is an `MPI_Type_indexed` type selecting three contiguous regions from a byte array. The third type is a complex derived datatype we captured from the HDF5 [3] storage library. This type was created to describe the on-disk access pattern used when accessing 5000 bytes of a dataset stored in the HDF5 file.[5] This particular type is 12 type constructors deep.

As expected, more complicated types take additional space and time to marshal. The named and indexed types are cheaply serialized in the "external" format. In the case of the complex type, the "compressed" format consumes ≈ 8.6 times less space but takes ≈ 6.4 times as long to marshal.

Our marshalling implementation is currently a prototype and has not been extensively optimized. We expect that marshalling and unmarshalling times could be reduced further with additional effort. The data presented in Table 1 are intended to provide a rough idea of marshalling performance.

---

[5] The exact type can be found in the MPITypes distribution [7] as `test/very_deep.c`.

Table 1. Evaluation of marshalling and unmarshalling time and space consumption in the prototype implementation

Type	External			Compressed		
	Size (B)	Time (μs)		Size (B)	Time (μs)	
		Marshal	Unmarshal		Marshal	Unmarshal
Named	8	< 1	< 1	14	76	1
Indexed	40	1	4	30	79	3
Complex	824	23	33	95	147	34

5   Related Work

One focus of research in MPI datatypes has been the detection of mismatched datatypes passed to MPI communication functions. Gropp introduced the idea of using hashes for this purpose and defined a hashing function that maps the type signature into an integer tuple [2]. Langou et al. extended this idea, proposing alternative hashing schemes and examining the performance of these approaches [4]. Falzone et al used this approach in a library for detecting user errors in the use of MPI collective communication calls [1], building on a simpler scheme first presented by Träff et al. [11]. These hashes can be used, for instance, to reference a datatype in a local cache, allowing a remote entity to query if a datatype is already represented in the cache.

Another focus has been in the efficient processing of MPI datatypes. Ross et al. describe the implementation of datatype processing used in MPICH2 [9] and an external library for processing MPI datatypes [8]. The approach used is similar to the one first described by Träff et al. [12]. Recently, Mir and Träff studied *transpacking*, or moving data between datatype representations [6]. These approaches generally rely on a simplified, underlying datatype representation with a type signature identical to the original user datatype. When these representations are available, they allow marshalling of a simplified description of a datatype, if envelope and contents information is not needed.

The Hierarchical Data Format version 5 (HDF5) [3] provides functionality similar to the MPI datatypes (called *datasets* in HDF5), splitting the definition of a dataset into a *dataspace* that describes the organization of elements and a datatype that describes a single element, similar to an MPI struct. HDF5 stores these descriptions persistently in HDF5 files, but it does not present an interface for marshalling these descriptions to users.

6   Conclusions and Future Work

In this paper we have discussed the notion of MPI datatype equivalence, arguing that the definition put forth in the standard is appropriate only for a certain set of use cases. We have identified a number of other interpretations, and we have provided an API and a library of functions that enable marshalling of MPI datatypes in order to meet various levels of equivalence.

We intend to release this functionality for general use in the MPITypes library [8,7]. We also plan to investigate using a combination of MPI attribute caching and datatype hashing techniques [4] to optimize the case when types are repeatedly serialized. Issues in the design and implementation of "internal" marshalling schemes also merit further study. We intend to propose this interface for the MPI-3 standardization process.

Acknowledgments. This work was supported by the U.S. Department of Energy, under Contract DE-AC02-06CH11357.

References

1. Falzone, C., Chan, A., Lusk, E., Gropp, W.: A portable method for finding user errors in the usage of MPI collective operations. International Journal of High Performance Computing Applications 21(2), 155–165 (2007)
2. Gropp, W.: Runtime checking of datatype signatures in MPI. In: Dongarra, J., Kacsuk, P., Podhorszki, N. (eds.) PVM/MPI 2000. LNCS, vol. 1908, pp. 160–167. Springer, Heidelberg (2000)
3. HDF5, http://hdf.ncsa.uiuc.edu/HDF5/
4. Langou, J., Bosilca, G., Fagg, G., Dongarra, J.: Hash functions for datatype signatures in MPI. In: Di Martino, B., Kranzlmüller, D., Dongarra, J. (eds.) EuroPVM/MPI 2005. LNCS, vol. 3666, pp. 76–83. Springer, Heidelberg (2005)
5. Message Passing Interface Forum: MPI: A Message-Passing Interface Standard Version 2.2 (September 2009), http://www.mpi-forum.org/docs/docs.html
6. Mir, F.G., Träff, J.L.: Constructing MPI input-output datatypes for efficient transpacking. In: Lastovetsky, A., Kechadi, T., Dongarra, J. (eds.) EuroPVM/MPI 2008. LNCS, vol. 5205, pp. 141–150. Springer, Heidelberg (2008)
7. MPITypes library, http://www.mcs.anl.gov/mpitypes/
8. Ross, R., Latham, R., Gropp, W., Lusk, E., Thakur, R.: Processing MPI datatypes outside MPI. In: Ropo, M., Westerholm, J., Dongarra, J. (eds.) Recent Advances in Parallel Virtual Machine and Message Passing Interface. LNCS, vol. 5759, pp. 42–53. Springer, Heidelberg (2009)
9. Ross, R., Miller, N., Gropp, W.: Implementing fast and reusable datatype processing. In: Dongarra, J., Laforenza, D., Orlando, S. (eds.) EuroPVM/MPI 2003. LNCS, vol. 2840, pp. 404–413. Springer, Heidelberg (2003)
10. Srinivasan, R.: XDR: External data representation standard (1995)
11. Träff, J., Worringen, J.: Verifying collective MPI calls. In: Kranzlmüller, D., Kacsuk, P., Dongarra, J. (eds.) EuroPVM/MPI 2004. LNCS, vol. 3241, pp. 18–27. Springer, Heidelberg (2004)
12. Träff, J.L., Hempel, R., Ritzdorf, H., Zimmermann, F.: Flattening on the fly: Efficient handling of MPI derived datatypes. In: Margalef, T., Dongarra, J., Luque, E. (eds.) PVM/MPI 1999. LNCS, vol. 1697, pp. 109–116. Springer, Heidelberg (1999)

Design of Kernel-Level Asynchronous Collective Communication

Akihiro Nomura and Yutaka Ishikawa

Dept. of Computer Science, Graduate School of Information Science and Technology,
The University of Tokyo
7-3-1 Hongo, Bunkyo-ku, Tokyo, Japan
nomura@il.is.s.u-tokyo.ac.jp,
ishikawa@is.s.u-tokyo.ac.jp

Abstract. Overlapping computation and communication, not only point-to-point but also collective communications, is an important technique to improve the performance of parallel programs. Since the current non-blocking collective communications have been mostly implemented using an extra thread to progress communication, they have extra overhead due to thread scheduling and context switching. In this paper, a new non- blocking communication facility, called KACC is proposed to provide fast asynchronous collective communications. KACC is implemented in the OS kernel interrupt context to perform non-blocking asynchronous collective operations without an extra thread. The experimental results show that the CPU time cost of this method is sufficiently small.

Keywords: Non-blocking collective communication, Linux kernel.

1 Introduction

In parallel applications, the performance and efficiency of communications often dominate the performance of the whole calculation. In addition to blocking point-to-point communication APIs in the MPI (Message Passing Interface) [5], some APIs for non-blocking communication, such as `MPI_Isend` and `MPI_Irecv`, are defined. Non-blocking communication allows calculations to continue during communication. This enables the MPI processes to overlap between calculation and communication.

MPI also defines collective communication APIs, such as `MPI_Reduce` and `MPI_Bcast`, to perform the conventional sets of communications easily and efficiently. The users of the MPI library do not need to know what is going on during the collective communication. The MPI library offers the most efficient algorithms for the requested collective communication with regard to the communication size, topology, and other information. Both APIs are used to efficiently perform the communication.

In the current version of MPI, due to the lack of non-blocking collective communication APIs, users must implement non-blocking communications in order to perform the collective communications asynchronously. For example, in the

R. Keller et al. (Eds.): EuroMPI 2010, LNCS 6305, pp. 92–101, 2010.

HPL [8] implementation, a non-blocking version of MPI_Bcast was implemented, but it is hard to maintain the code due to the complexity of the collective algorithms and a mixture of communication and computation routines. Furthermore, the code might be inefficient in some topologies because the broadcast algorithm is based on some assumed network topology. Thus, the introduction of non-blocking collective communication APIs to the MPI standard has been discussed.

In the next version of the MPI standard, MPI 3.0, non-blocking collective communication APIs are to be introduced. There is a reference implementation of those APIs, LibNBC [2, 3]. In the implementation, non-blocking communication operations are implemented using threads for communication progress. The thread implementation has two limitations. Firstly, if an extra thread that performs communications is introduced, it consumes CPU resources due to the overhead of both task scheduling and context switching. For example, if an MPI application runs on an eight-core cluster in which each process runs on each CPU core, sixteen threads are created. Eight threads are for processes, and the other eight threads are for communication progress. This means that the execution of those threads is multiplexed. Secondly, since the timing of communication progress depends on the task scheduling in the operating system, it is not guaranteed that the progress thread runs immediately when the communication processing is ready when a message arrives.

In this paper, a new non-blocking collective communication facility, called KACC, is designed and implemented to overcome the limitations described above. KACC is implemented in the OS kernel interrupt context in order to perform the non-blocking collective operations without an extra thread. Since the communication progress is handled when a message arrives, there is no delay in the progress, and no extra context switching overhead is introduced. The facility has been implemented as a kernel module with a user-level library in the Linux kernel. KACC is evaluated by a benchmark which uses non-blocking broadcast algorithm. The benchmark reveals how much the non-blocking broadcast operation contributes to overlapping communication and computation. Four implementations of the non-blocking broadcast operation are considered: a tree-based broadcast operation written in MPI, a non-blocking point-to-point operation; a tree-based broadcast operation written in KACC; a pipeline-based broadcast operation written in the threads; and a pipeline-based broadcast operation written in KACC.

The CPU waste time depends how often the application program examines the completion of the non-blocking operations. The results of the benchmarks show that 97 % of CPU time is lost in LibNBC, while only 31.8 % of CPU time is lost in KACC during a high frequency of examinations. The total execution time of non-blocking collective operation is also improved. The result shows that the execution time in KACC is 79 to 101% of that in LibNBC.

2   Issues

Implementation of non-blocking collective communications is not trivial due to progressions. Most of the implementations of collective communications consist of the set of point-to-point communications that are connected by data dependencies. Progression is the procedure to connect these point-to-point communications. The MPI library must issue the communication when all of the dependent communications are completed in order to continue collective communication. Two implementations have been introduced so far: thread implementation and explicit progression.

Thread Implementation. The straightforward solution to this problem is creating a thread for communication and performing progression in this thread. An example of this method is used in the LibNBC implementation [3]. The advantage of this method is that the communication will execute asynchronously to computation that runs on another thread. Theoretically, the communication thread runs independently from the computation thread, and the progression is always executed at the appropriate timing.

However, the real situation is different due to the limitation of the number of CPU cores. The user usually spawns the same number of MPI processes as the number of CPU cores, because the user often assumes that all cores can be used for computation. In this situation, if the MPI library makes the communication thread, the number of active threads exceeds the number of cores. This results in frequent context switches among these threads. If the context switches are performed by the operating system, and the OS does not know about the dependencies among these threads, then the timing of context switches might not be optimal. This will result in waste of CPU time and delay of communications.

Explicit Continuation. Another way to implement non-blocking communications instead of creating communication threads is to implement the progressions in the MPI library, that is, the progression of collective communications is only done within the MPI functions, such as `MPI_Test` and `MPI_Wait`, when invoked by the application program. This method does not create any threads, and thus the context switching problem does not happen.

On the other hand, the progression is not processed if the application does not call any MPI functions. This results in no overlapping computation and communication, that is, although a non-blocking collective communication has been posted, the communication does not progress during the computation. If the program waits for the completion of the non-blocking collective communication by issuing `MPI_Wait` when computation is completed, the progression starts. If the user calls the progression too frequently to avoid missing progression timing, this results in a loss of CPU time.

This kind of explicit progression method is used in some MPI applications. For example, in the Linpack benchmark program, non-blocking broadcast is implemented using non-blocking send and receive primitives. For example, during local computation, Linpack polls whether the broadcast has been completed using the `MPI_Test` primitive.

3 Design

In order to solve both the frequent context switching and false asynchronization problems at the same time, the KACC facility is designed and implemented in this paper. In KACC, the progression routine is implemented as an OS kernel's soft-interrupt handler. In this method, the number of threads does not increase, because the progression routine does not have a thread context, and it is called at the appropriate time by the kernel interrupt handler instead of the OS scheduler.

3.1 Collective Algorithm Design

Splitting the collective algorithm from the kernel module is important to design new collective algorithms. However, it becomes a security hole if the program binaries described in the user-level program can be passed to kernel space directly. Instead of sending binary, a data structure, called a CAD (Collective Algorithm Design) structure, is introduced to describe the collective algorithms. The MPI library creates the CAD structure and passes the structure to the kernel module.

Collective communication algorithms can be mapped to directed acyclic graphs which shows dependencies among the point-to-point and reductive operations [6, 3]. In the CAD structure, collective algorithms are expressed using these graph structures instead of the binary program. There are three types of nodes: SEND, RECV, and CALC. The SEND and RECV nodes represent communication. These nodes contain information required for communication: address of data, data size, rank of sender or receiver, and tag information to match the messages. The CALC node represents calculation in the MPI reduction function, such as MPI_Sum and MPI_Max. This node contains information about the reductive calculation operator, memory address, size, and data type. The edges between nodes denote dependencies between each operation.

A CAD, describing a collective algorithm, is created and executed using the following API:

- InitCAD(): creates new CAD
- MakeSendNode(), MakeRecvNode(), MakeCalcNode(): creates CAD nodes for SEND, RECV and CALC, respectively
- ConnectNode(A, B): marks the dependency edges from node A to node B START and END nodes are pre-defined
- IssueCAD(): tells the system to start communication
- QueryCAD(): queries to system whether the operation has been completed

An example of a CAD structure, representing a non-blocking broadcast algorithm, is shown in Figure 1. This structure is generated at rank 1 by the user-level code shown in Figure 2. Note that in this broadcast implementation, each SEND/RECV node sends/receives a fragment of the message. The message is stored in the addr memory area that is defined as an array for simplicity. In the code shown in Figure 2, a data structure to store CAD tree is allocated by InitCAD in line 2 at first. Then, RECV and SEND nodes for each fragment are created by MakeRecvNode and MakeSendNode. Each RECV/SEND node has

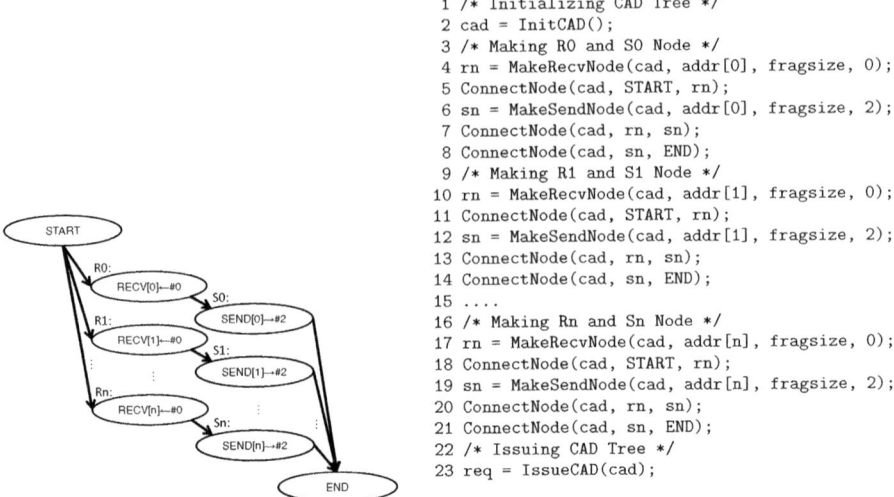

```
 1 /* Initializing CAD Tree */
 2 cad = InitCAD();
 3 /* Making R0 and S0 Node */
 4 rn = MakeRecvNode(cad, addr[0], fragsize, 0);
 5 ConnectNode(cad, START, rn);
 6 sn = MakeSendNode(cad, addr[0], fragsize, 2);
 7 ConnectNode(cad, rn, sn);
 8 ConnectNode(cad, sn, END);
 9 /* Making R1 and S1 Node */
10 rn = MakeRecvNode(cad, addr[1], fragsize, 0);
11 ConnectNode(cad, START, rn);
12 sn = MakeSendNode(cad, addr[1], fragsize, 2);
13 ConnectNode(cad, rn, sn);
14 ConnectNode(cad, sn, END);
15 ....
16 /* Making Rn and Sn Node */
17 rn = MakeRecvNode(cad, addr[n], fragsize, 0);
18 ConnectNode(cad, START, rn);
19 sn = MakeSendNode(cad, addr[n], fragsize, 2);
20 ConnectNode(cad, rn, sn);
21 ConnectNode(cad, sn, END);
22 /* Issuing CAD Tree */
23 req = IssueCAD(cad);
```

Fig. 1. CAD tree example   **Fig. 2.** Code generating CAD Tree

corresponding SEND/RECV node which is created by the code in corresponding rank. After that, each SEND/RECV nodes and special START and END nodes are connected using ConnectNode to form dependencies shown in Figure 1. Finally, the CAD tree is fixed and sent to the KACC system using IssueCAD in line 23. The progression routine, which will be introduced in the following section, starts communication in CAD tree at this time. The user can query the completion of issued CAD using QueryCAD.

4   Implementation

4.1   Structure of KACC System

As shown in Figure 3, the KACC facility consists of three layers: the CAD API, the Progress Engine (PE), and the point-to-point (P2P) communication interface. The latter two layers are implemented inside the Linux kernel as a kernel module. The CAD API described in the previous section is implemented as a user-level library. Using the API, the CAD structure is created in a special memory area shared by MPI processes and the kernel module, so that no structure copy between the user and kernel memory spaces is required.

The Progress Engine (PE) plays two roles. The first role, invoked at the user-level, is to start communication between the CAD structures and to report its progress. The second role, invoked by the P2P communication interface, is to perform the communication algorithm on the CAD structure. The latter role is implemented as a *tasklet* in the Linux kernel, that is, it runs under the kernel context triggered by the interrupt routine. Thus, the implementation of KACC does not have threads.

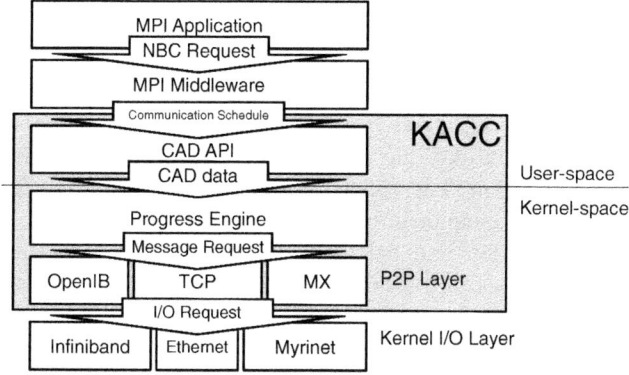

Fig. 3. Structure of KACC facility

The Linux tasklet runs on the same core as the tasklet is scheduled on. If the network interrupts always trigger one specific core, all of the PE routine is executed on that core and fully serialized. In order to avoid this serialization and balance the PE tasklets among CPU cores, we use inter-processor interrupts(IPI) to schedule them on arbitrary core. This method enables each PE event to be executed simultaneously and reduces total execution time. On the other hand, CPU time loss increases due to IPI costs.

The point-to-point (P2P) layer offers communication APIs to the PE layer. The P2P APIs are independent from the network devices and are similar to MPI's non-blocking point-to-point communication APIs. The difference between them is the notification method of completion. The P2P layer invokes PE's callback routine immediately after completion, instead of offering a polling interface of completion. Thus, the PE's routine is always called at the appropriate timing.

In the current implementation, only the kernel-mode non-blocking TCP has been implemented in the P2P layer, since the recent interconnect devices often have an IP interface [7,4]. It is possible to implement the P2P layer using native APIs for Myrinet or InfiniBand instead of using IP compatibility layer of these interfaces. The interface between PE and P2P layer is message-oriented, it will fit nicely to the message-oriented communication in Myrinet MX. If we use the native implementation, the communication will be faster and the CPU time loss will be smaller.

Currently, the network connection used in P2P layer is established independently from connection of MPI library in order to distinguish KACC's traffic from other traffics. We are planning to implement communication device interface for MPICH2 or other MPI implementations. This interface also provides point-to-point non-blocking communication API and all communications in MPI programs are executed by KACC facility.

5 Evaluation

The performance of KACC facility is compared to the other implementation, LibNBC, using a non-blocking broadcast communication in this section. A benchmark program is designed based on the HPL benchmark [8] in order to show how much the non-blocking operation contributes to overlapping communication and computation. The benchmark is named the HPL codelet because it is not a real Linpack benchmark, but is a code snippet from HPL that performs broadcast communication and calculation simultaneously. In this benchmark, after issuing a non-blocking broadcast operation, the fixed amount of calculation is computed repeatedly until the broadcast operation is completed, that is, at every end of calculation, the completion of the operation is examined. The calculation is the matrix multiplication whose size is specified at the run time so that the examination frequency of completion is programmable.

Four implementations of the non-blocking broadcast communication have been carried out. The first implementation is the original HPL non-blocking broadcast communication implementation, using MPI point-to-point operations, whose communication algorithm is based on a binary tree, which will be denoted as *MPI Tree*. The second implementation is the same algorithm written using KACC, which will be denoted as *KACC Tree*. The third implementation is the version used in the LibNBC's library, that is based on pipeline processing, which will be denoted as *LibNBC Pipeline*. The fourth implementation is the same algorithm as LibNBC but written in KACC, which will be denoted as *KACC Pipeline*.

The execution time of the benchmark was measured for the four implementations. The experimental environment consists of eight computing nodes connected by a 1Gbps Ethernet. Each computing node has two dual-core 2GHz Opteron CPUs; thus, there are 32 cores in this cluster. MPICH2/TCP 1.0.6 [1] is used as the base MPI environment.

The percentage of CPU time spent for communication is estimated by comparing the number of calculation to the ideal number of calculations without communication. The effect of the frequency of examining the completion of the communication operation is revealed by varying both computation and communication lengths. For the granularity of the computation, two different matrix sizes for calculation are considered: 40×40 matrices for coarse-grained testing and 4×4 matrices for fine-grained testing.

The results of CPU time loss are shown in Figures 4 and 5. In order to show the cost of tasklet load-balancing by IPI, the CPU time loss in KACC without load-balancing is also shown in graph (b) of each figures. In the coarse-grained workload shown in Figure 4, the frequency of polling is less, and thus, in this case, the CPU is not consumed by polling communication activity. However, in LibNBC, 28.3 to 57.3% of CPU time is always spent for the communication thread, and, therefore, the effect from the communication computation overlap is relatively small. On the other hand, in the same pipeline algorithm with KACC facility, the CPU time spent for communication is limited to about half of the case with LibNBC, and about half of the CPU time is due to the cost from IPI handling.

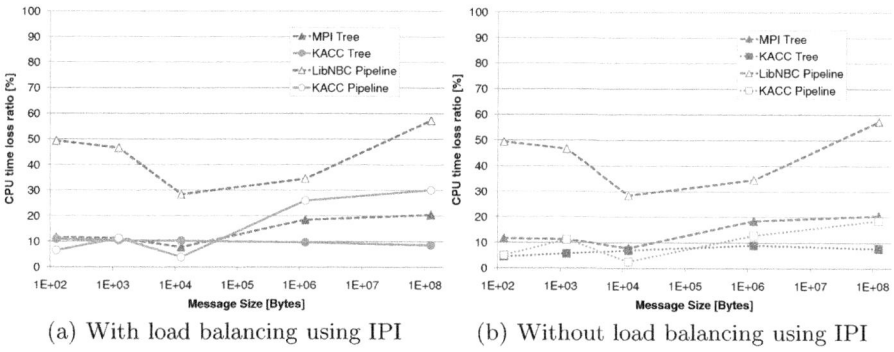

(a) With load balancing using IPI   (b) Without load balancing using IPI

Fig. 4. CPU time loss during broadcast with coarse-grained workload

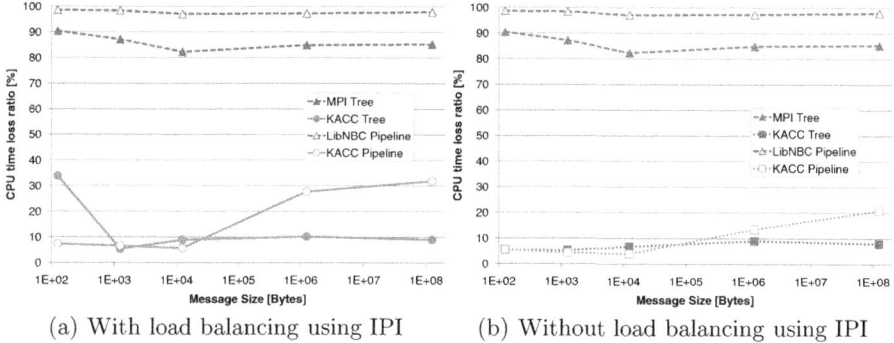

(a) With load balancing using IPI   (b) Without load balancing using IPI

Fig. 5. CPU time loss during broadcast with fine-grained workload

When the granularity of computation is fine as shown in Figure 5, the frequency of tests for the communication's completion is high. In this case, 97.0% of the CPU time is lost to the communication thread in LibNBC. On the other hand, KACC's CPU loss rate is limited to 31.8% even if load-balancing using IPI is enabled. This result shows that the users can continue calculation effectively during collective communication under the KACC facility.

The ratio of total execution time in broadcast on KACC compared to normal MPI tree and LibNBC pipeline implementations is shown in Figure 6. If the ratio is smaller than 1, corresponding method is faster than the compared case, MPI Tree or LibNBC Pipeline. In the best case, it took 2.15 milliseconds for 12.5kB broadcast in MPI and 1.19 milliseconds in the same algorithms with KACC facility and thus execution time ratio is calculated as 0.55. Similarly, it took 22.6 milliseconds for 1.25MB broadcast in LibNBC and 17.9 milliseconds in the same algorithms with KACC facility and execution time ratio is calculated as 0.79. For various communication size and workload granularity, the execution time in the KACC is 55 to 98% of that in MPI implementation and 79 to 101% of that in LibNBC except for small message size. However, if the message size is small,

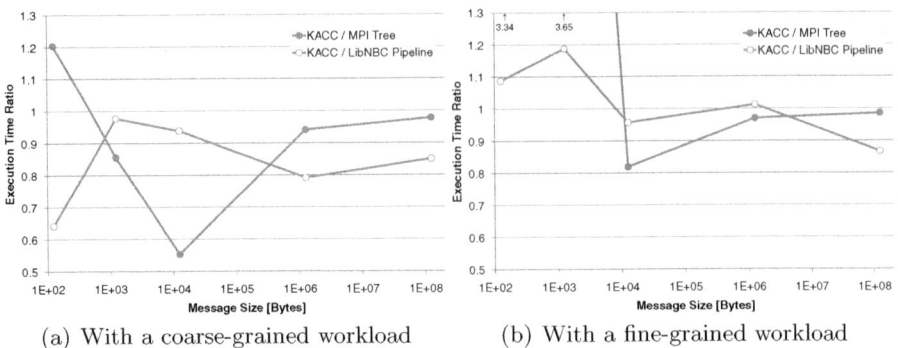

(a) With a coarse-grained workload   (b) With a fine-grained workload

Fig. 6. Total execution time ratio in non-blocking broadcast

KACC implementation is slower than normal MPI or LibNBC implementation. We have to improve the performance of small messages in the future.

6   Conclusions

Non-blocking collective operations have been proposed and implemented. If non-blocking collective operations are performed asynchronously with the main computation, the communication latency can be hidden, and more time can be made available for computation. However, these implementations are not always scaled due to the extra cost of asynchronous communication management using threads.

This paper proposed the KACC facility, which performs collective communications in the OS kernel's interrupt context. Since non-blocking collective operations are performed in the kernel interrupt context, no extra threads are introduced. Furthermore, a collective operation is progressed immediately when a message for the operation arrives. These two features contribute to the application program spending more CPU time in computation.

A benchmark was used to test four different implementations, including the thread implementation in LibNBC [2,3], of the non-blocking broadcast operation. The results show that the thread implementation, LibNBC, consumes almost half of the CPU when less frequent polling of completion is requested or 97% of CPU time when frequent polling is performed, while KACC consumes only 10 to 30% of CPU time for both cases while keeping its execution time up to 79 to 101% of that in LibNBC.

There are four limitations in the current implementation. Firstly, the user-defined operations have not yet been considered. The execution of operations must be safe in terms of security. There are two potential approaches to this problem: interpreter and thread approaches. In the interpreter approach, the function is compiled to virtual machine code, and this code is interpreted in the kernel. Another approach is to introduce a thread for executing the operations. In this approach, an extra thread is created. However, unlike the thread for communication progress, the extra thread is only activated during the operation of collectives, and the overhead is expected to be small.

Secondly, the current implementation has not yet been optimized for the intra-node communication, since intra-node communications are handled by the TCP connections. If messages are directly copied by the memory copy operation, performance is improved.

Thirdly, the P2P layer supports only TCP in the current implementation. Porting to other hardware, such as InfiniBand and Myrinet, is being considered.

Fourthly, the current implementation is slow if the messages are small. Improving the performance for small messages is strongly needed. We may recommend not to use non-blocking collective communications if the message size is small. In this case, we should show reasonable method to determine the threshold of message size to use non-blocking collective communications.

References

1. Argonne National Laboratory: MPICH2: High-performance and widely portable MPI, http://www.mcs.anl.gov/research/projects/mpich2/
2. Hoefler, T., Lumsdaine, A.: Design, Implementation, and Usage of LibNBC. Tech. rep., Open Systems Lab, Indiana University (August 2006)
3. Hoefler, T., Lumsdaine, A., Rehm, W.: Implementation and Performance Analysis of Non-Blocking Collective Operations for MPI. In: Proceedings of the 2007 International Conference on High Performance Computing, Networking, Storage and Analysis, SC 2007. IEEE Computer Society/ACM (November 2007)
4. Kashyap, V.: IP over InfiniBand (IPoIB) Architecture. RFC 4392 (Informational) (April 2006), http://www.ietf.org/rfc/rfc4392.txt
5. MPI Forum: Message passing interface, http://www.mpi-forum.org/
6. MPI Forum: MPIplans - an alternative for all other collectives proposals? https://svn.mpi-forum.org/trac/mpi-forum-web/wiki/MPIplans
7. Myricom, Inc.: IP over myrinet, http://www.myri.com/scs/documentation/mug/ip/
8. Petitet, A., Whaley, R.C., Dongarra, J., Cleary, A.: HPL - a portable implementation of the high-performance linpack benchmark for distributed-memory computers, http://www.netlib.org/benchmark/hpl/

Network Offloaded Hierarchical Collectives Using ConnectX-2's CORE-*Direct* Capabilities

Ishai Rabinovitz[1], Pavel Shamis[1], Richard L. Graham[2],
Noam Bloch[1], and Gilad Shainer[3]

[1] Mellanox Technologies, Inc.
{ishai,pasha,noam}@mellanox.co.il
[2] Oak Ridge National Laboratory (ORNL)
rlgraham@ornl.gov*
[3] Mellanox Technologies, Inc.
shainer@mellanox.com

Abstract. [1]As the scale of High Performance Computing (HPC) systems continues to increase, demanding that we extract even more parallelism from applications, the need to move communication management away from the Central Processing Unit (CPU) becomes even greater. Moving this management to the network, frees up CPU cycles for computation, making it possible to overlap computation and communication. In this paper we continue to investigate how to best use the new CORE-*Direct* support added in the ConnectX-2 Host Channel Adapter (HCA) for creating high performance, asynchronous collective operations that are managed by the HCA. Specifically we consider the network topology, creating a two-level communication hierarchy, reducing the MPI_Barrier completion time by 45%, from 26.59 microseconds, when not considering network topology, to 14.72 microseconds, with the CPU based collective barrier operation completing in 19.04 microseconds. The nonblocking barrier algorithm has similar performance, with about 50% of that time available for computation.

Keywords: InfiniBand, Offload, Collectives, Hierarchy.

1 Introduction

System-area network characteristics on any HPC system have a large impact on the performance of applications running on them, equally important to that of

* Research sponsored by the Office of Advanced Scientific Computing Research's FASTOS program; U.S. Department of Energy, and performed at ORNL, which is managed by UT-Battelle, LLC under Contract No. DE-AC05-00OR22725. The HPC Advisory Council (http://www.hpcadvisorycouncil.com) provided computational resource for testing and data gathering.

[1] Research sponsored by the Office of Advanced Scientific Computing Research's FASTOS program; U.S. Department of Energy, and performed at ORNL, which is managed by UT-Battelle, LLC under Contract No. DE-AC05-00OR22725. The HPC Advisory Council (http://www.hpcadvisorycouncil.com) provided computational resource for testing and data gathering.

the CPU and the attached storage. However, much more attention has been focused on improving CPU capabilities, with a relatively modest effort put into improving network hardware capabilities. Since the need for computational power continues to grow and CPU speeds are remaining essentially constant, top end system core counts are expected to continue their rapid growth, and even accelerate given energy consumption constraints, as we move towards exascale class computing [16]. Core counts of the top-ten systems on the Top500 list [5] in November 2009 reached an average of 134,893, and are expected to reach about 10,000,000 cores to provide an exascale computing environment [16]. To make effective use of such systems, a more balanced system design is needed. Enhancing network capabilities, performance, and scalability is required.

Improving network performance, the performance of point-to-point (PTP) communications and collective communications needs be addressed. These include addressing network specific issues such as reducing latency (PTP and collective), increasing network bandwidth, and increasing message rates. Because of their absolute performance and scalability, collective communications are often the limiting factor for achieving good application scalability. The ordered communication patterns used by high performance implementation of collective algorithms present an application scalability challenge, as a delay in entering a particular stage of the algorithm propagates through the rest of that particular collective operation. The reason for such delays may be classified as application or system effects. Application load imbalance, with different application processes entering a collective operation at different times are a cause for such delays. System activity, also known as system noise [21], which displaces application processes from the CPU also delays the participation of the displaced processes in the collective operation, further posing a scalability challenge. In addition, because CPU cycles are used in managing network communications, taking away compute CPU cycles from the application, progressing network communication also tends to hamper application scalability.

CORE-*Direct* functionality, recently added to the InfiniBand ConnectX-2 HCAs by Mellanox Technologies [2], provides comprehensive hardware support for offloading a sequence of data-dependent communications to the network. This allows moving some of the network processing from the CPU to the network, freeing up CPU cycles for computation. It also supports asynchronous collective communications, such as the pending MPI [19] nonblocking collective operations, providing some of the support needed from the network for a more balanced system design.

The major contributions described in this paper include the first implementation of asynchronous hierarchical InfiniBand blocking and nonblocking barrier algorithms, which can be fully offloaded to the network. We briefly describe the new InfiniBand CORE-*Direct* capabilities, and a hierarchical design used to improve collective (barrier) performance. We present the results of numerical studies, designed to understand the characteristics of this new approach, as applied to blocking and nonblocking barrier algorithm, presenting both timing data, as well as the results of computation and communication overlap experiments. We finish with a discussion of these results.

2 Related Work

Several authors have investigated using the HCA to offload collective communication management from the CPU. Analyses of HCA-based broadcast algorithms are available in references [6,26]. Generally, these all tend to use HCA-based packet forwarding as a means of improving performance of the broadcast operation. Some of the benefits of offloading *barrier, reduce,* and *broadcast* operations to the HCA are described in references [8,7,20], and [26,25]. These show that *barrier* and *reduce* operations can benefit from reduced host involvement, efficient communication processing, and are more tolerant to process skew. Keeping the data transfer paths relatively short in multi-stage communication patterns is appealing, as it offers a favorable payback for moving this work to the network. Our recent work [12,13] has started to investigate the new support for offloading collective operation management to the network using the CORE-*Direct* technology introduced in ConnectX-2. We have shown good blocking and nonblocking barrier performance, and about 80% CPU availability while executing a non-blocking barrier. Amongst all the previous HCA-based collective implementations, only the efforts from Quadrics [4] for Elan 3/4 and IBM [11] for Blue Gene have been widely used.

Others studies have shown that the CPU may be used in support of asynchronous collective operation progress. Sancho et. al [22] have dedicated several CPU's to processing collective communications to improve their performance and scalability. Hoefler et. al [15,14] have implemented nonblocking collectives, and investigated ways to minimize CPU overhead for progressing these collectives. Our approach for supporting both blocking and non-blocking collective operations is aimed at avoiding CPU involvement altogether in progressing these operations.

Much of the work on collective communication hierarchies occurs in the context of grid computing, aimed primarily at handling clusters connected by a relatively low-performance network, with limited connectivity, which is typified by the work described in [17,23]. In the context of HPC, most of the work on hierarchical collectives has been exploiting the low-latency shared memory communication available between processes on a given host [24]. For example, LA-MPI [10], Open MPI [9], MVAPICH [3], and MPICH2 [27] have such support. The key idea behind these implementations is the use of a two-level communication hierarchy, to maximize the utilization of the higher performing shared memory communication at one level, and minimize network communications at another level, by restricting it to one rank within the shared memory domain, also known as the *local leader*.

Since our goal it to develop high-performance asynchronous collective operations, we will explore the characteristics of hierarchical collectives whose progression is fully offloaded to the HCA. As such, we will take advantage of the

the on-host hierarchy, but will use the CORE-*Direct* capabilities, rather than shared memory communication.

3   Overview of Technical Approach

A detailed description of ConnectX-2's CORE-*Direct* design and how this is used to implement collective operations is provided elsewhere [12,13], so we will provide only a brief description of these here, and then describe the new design features studied in this paper. We describe a hierarchical approach for offloading the collective operations, both blocking and nonblocking, that provides full network hardware support for collective communication management. This also provides a large performance boost relative to previous CORE-*Direct* algorithms.

The CORE-*Direct* functionality is an extension of the InfiniBand Architecture (IBA) [1] specification. IBA is designed for interconnecting compute nodes, Input/Output (I/O) nodes and storage devices in a system area network. It defines a communication architecture from the switch-based network fabric to transport layer communication interface for inter-processor communication. Processing nodes and I/O nodes are connected as end-nodes to the fabric by two kinds of channel adapters: HCA and Target Channel Adapters.

Send Work Queue Entries (WQEs) are interpreted and processed by the HCA, which sends data on behalf of this WQE using either send/receive or Remote Direct Memory Access (RDMA) capabilities. Received data is delivered in the order in which it is sent. Send/receive packets are delivered to the appropriate Queue Pair (QP), and are matched up with the appropriate receive WQE. RDMA data is delivered directly to the address specified in the send operation producing a receive completion queue entry if immediate data mode is used. Completion of a WQE results in a Completion Queue Entry being posted to a Completion Queue (CQ), where the CQ can be polled for completions.

The IBA defines several communication tasks, these include send, receive, read, write, and atomic tasks. CORE-*Direct* adds hardware support for cross QP synchronization operations - wait, send_enable, and receive_enable tasks. Wait takes as an argument a completion queue and the number of completion tasks to wait for, and can be used to order communications taking place using different QP's. Information about completed tasks consumed by a wait task may not be obtained from a completion queue, and must be inferred from QP completion ordering, with receive buffers being consumed in the order in which they are posted. If a single completion queue was used for more than a single receive queue, there is no way to correctly identify the source of the arriving data, as there is no ordering of completion events from different sources. Send_enable and receive_enable tasks activate an already posted send or receive task, respectively, allowing the HCA to process these tasks. This provides the ability to delay activation of the send or receive tasks until after the tasks preceding the enable tasks in their queues have been executed. The Multiple Work Request (MWR) is a list of InfiniBand communication tasks which the driver posts, in order, to the queues specified by the individual work requests. These tasks include the send, receive, RDMA write and synchronization tasks. An MWR completion entry is posted after the task that is marked with the flag MQE_WR_FLAG_SIGNAL is processed by the HCA. The MWR may be used to chain a series of network tasks, and, once posted, the HCAs progress the communication, without using the CPU.

The Management Queue (MQ) is set up to handle MWRs. When an MQ is created, a completion queue is also created, thus imposing a one-to-one mapping of MQ onto MQ Completion queue. When an MWR is posted to the MQ, the driver posts the individual work requests in-order to the specified QP's with no interleaving of individual tasks from different MWR's. Wait tasks are posted either to the QP specified in the task, or if a NULL queue is specified, the wait task is posted to the MQ. The send/receive tasks are posted to the specified QPs. The driver will also generate additional tasks, based on the structure of the user's MWR. When a send/receive task in the MWR follows a wait task that is posted to the MQ the send/receive task is posted to the specified QP, but is not enabled for send/receive, and it cannot be processed by the HCA until it is enabled. In addition, a send/receive-enable task is posted to the MQ after the wait task, which will cause the corresponding send/receive task to be enabled after the wait task completes.

Collective operations can employ the hardware defined tasks and the multiple work request functionality. To achieve this, each process in the communicator creates an MWR that describes its local portion of the collective communication pattern, with the necessary synchronization operations (e.g., wait or enable tasks). The MWR is posted to the management queue to initiate the collective operations, with the completion-queue being polled for completion. For example, the MWR task list for a four process recursive doubling barrier operations is provided in Table 1.

The initial implementation of the collective operations showed the benefit of reducing the number of queue pairs used to implement these collective operations. However, they did not consider cluster topology in the algorithms, that is it did not take into account optimizations for ranks sharing the services of the same HCA. We have developed network topology aware collective operations taking into account processes that share the use of a given HCA. Processes sharing an HCA are grouped in one hierarchy, with a local-leader from each group participating in the second-level inter-host hierarchy. The local traffic uses a small number of QP's, reducing pressure on HCA resources, and also avoids

Table 1. MWR task list for each rank participating in a four process recursive doubling barrier

proc 0	proc 1	proc 2	proc 3
send to 1	send to 0	send to 3	send to 2
recv wait from 1	recv wait from 0	recv wait from 3	recv wait from 2
send to 2	send to 3	send to 0	send to 1
recv wait from 2	recv wait from 3	recv wait from 0	recv wait from 1

generating traffic to the switch, while using the CORE-*Direct* capabilities. Also, the number of inter-host messages is reduced, putting less pressure on the network. In many implementations of hierarchical collective operations, on-host communication use shared memory, using CPU cycles to progress the shared memory phase of the algorithm. However, in this approach, all communications take place using the ConnectX-2 CORE-*Direct* capabilities, thus still making it possible to offload all phases of a given collective operation, freeing the CPU for computation. For example, a barrier collective operation may be implemented as: 1) on-host communication, with fan-in to the local-leader, 2) off-host recursive doubling algorithm amongst the local-leaders, and 3) on-host, fan-out from the local leader. In the current implementation, we use linear all-to-one fan-in, and one-to-all fan-out communication patterns. For a non-power-of-two local-leader recursive doubling algorithm we pair the "excess" ranks with a single rank in the set of ranks that is the largest power-of-two smaller than the size of the off-host group. This is described in detail in [12,13].

4   Experimental Setup

The performance was measured on an eight node, dual socket, quad-core, 3.00 Gigahertz Intel Xeon Quad-core X5472 with 32 gigabytes of memory. Red Hat Enterprise Linux Server 5.1 is used, kernel version 2.6.18-53.el5, a dual port quad data rate ConnectX-2 HCA, and a 36-port QDR switch running Mellanox firmware version 2.7.650. This is pre-release firmware, and provides the first working implementation of the new MQ capability. Platform availability constrained the scale of the experiments.

The prototype offloaded IB collectives are implemented within version 1.5 of the Open MPI code base, as a new collective module. To measure raw barrier time, we measure the completion time of a tight loop over barrier calls, and report the average time for the MPI rank 0. Similarly, for the nonblocking collectives we loop over nonblocking barrier initiation and barrier completion. To measure the overlap characteristics of these collective operations, we modify the ideas introduced in the COMB [18] benchmark, adapting them for collective operations. We measure communication-computation overlap by initiating the collective operation, executing a work loop, and then waiting for collective operation completion. The work loop starts at about 10 percent of the raw completion time, and is incremented in steps of about 10 percent up to about approximately 100 percent of raw completion time. The work loop is created by looping over the "nop" assembler instruction. The later is used to consume processor time, thus simulating the CPU being utilized for computation.

5   Benchmark Results

We have measured the performance of a number of different barrier operations in the range of eight to 64 processes on an eight node cluster system, with each host having the same number of MPI ranks in each measurement. Figure 1

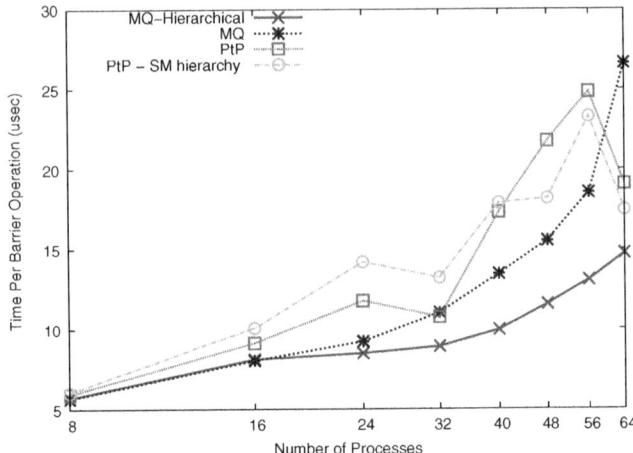

Fig. 1. MPI_Barrier performance, in micro-seconds per call, as a function of number of processes, and algorithm type. The algorithms include IB network offload with local host optimization, generic IB network offload, IB point-to-point RDMA based, and IB point-to-point with shared-memory local host optimization.

shows the results of measuring the performance of MPI_Barrier() with the original recursive-doubling offload algorithm [12](labeled MQ), in which each rank is treated identically. The performance ranges from 5.65 to 26.59 microseconds per iterations over the range of measurements. The performance of the hierarchical offloaded barrier (labeled MQ-Hierarchical), the new algorithm being described in this paper, is in the range of 5.68 to 14.72 microseconds. We also measured the performance of the barrier algorithm using a point-to-point implementation

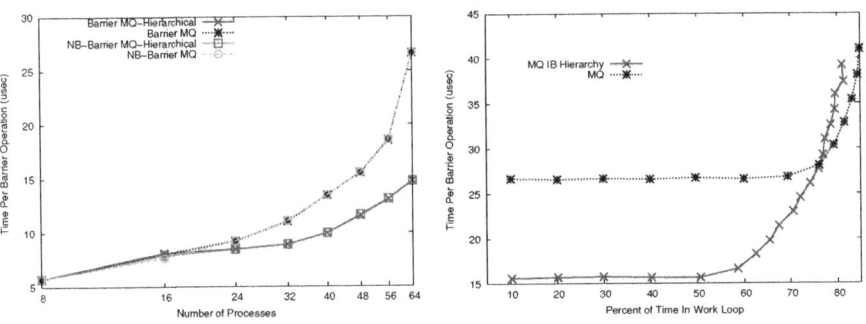

Fig. 2. Barrier performance data. **Left:** Blocking and nonblocking barrier latency (in micro-seconds) as a function of number of processes, and algorithm type. The algorithms used include IB network offload with IB offload local host optimization. **Right:** Nonblocking barrier completion time available for computation. Hierarchical IB and "flat" algorithms used. Available work time is reported as the percent of iteration time.

being progressed by the CPU (labeled PtP), with the performance ranging from 5.90 to 24.80 microseconds. Finally, we used a shared memory optimization for point-to-point based technique (labeled PtP - SM hierarchy), and measured the performance of the MPI_Barrier() algorithm in the range of 6.02 to 23.26 microseconds.

Figure 2 compares the performance of the blocking MPI barrier, to that of a nonblocking barrier implementation, for both the hierarchical and the uniform approach. As this figure shows, the performance of the blocking and nonblocking barriers are essentially the same.

Figure 2 shows the overlap capabilities of the nonblocking algorithms at a count of 64 processes. As the figure shows, about 50% of the 14.72 microseconds to run the hierarchical collective barrier algorithm can be used for computation without impacting the performance of the nonblocking barrier. The uniform barrier algorithm completes in 26.59 microseconds with about 70% of the time available for computation.

6   Discussion and Summary

As the results in Figure 1 show, using the network hierarchy reduces the barrier at 64 nodes by almost 12 microseconds, or about 45%, which is a large reduction in latency. The main reason for this reduction is the reduced number of queues per process, reducing the pressure on HCA resources. With the original uniform algorithm, each process opens six QP's, one MQ, one send completion queue, six receive completion queues, and one MQ completion queue, for a total of 168 queues per host. However in the hierarchical queue approach, each local leader opens three QP's for inter-node communication, seven QP's for intra-node communications, one MQ, one send completion queue, 10 receive completion queues, and one MQ completion queue. All other processes have only one QP, one MQ, one send CQ, one receive CQ, and one MQ completion queue. This gives a total of 67 queues per host, or a reduction of about 60% in the number of queues per host. This is believed to be the primary reason for the improved performance of the barrier algorithm, based on earlier work [12] studying the performance of the MPI_Barrier() collective operation as a function of QPs. We also experimented with rank layout, with the ranks involved in the first three levels of the recursive doubling algorithm sharing the same HCA, and by distributing the ranks across nodes in a round-robin manner, where the amount of local communication varies from rank to rank, but observed very little impact on performance. This is not surprising given the relatively small difference in performance of the off-host and on host communication using InfiniBand. We measured the two process on-host recursive doubling barrier to be 1.40 microseconds, and 1.78 microseconds for the off-host barrier, or about four tenths of a microsecond difference. This leads us to believe that this performance has little impact on the overall measured barrier function performance.

We further studied the barrier algorithm to understand how the time is distributed between the off-host communications and the on-host communication for the 64 rank barrier operation. The eight rank barrier operation completes in 5.68 microseconds, and uses the same inter-host communication pattern used by the 64 rank barrier, implying that the on-host portion of the algorithm, the initial fan-in to the local leader and the final fan-out from the local leader, account for about nine microseconds. We believe the on-host portion of the barrier algorithm, which is independent of the number of nodes being used, can be improved, and be closer to one microsecond that one can obtain with a CPU based algorithm using shared memory communications. However, since this hierarchical approach fully offloads collective management to the HCA, it is able to proceed while the CPU is being used for computation, where high-performance shared memory implementations are polling based, and do not allow for effective communication and computation overlap.

This particular algorithm scales very well, as we only need to add three queues per node, one QP and one receive completion queue, to double the size of the cluster, compared with three queues per process, or 24 queues per node for this particular host configuration. Similarly, scaling this system out to sixteen thousand processes requires only 101 queues. This algorithmic approach and the CORE-*Direct* technology provide needed system level capabilities for good application scalability.

As the results in Figure 2 show, the performance of the blocking and non-blocking barrier algorithms are essentially identical. This is not surprising given that they differ slightly in completion implementation, with the blocking barrier polling for completion as part of the barrier function, but the nonblocking barrier implementation polls for completion through the request object.

We studied the overlap capabilities of only the fully offloaded nonblocking barrier algorithms, as we have already shown [13] that those implementations which rely on the CPU for collective management are not well suited to overlapping communication with computation. We see from Figure 2 the hierarchical algorithm allows for about 50% overlap of computation with communication, which is less than the roughly 75% that could be overlapped with the original nonblocking barrier implementation. However, when converting this to time lost to overhead, the implementation of the hierarchical algorithm uses a bit more than seven microseconds of processing time, as does that of the original implementation. The nonblocking barrier is the collective operations equivalent of zero byte point-to-point ping-pong measurements, as no user data is involved, and the performance is determined by hardware and software latency effects. As such, this is the collective operation that is expected to provide the smallest overlap capabilities. While these overlap capabilities can be very helpful for applications, we will continue to investigate what may be done to lower the CPU overhead used in setting up the nonblocking collective operation, and then completing it.

To summarize, we have developed new hierarchical blocking and nonblocking barrier algorithms that use the CORE-*Direct* capabilities introduced in ConnectX-2, and have greatly improved the performance of the barrier algorithms using these capabilities, while maintaining the ability to fully offload the management of these collectives to the HCA. The new barrier algorithm also outperforms the point-to-point algorithm using InfiniBand with the CPU progressing the collective operation. While there is significant improvement in performance using these techniques, more work is needed to further reduce the overhead, thus continuing to improve the barrier implementations.

References

1. InfiniBand Trade Association, http://www.infinibandta.org/specs
2. Mellanox Technologies, http://www.mellanox.com/
3. Mvapich, http://mvapich.cse.ohio-state.edu/
4. Quadrics, http://www.quadrics.com/
5. Top 500 Super Computer Sites, http://www.top500.org/
6. Bhoedjang, R.A.F., Ruhl, T., Bal, H.E.: Efficient Multicast on Myrinet Using Link-Level Flow Control. In: 27th ICPP (1998)
7. Buntinas, D., Panda, D.K.: NIC-Based Reduction in Myrinet Clusters: Is It Beneficial. In: SAN-2002 Workshop (in conjunction with HPCA) (February 2003)
8. Buntinas, D., Panda, D.K., Sadayappan, P.: Fast NIC-Level Barrier over Myrinet/GM. In: Proceedings of IPDPS (2001)
9. Garbriel, E., et al: Open MPI: Goals, Concept, and Design of a Next Generation MPI Implementation. In: Proceedings, 11th European PVM/MPI Users' Group Meeting (2004)
10. Graham, R.L., et al.: A Network-Failure-tolerant Message-Passing System for Terascale Clusters. In: Proceedings of ICS (June 2002)
11. Kumar, S., et al: The deep computing messaging framework: generalized scalable message passing on the blue gene/P supercomputer. In: ICS 2008: Proceedings of the 22nd annual international conference on Supercomputing, pp. 94–103. ACM, New York (2008)
12. Graham, R.L., Poole, S., Shamis, P., Bloch, G., Bloch, N., Chapman, H., Kagan, M., Shahar, A., Rabinovitz, I., Shainer, G.: ConnectX-2 InfiniBand Management Queues: First investigation of the new support for network offloaded collective operations. Accepted for the 10th IEEE/ACM International Symposium CCGrid (2010)
13. Graham, R.L., Poole, S., Shamis, P., Bloch, G., Bloch, N., Chapman, H., Kagan, M., Shahar, A., Rabinovitz, I., Shainer, G.: Overlapping Computation and Communication: Barrier Algorithms and ConnectX-2 Core-DIRECT Capabilities. Accepted to CAC (2010)
14. Hoefler, T., Lumsdaine, A.: Optimizing non-blocking Collective Operations for InfiniBand. In: Proceedings of the 22nd IPDPS (April 2008)
15. Hoefler, T., Lumsdaine, A., Rehm, W.: Implementation and Performance Analysis of Non-Blocking Collective Operations for MPI. In: SC 2007: Proceedings of the SC 2007, pp. 1–10. ACM, New York (2007)
16. Dongarra, J., et al.: The International Exascale Software Project: a Call To Cooperative Action By the Global High-Performance Community. Int. J. High Perform. Comput. Appl. 23(4), 309–322 (2009)
17. Kielmann, T., Hofman, R.F.H., Bal, H.E., Plaat, A., Bhoedjang, R.A.F.: MagPIe: MPI's collective communication operations for clustered wide area systems. SIGPLAN Not. 34, 131–140 (1999)
18. Lawry, W., Wilson, C., Maccabe, A.B., Brightwell, R.: Comb: a portable benchmark suite for assessing mpi overlap. In: 2002 IEEE International Conference on Cluster Computing, pp. 472–475 (2002)

19. Message Passing Interface Forum. MPI: A Message-Passing Standard (June 2008)
20. Moody, A., Fernandez, J., Petrini, F., Panda, D.: Scalable NIC-based Reduction on Large-Scale Clusters. In: SC 2003 (November 2003)
21. Mraz, R.: Reducing the Variance of Point to Point Transfers in the IBM 9076 Parallel Computer. In: Proceedings of the 1994 ACM/IEEE conference on Supercomputing, pp. 620–629 (November 1994)
22. Sancho, J.C., Kerbyson, D.J., Barker, K.J.: Efficient Offloading of Collective Communications in Large-Scale Systems. In: IEEE International Conference on Cluster Computing, pp. 169–178 (2007)
23. Steffenel, L.A., Mounié, G.: A Framework for Adaptive Collective Communications for Heterogeneous Hierarchical Computing Systems. J. Comput. Syst. Sci. 74(6), 1082–1093 (2008)
24. Tipparaju, V., Nieplocha, J., Panda, D.: Fast Collective Operations Using Shared and Remote Memory Access Protocols on Clusters. In: Proceedings of the IPDPS (2003)
25. Yu, W., Buntinas, D., Graham, R.L., Panda, D.K.: Efficient and Scalable Barrier over Quadrics and Myrinet with a New NIC-Based Collective Message Passing Protocol. In: CAC Workshop, in Conjunction IPDPS 2004 (April 2004)
26. Yu, W., Buntinas, D., Panda, D.K.: High Performance and Reliable NIC-Based Multicast over Myrinet/GM-2. In: Proceedings of the IPDPS 2003 (October 2003)
27. Zhu, H., Goodell, D., Gropp, W., Thakur, R.: Hierarchical Collectives in MPICH2. In: Proceedings of the 16th European PVM/MPI Users' Group Meeting on Recent Advances in Parallel Virtual Machine and Message Passing Interface, pp. 325–326. Springer, Heidelberg (2009)

An In-Place Algorithm for Irregular All-to-All Communication with Limited Memory

Michael Hofmann⋆ and Gudula Rünger

Department of Computer Science
Chemnitz University of Technology, Germany
{mhofma,ruenger}@cs.tu-chemnitz.de

Abstract. In this article, we propose an in-place algorithm for irregular all-to-all communication corresponding to the `MPI_Alltoallv` operation. This in-place algorithm uses a single message buffer and replaces the outgoing messages with the incoming messages. In comparison to existing support for in-place communication in MPI, the proposed algorithm for `MPI_Alltoallv` has no restriction on the message sizes and displacements. The algorithm requires memory whose size does not depend on the message sizes. Additional memory of arbitrary size can be used to improve its performance. Performance results for a Blue Gene/P system are shown to demonstrate the performance of the approach.

Keywords: All-to-all, Irregular communication, In-place, Limited memory, MPI.

1 Introduction

The amount of the memory required to solve a given problem can be one of the most important properties for algorithms and applications in parallel scientific computing. Since main memory is a limited resource, even for distributed memory parallel computers, the memory footprint of an application decides whether a certain problem size can be processed or not. Examples are parallel applications that use domain-decomposition techniques, e.g. mesh-based algorithms or particle codes. Adaptive or time-dependent solutions often require periodical redistributions of the workload and its associated data. This may require irregular communication based on `MPI_Alltoallv` where individual messages of arbitrary size are exchanged between processes. Even though the redistribution step may require only a small part of the runtime, it can significantly reduce the maximum problem size if a second fully-sized buffer has to be kept available only for receiving data during this step.

MPI communication operations commonly use separate send and receive buffers. MPI version 2.0 has introduced "in place" buffers for many intracommunicator collective operations using the `MPI_IN_PLACE` keyword. The resulting *in-place* communication operations use only a single message buffer and

---

⋆ Supported by Deutsche Forschungsgemeinschaft (DFG).

R. Keller et al. (Eds.): EuroMPI 2010, LNCS 6305, pp. 113–121, 2010.

replace the outgoing messages with the incoming messages. Support for an in-place MPI_Alltoallv operation was introduced in the MPI version 2.2 [1], but only with the restriction that counts and displacements of the messages to be sent and received are equal. In this article, we present an algorithm for an in-place MPI_Alltoallv operation with arbitrary counts and displacements. This in-place algorithm requires memory whose size does not depend on the message sizes. If additional memory is available, it can be used to speed up the in-place algorithm. The algorithm is described in the context of all-to-all communication, but can directly be adapted to many-to-many or sparse communication. We have implemented the algorithm and present performance results for a Blue Gene/P system using up to 4096 processes to demonstrate the efficiency of our approach.

The rest of this paper is organized as follows. Section 2 presents related work. Section 3 introduces the in-place algorithm for the MPI_Alltoallv operation. Section 4 shows performance results and Section 5 concludes the paper.

2   Related Work

Optimizations of MPI communication operations usually address latency and bandwidth results. Interconnection topologies and other architecture-specific properties are the subject of performance improvements, especially for high scaling parallel platforms [2]. Specific MPI implementations as well as the MPI specification itself are analyzed with respect to future scalability requirements [3]. Here, the memory footprint of an MPI implementation becomes an important optimization target.

Efficient data redistribution is a common problem in parallel computing. Especially data parallel programming models like High Performance Fortran provide support for flexible distributions of regular data structures to different processors. In this context, numerous algorithms for efficient redistributions of block-cyclic data distributions have been proposed (e.g. [4,5,6]). Algorithms for irregular data redistribution with limited memory have received less attention. Pinar et al. have proposed algorithms for data migration in the case of limited memory based on sequences of communication phases [7]. Siegel et al. have implemented various algorithms for data redistribution with limited memory in the MADRE library [8]. In [9], they have provided a modification to prevent livelocks in the basic algorithm of Pinar et al. In [10], we have proposed a fine-grained data distribution operation in MPI and provided several implementation variants including an in-place implementation based on parallel sorting.

3   An In-Place Algorithm for MPI_Alltoallv

The MPI_Alltoallv operation is one of the most general collective communication operations in MPI. With p participating processes, each process sends p individual messages to the other processes and receives p messages from them. For each process, the sizes of the messages to be sent and received are given by arrays $scounts[1\ldots p]$ and $rcounts[1\ldots p]$. Additional arrays $sdispls[1\ldots p]$ and

$rdispls[1...p]$ specify displacements that determine the locations of the messages. The standard MPI_Alltoallv operation uses separate send and receive buffers. Therefore, the locations of all messages are non-overlapping and all messages can be sent and received independently from each other.

The in-place MPI_Alltoallv operation uses a single buffer for storing the messages to be sent and received. This leads to additional dependencies for sending and receiving the messages. A message cannot be received until there is enough free space available at the destination process. Furthermore, a message cannot be stored at their target location if this location is occupied by a message that has to be sent (in advance). A trivial solution for this problem uses an intermediate buffer to receive the messages. This requires additional memory whose size depends on the size of the messages. The proposed in-place algorithm solves this problem using additional memory of a size independent from the message sizes.

3.1   Basic Algorithm

The algorithm is described from the perspective of a single process. We assume that all messages consist of data items of the same type and that the buffers used to store the messages are arrays of this type. Let S_i denote the set of indices belonging to the data items of the message to be sent to process i for $i = 1, ..., p$. Let R_i denote the set of indices of the locations where the incoming data items from process i should be stored. The initial index sets can be calculated from the given counts and displacements. We assume that the initial send index sets S_i are disjoint. The same applies to the initial receive index sets R_i. The messages are sent and received in several partial submessages and the index sets are updated as the algorithm proceeds. Even though the buffer can be seen as a large array, only the locations given by the initial index sets are accessible.

The in-place algorithm is based on the basic algorithm of Pinar et al. [7,9] and consists of a sequence of communication phases. In each phase the following steps are performed. (1) The number of data items that can be received in free space from every other process is determined. (2) These numbers are sent to the corresponding source processes. This represents an exchange of request messages between all processes that still have data items left to be exchanged with each other. (3) The data items are transferred. The algorithm terminates when all data items are exchanged. Algorithm 1 shows the basic algorithm adapted to our notation. Additionally, our algorithm includes procedures for the initialization of the index sets (line 2), for determining the items that can be received in free space (line 4), and for updating the index sets at the end of each communication phase (line 11). A description of these procedures is given in the following subsections. Exchanging the request messages (lines 5–7) and the data items (lines 8–10) can be implemented with non-blocking communication.

Siegel et al. have shown that the basic algorithm will neither deadlock nor livelock, provided there is additional free space on every process used to receive data items [9]. This is independent from the actual size of the additional free space and from the particular strategy that determines how free space is used

Algorithm 1. Basic algorithm of the in-place `MPI_Alltoallv` operation.

1: let $recv[1 \ldots p]$ and $send[1 \ldots p]$ be arrays of integers
2: init the index sets S_i and R_i for $i = 1, \ldots, p$ (see Sect. 3.2)
3: **while** $(\sum_i |S_i| + \sum_i |R_i| > 0)$ **do**
4:     $recv[1 \ldots p]$ = determine the number of data items to be received (see Sect. 3.3)
5:     exchange requests
6:         → send $recv[i]$ to process i for all i with $|R_i| > 0$
7:         → receive $send[i]$ from process i for all i with $|S_i| > 0$
8:     exchange data items
9:         → send $send[i]$ items at indices S_i to process i for all i with $send[i] > 0$
10:         → receive $recv[i]$ items at indices R_i from process i for all i with $recv[i] > 0$
11:     update the index sets (see Sect. 3.4)
12: **end**

to receive data items (line 4). The proof for deadlock-freedom applies to the basic algorithm and is independent from our modifications. The original proof for livelock-freedom assumes that all locations that become free (during the algorithm) can be used as free space. However, for the `MPI_Alltoallv` operation, only locations given by the initial index sets R_i can be used as free space. All locations that become free and do not belong to the initial index sets R_i cannot be used as free space. The original proof can be modified to distinguish between the usable and not-usable free locations, leading to the same result.

The additional available memory is used to create auxiliary buffers that are independent from the input buffer. These auxiliary buffers provide the additional free space that is required for the successful termination of the basic algorithm. The usage of the auxiliary buffers is independent from the rest of the algorithm and described in Sect. 3.5.

3.2   Initializing the Index Sets

The initial send index sets S_i are defined according to the given send counts and displacements: $S_i = \{sdispls[i], \ldots, sdispls[i] + scounts[i] - 1\}$. We assume that the indices of S_i are lower than the indices of S_{i+1} for $i = 1, \ldots, p-1$. Otherwise, a local reordering of the index sets is necessary. The receive index sets R_i are initialized analogously. Each send index set S_i is split into a finite number of disjoint subsets $S_i^1, \ldots, S_i^{n_i}$ with $S_i = S_i^1 \cup \ldots \cup S_i^{n_i}$. The splitting is performed at the positions given by the lowest and highest indices of the receive index sets. Each subset created contains contiguous indices and is either disjoint from all receive index sets or completely overlapped by one of the receive index sets. There are at most $2p$ possible splitting positions, since each receive index set provides two splitting positions. Thus the p initial send index sets can be split in at most $p + 2p$ subsets. We define two functions to specify how the send index subsets and the receive index sets overlap each other. A send subset is assigned a matching receive set and vice versa using the functions $rmatch$ and $smatch$. For a send subset $S_i^k \neq \varnothing$, $k \in \{1, \ldots, n_i\}$, the matching receive set $rmatch(S_i^k)$ corresponds to the receive set R_j that overlaps with S_i^k. If there exists no such

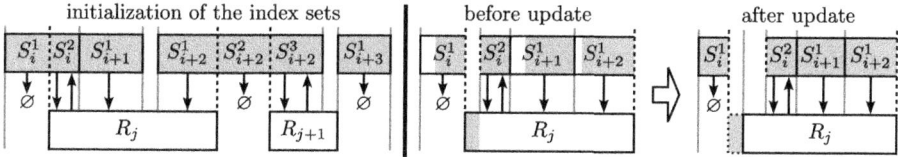

Fig. 1. Examples for initialization (left) and update (right) of send and receive index sets. Arrows indicate matching sets, e.g. $\boxed{S_i^2} \rightarrow \boxed{R_j}$ corresponds to $rmatch(S_i^2) = R_j$.

receive set then $rmatch(S_i^k) = \varnothing$. For a given receive set R_j, the matching send subset $smatch(R_j)$ corresponds to the send subset $S_i^k \neq \varnothing$ such that i and k are minimal and S_i^k is overlapped by R_j. If there exists no such send subset then $smatch(R_j) = \varnothing$. Figure 1 (left) shows an example for this initialization.

3.3 Determine the Number of Data Items to Be Received

For each message, the data items are received from the lowest to the highest indices. The number of data items that can be received from process j in free space at locations given by R_j is determined from the matching send subset of R_j and is stored in $recv[j]$.

$$
recv[j] = \begin{cases} |R_j| & \text{if } smatch(R_j) = \varnothing \\ \min\{x|x \in smatch(R_j)\} - \min\{x|x \in R_j\} & \text{otherwise} \end{cases}
$$

If no matching send subset exists, then all remaining data items from process j can be received. Otherwise, the number of data items that can be received is limited by the matching send subset $smatch(R_j)$. The lowest index of this subset corresponds to the lowest location of R_j that is not free. In addition to the data items that can be received in the input buffer, the free space that is available in the auxiliary buffers is used to receive additional data items.

3.4 Updating the Index Sets

The data items of the messages are sent and received from the lowest to the highest indices. All index sets are updated when the data items of the current phase are sent and received. Each receive set R_j is updated by removing the $recv[j]$ lowest indices, since they correspond to the data items that were previously received. Similarly, the $send[i]$ lowest indices are removed from the send subsets $S_i^1, \ldots, S_i^{n_i}$. Sending data items to other processes creates free space at the local process. However, only contiguous free space available at the lowest indices of a receive set can be used to receive data items in the next communication phase. For each receive set R_j, the free space corresponding to its indices is joined at its lowest indices. This is achieved by moving all data items that correspond to send subsets S_i^k with $rmatch(S_i^k) = R_j$ towards the highest indices of R_j (the index values of S_i^k are shifted accordingly). After that, data items from process

j that are stored in the auxiliary buffers are moved to the freed space and R_j is updated again. Finally, the functions $rmatch$ and $smatch$ are adapted and for each send set S_i the value of $|S_i|$ is computed according to its updated subsets (only $|S_i|$ is required in Algorithm 1). Figure 1 (right) shows an example for this update procedure. The costs for updating the index sets depend on the number of data items that have to be moved. In the worst case, all data items of the remaining send subsets have to be moved during the update procedure in every communication phase.

3.5   Using Auxiliary Buffers

Additional memory of arbitrary size a is used to create auxiliary buffers on each process. These buffers are used to receive additional data items while their target locations are still occupied. The efficient management of the auxiliary buffers can have a significant influence on the performance. We use a static approach to create a fixed number of b auxiliary buffers, each of size $\frac{a}{b}$. Data items from process j can be stored in the $(j \bmod b)$-th auxiliary buffer using a *first-come, first-served* policy. This static partitioning of the additional memory allows a more flexible utilization in comparison to a single auxiliary buffer, but prevents fragmentation and overhead costs (e.g., for searching for free space). More advanced auxiliary buffer strategies (e.g., with dynamic heap-like allocations) can be subject of further optimizations.

4   Performance Results

We have performed experimental results for a Blue Gene/P system to investigate the performance of the proposed in-place algorithm for MPI_Alltoallv. The implementation uses the standard MPI_Alltoallv operation for exchanging the request messages, because using non-blocking communication for this exchange has caused performance problems for large numbers of processes (≥ 1024). In-place communication usually involves large messages that occupy a significant amount of main memory. Unless otherwise specified, each process uses 100 MB data that is randomly partitioned into blocks and sent to other processes. Results for the platform-specific MPI_Alltoallv operation are obtained using a separate (100 MB) receive buffer.

Figure 2 (left) shows communication times for different numbers of processes p depending on the number of auxiliary buffers b. The total size of additional memory used for the auxiliary buffers is 1 MB. Increasing the number of auxiliary buffers leads to a significant reduction in communication time, especially for large numbers of processes. Choosing the number of auxiliary buffers depending on the total number of processes shows good results for various values of p. For the following results we continue to use $b = \frac{p}{8}$. Figure 2 (right) shows communication times for different sizes of additional memory (in % with respect to the total message size of 100 MB) depending on the number processes. Increasing the additional memory up to 10 % leads to a significant performance improvement. A further increase up to 100% shows only small differences. This can be

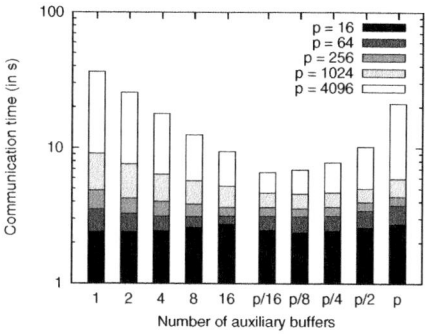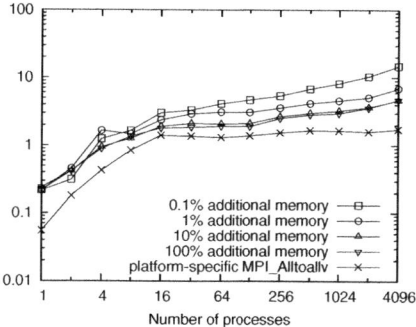

Fig. 2. Communication times for in-place `MPI_Alltoallv` depending on the number of auxiliary buffers (left) and with different sizes of additional memory (right)

attributed to the static auxiliary buffer strategy. With 100 % additional memory, the sizes of the auxiliary buffers exceed the sizes of the messages to be received. This leads to an insufficient utilization of the additional memory. However, even with an optimal auxiliary buffer strategy there can be differences in performance in comparison to a platform-specific `MPI_Alltoallv` operation that includes optimizations for the specific system architecture [2]. For the following results we continue to use 1 % additional memory (1 MB).

Figure 3 (left) shows communication times for different total message sizes depending on the number of processes. The communication time with small messages strongly depends on the number of processes, while for large messages it increases more slowly. Figure 3 (right) shows the time spend on different parts of the in-place algorithm depending on the number of processes. The major part of the communication time is spent for exchanging the data items. The costs for exchanging the request messages are comparably small, but they increase with the number of processes. The costs for updating the index sets are also rather small. However, these costs strongly depend on the actual data redistribution

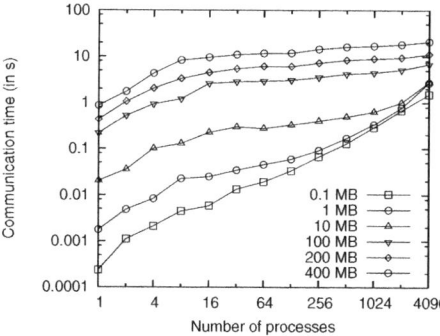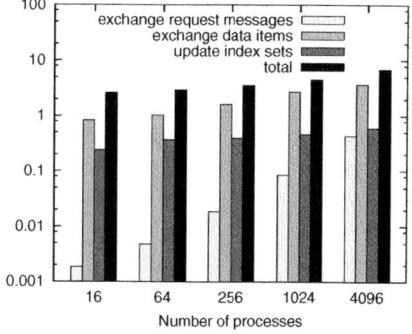

Fig. 3. Communication times for in-place `MPI_Alltoallv` with different total message sizes (left). Times spend in the different parts of in-place `MPI_Alltoallv` (right).

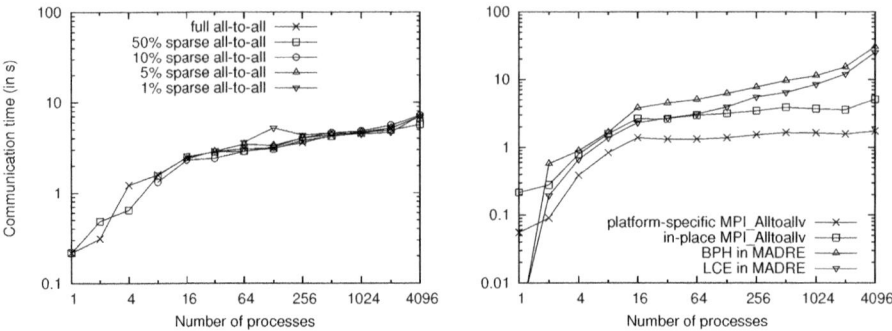

Fig. 4. Communication times for in-place `MPI_Alltoallv` for sparse data redistribution schemes (left). Comparison of communication times for platform-specific `MPI_Alltoallv`, in-place `MPI_Alltoallv` and two MADRE algorithms (right).

problem. If the number of communication phases increases and a large number of data items need to be moved for each update, the total time of the in-place `MPI_Alltoallv` operation can be dominated by the local data movements.

Figure 4 (left) shows communication times for different sparse data redistribution schemes depending on the number of processes. For the sparse all-to-all communication, each process sends messages only to a limited number of random processes. The results show that the performance of the in-place algorithm is almost independent from the actual number of messages. Figure 4 (right) shows communication times for the platform-specific `MPI_Alltoallv` operation, the in-place algorithm, and for the Basic Pinar-Hendrickson algorithm (BPH) and the Local Copy Efficient algorithm (LCE) from the MADRE library [11] depending on the number of processes. MADRE and its in-place algorithms are designed to redistribute an arbitrary number of equal-sized (large) blocks according to a given destination rank and index for each block. To compare these algorithms to the `MPI_Alltoallv` operation, we increase the data item size up to 16 KB and treat each data item as a separate block in MADRE. Additional free blocks are used to provide the additional memory to the MADRE algorithms. There is a general increase in communication time for the MADRE algorithms depending on the number of processes. In comparison to that, the results for the platform-specific `MPI_Alltoallv` operation and the in-place algorithm are more stable. The communication time of the in-place algorithm is within a factor of three of the platform-specific `MPI_Alltoallv` operation.

5   Summary

In this paper, we have proposed an in-place algorithm for `MPI_Alltoallv` that performs data redistribution with limited memory. The size of required memory is independent from the message sizes and depends only linearly on the number of processes a single process has messages to exchange with. Additional memory can be used to improve the performance of the implementation. It is shown that the size of the additional memory and its efficient usage has a significant influence on the performance, especially for large numbers of processes. Performance results with large messages demonstrate the good performance of our approach.

Acknowledgments. The measurements are performed at the John von Neumann Institute for Computing, Jülich, Germany.
http://www.fz-juelich.de/nic

References

1. MPI Forum: MPI: A Message-Passing Interface Standard Version 2.2. (2009)
2. Almási, G., Heidelberger, P., Archer, C.J., Martorell, X., Erway, C.C., Moreira, J.E., Steinmacher-Burow, B., Zheng, Y.: Optimization of MPI collective communication on BlueGene/L systems. In: Proc. of the 19th annual Int. Conf. on Supercomputing, pp. 253–262. ACM Press, New York (2005)
3. Balaji, P., Buntinas, D., Goodell, D., Gropp, W., Kumar, S., Lusk, E., Thakur, R., Träff, J.L.: MPI on a Million Processors. In: Ropo, M., Westerholm, J., Dongarra, J. (eds.) Recent Advances in Parallel Virtual Machine and Message Passing Interface. LNCS, vol. 5759, pp. 20–30. Springer, Heidelberg (2009)
4. Thakur, R., Choudhary, A., Ramanujam, J.: Efficient Algorithms for Array Redistribution. IEEE Trans. Parallel Distrib. Syst. 7(6), 587–594 (1996)
5. Walker, D.W., Otto, S.W.: Redistribution of block-cyclic data distributions using MPI. Concurrency - Practice and Experience 8(9), 707–728 (1996)
6. Lim, Y., Bhat, P., Prasanna, V.: Efficient Algorithms for Block-Cyclic Redistribution of Arrays. Algorithmica 24, 298–330 (1999)
7. Pinar, A., Hendrickson, B.: Interprocessor Communication with Limited Memory. IEEE Trans. Parallel Distrib. Syst. 15(7), 606–616 (2004)
8. Siegel, S.F., Siegel, A.R.: MADRE: The Memory-Aware Data Redistribution Engine. Int. J. of High Performance Computing Applications 24, 93–104 (2010)
9. Siegel, S.F., Siegel, A.R.: A Memory-Efficient Data Redistribution Algorithm. In: Ropo, M., Westerholm, J., Dongarra, J. (eds.) Recent Advances in Parallel Virtual Machine and Message Passing Interface. LNCS, vol. 5759, pp. 219–229. Springer, Heidelberg (2009)
10. Hofmann, M., Rünger, G.: Fine-Grained Data Distribution Operations for Particle Codes. In: Ropo, M., Westerholm, J., Dongarra, J. (eds.) Recent Advances in Parallel Virtual Machine and Message Passing Interface. LNCS, vol. 5759, pp. 54–63. Springer, Heidelberg (2009)
11. Siegel, S.F., Siegel, A.R.: MADRE: The Memory-Aware Data Redistribution Engine, Version 0.4 (2010), http://vsl.cis.udel.edu/madre/

Massively Parallel Finite Element Programming

Timo Heister[1], Martin Kronbichler[2], and Wolfgang Bangerth[3]

[1] NAM, University of Göttingen, Germany
heister@math.uni-goettingen.de
[2] Department of Information Technology, Uppsala University, Sweden
martin.kronbichler@it.uu.se
[3] Department of Mathematics, Texas A&M University
bangerth@math.tamu.edu

Abstract. Today's large finite element simulations require parallel algorithms to scale on clusters with thousands or tens of thousands of processor cores. We present data structures and algorithms to take advantage of the power of high performance computers in generic finite element codes.

Existing generic finite element libraries often restrict the parallelization to parallel linear algebra routines. This is a limiting factor when solving on more than a few hundreds of cores. We describe routines for distributed storage of all major components coupled with efficient, scalable algorithms. We give an overview of our effort to enable the modern and generic finite element library deal.II to take advantage of the power of large clusters. In particular, we describe the construction of a distributed mesh and develop algorithms to fully parallelize the finite element calculation. Numerical results demonstrate good scalability.

Keywords: Finite Element Software, Parallel Algorithms, Massively Parallel Scalability.

1 Introduction

Modern computer clusters have up to tens of thousands of cores and are the foundation to deal with large numerical problems in finite element calculations. The hardware architecture requires software libraries to be specifically designed.

This has led to a significant disparity between the capabilities of current hardware and the software infrastructure that underlies many finite element codes for the numerical simulation of partial differential equations: there is a large gap in parallel scalability between the specialized codes designed to run on those large clusters and general libraries. The former are hand-tailored to the numerical problem to be solved and often only feature basic numerical algorithms such as low order time and spatial discretizations on uniform meshes. On the other hand, most general purpose finite element libraries like deal.II [4,5] presently do not scale to large clusters but provide more features, such as higher order finite elements, mesh adaptivity, and flexible coupling of different elements and equations, and more.

R. Keller et al. (Eds.): EuroMPI 2010, LNCS 6305, pp. 122–131, 2010.
© Springer-Verlag Berlin Heidelberg 2010

By developing parallel data structures and algorithms we enable deal.II to perform numerical simulations on massively parallel machines with a distributed memory architecture. At the same time, by implementing these algorithms at a generic level, we maintain the advanced features of deal.II.

While we describe our progress with deal.II in [3], the generic algorithms developed here are applicable to nearly any finite element code. The modifications to deal.II discussed in this paper are a work in progress but will become available soon as open source. In Section 3 we describe the data structures and algorithms for the parallelization of finite element software. We conclude with numerical results in Section 4. The parallel scalability is shown using a Poisson problem and we present results for a more involved mantle convection problem.

2   Related Work

Finite element software has long been packaged in the form of libraries. A cursory search of the internet will yield several dozen libraries that support the creation of finite element applications to various degrees. While most of these are poorly documented, poorly maintained, or both, there are several widely used and professionally developed libraries. Most of these, such as DiffPack [6,14], libMesh [12], Getfem++ [17], OOFEM [16], or FEniCS [15] are smaller than the deal.II library but have similar approaches and features.

To the best of our knowledge, none of these libraries currently support massive parallel computations. What parallel computation they support is similar to what is available in publicly available releases of deal.II: meshes are either statically partitioned or need to be replicated on every processor, only linear solvers are fully distributed. While this allows for good scaling of solvers, the replication of meshes on all processors is a bottleneck that limits overall scalability of parallel adaptive codes to a few dozen processors. The reason for this lack of functionality is the generally acknowledged difficulty of fully distributing the dynamically changing, complex data structures used to describe adaptive finite element meshes. A particular complication is the fact that all of the widely used libraries originate from software that predates massively parallel computations, and retrofitting the basic data structures in existing software to new requirements is nontrivial in all areas of software design.

The only general framework for unstructured, fully parallel adaptive finite element codes we are aware of that scales to massive numbers of processors is ALPS, see [7]. Like the work described here, ALPS is based on the p4est library [8]. On the other hand, ALPS lacks the extensive support infrastructure of deal.II and is not publicly available.

3   Massively Parallel Finite Element Software Design

In order for finite element simulations to scale to a large number of processors, the compute time must scale linearly with respect to the number of processors and the problem size. Additionally, local memory consumption should only depend on the local, not the global, problem size. The former requires minimizing global communication. The latter requires distributed data structures, where only necessary data is stored locally.

Thus, our focus is on the primary bottlenecks to parallel scalability: the mesh handling, the distribution and global numbering of the degrees of freedom, and the numerical linear algebra. See [3] for the technical details concerning the implementation with `deal.II`.

3.1   Distributed Mesh Handling

The computational mesh is duplicated on each processor in most generic finite element libraries, if they support distributed parallel computing at all. This is not feasible for massively parallel computations because the generic description of a mesh involves a significant amount of data for each cell which is then replicated on each processor, resulting in huge memory overhead.

Each processor only needs to access a small subset of cells. Most parts of the replicated global mesh are unnecessary locally. We will say that the processor "owns" this required subset of cells. In addition, a processor also needs to store cells touching the cells it owns, but that are in fact owned by neighboring machines. These neighboring cells are called *ghost cells*. Information about ghost cells is needed for several reasons, most obviously because continuous finite elements share degrees of freedom on the lines and vertices connecting cells. Ghost cells are also needed for adaptive refinement, error estimation, and more.

Since `deal.II` only supports hexahedral cells, we restrict the discussion to that type of meshes. The active cells in h-adaptive finite element methods are attained from recursively refining a given "coarse" mesh. Thus, one also stores the hierarchy of refinement steps. We distinguish between three different kinds of information for mesh storage:

1. The *coarse mesh* consists of a number of coarse cells describing the domain. We assume that the coarse mesh only consists of a relatively small number of cells compared to the number of active cells in the parallel computation, say a few tens of thousands; it can be stored on each processor. The refinement process starts with the coarse mesh.
2. *Refinement information* can be stored in a (sparse) octree (a quadtree in two spatial dimensions) of refinement flags for each coarse cell. Each flag either states that this cells has been refined into eight (four in two dimensions) children or that it is an active cell if not.
3. *Active cells* are those on which the finite element calculation is done. Typical finite element programs attach a significant amount of information to each cell: vertex coordinates, connectivity information to faces, lines, corners, and neighboring cells, material indicators, boundary indicators, etc. We include the ghost cells here, but we set a flag to indicate that they to belong to a different machine.

We can store the coarse mesh (1.) on each machine without a problem. For the refinement information (2.) we interface to an external library called `p4est`, see

[8]. This library handles the abstract collection of octrees describing refinement from the coarse mesh and handles coarsening, refinement, and distribution of cells. Internally p4est indexes all terminal cells in a collection of octrees with a space filling curve. This allows rapid operations and scalability to billions of cells. For partitioning the space filling curve is cut into equally sized subsets, which ensures a well balanced workload even during adaptive refinement and coarsening. p4est allows queries about the local mesh and the ghost layer. With this information deal.II recreates only the active cells and the ghost layer. Because we chose to recreate a local triangulation on each machine, most of the finite element library works without modification, e.g. implementation of finite element spaces. However, we need a completely new method for creating a global enumeration of degrees of freedom. This is discussed next.

3.2 Handling of Degrees of Freedom

The finite element calculation requires a global enumeration of the degrees of freedom. The difficulty lies in the fact that every machine only knows about a small part of the mesh. The calculation of this numbering involves communication between processors. Additionally, the numbering on the ghost layer and the interface must be available on each machine. This is done in a second step.

Like all cells, each degree of freedom is owned by a single processor. All degrees of freedom inside a cell belong to the machine that owns the cell. The ownership of degrees of freedom on the interface between cells belonging to different machines is arbitrary, but processors that share such an interface need to deterministically agree who owns them. We assign such degrees of freedom to the processor with the smaller index.[1] The following algorithm describes the calculation and communication to acquire a global enumeration on the machines $p = 0, \ldots, P - 1$:

1. Mark all degrees of freedom as invalid (e.g. -1).
2. Loop over the locally owned cells and mark all degrees of freedom as valid (e.g. 0).
3. Loop over the ghost cells and reset the indices back to invalid if the cell is owned by a processor $q < p$. Now only indices that are owned locally are marked as valid.
4. Assign indices starting from 0 to all valid DoFs. This is done separately from the previous steps, because otherwise all neighbors sharing a degree of freedom would have to be checked for ownership. We denote the number of distributed DoFs on machine p with n_p.
5. Communicate the numbers n_p to all machines and shift the local indices by $\sum_{q=0}^{p-1} n_q$

Now all degrees of freedom are uniquely numbered with indices between 0 and $N = \sum_{q=0}^{P-1} n_q$. Next we must communicate the indices of degrees of freedom

---

[1] This rule is evaluated without communication. Assigning all degrees of freedom on one interface to the same processor also minimizes the coupling in the system matrix between the processors, see [3].

on the interface to the ghost layer and on the ghost layer itself. Each machine collects a packet of indices to send to its neighbors. Indices of a cell are sent to a neighbor if a ghost cell owned by that neighbor touches the cell. Indices on the interface may not be known at this point because they might belong to a third machine. As these communications are done concurrently there is no way to incorporate this information in this step. So we do this communication step twice: in the first round every machine receives all indices on its own cells, and after the second round every machine knows every index on the own cells and the ghost cells. There is no global communication required in these two steps.

3.3   Efficient Indexing

A subset $\mathcal{I}$ of the indices $\{0, \ldots, N\}$ is managed on each processor p. Each processor also needs to have the indices of degrees of freedom on the interface owned by another machine and indices on the ghost layer. Algebraic constraints induced by hanging nodes, solution vectors and other data structures need to access or store information for those indices. We typically look at three different subsets of indices:

1. The *locally owned* indices $\mathcal{I}_{l.o.}$ as described earlier. Following the algorithm outlined in the previous step, this is initially a contiguous range of n_p indices. However, we may later renumber indices, for example to in a block-wise way to reflect the structure of a partial differential equation in the linear system.
2. The *locally active* indices $\mathcal{I}_{l.a.}$ defined as the locally owned indices as well as the other indices on the interface. This is no longer a contiguous range.
3. The *locally relevant* indices $\mathcal{I}_{l.r.}$, which also includes the indices on the ghost cells.

We need an efficient data structure to define these subsets. If we store some information for each index in $\mathcal{I}_{l.r.}$, we would like to put that information into a contiguous memory location of $\#\mathcal{I}_{l.r.}$ elements. To access the information of an index $i \in \mathcal{I}_{l.r.}$, we need to find its position in the list $\mathcal{I}_{l.r.}$ (in other words, the number of indices $j \in \mathcal{I}_{l.r.}$ with $j < i$). This query is performed repeatedly, and thus it should be optimized.

We create a data structure of K sorted, disjoint, contiguous, intervals $[b_k, e_k)$ for defining the subset $\mathcal{I} \subset \{0, \ldots, N\}$ as $\tilde{\mathcal{I}} = \bigcup_{k=0}^{K}[b_k, e_k)$. Other libraries often go for a simpler description of this subset as a list of numbers, but this means more entries are stored because of the large contiguous subranges. Thus the important queries are slower. We also store the number $p_k = \sum_{\kappa=0}^{k-1}(e_\kappa - b_\kappa) = p_{k-1} + (e_{k-1} - b_{k-1})$ of indices in previous intervals with each interval $[b_k, e_k)$. This allows us to do queries like the one above in $\mathcal{O}(\log_2 K)$ operations.

3.4   Numerical Linear Algebra

The linear system can be stored with the global numbering of the degrees of freedom. There are existing, extensively tested, and widely used parallel libraries like PETSc, see [1,2] and Trilinos, see [11,10] to handle the linear system. They supply row-wise distributed matrices, vectors and algorithms, like iterative Krylov solvers, and preconditioners. `deal.II`, like most other finite element libraries, has interfaces to these libraries. We have tested both PETSc and Trilinos solvers up to many thousand cores and obtain excellent scaling results (see below).

The linear system is assembled on the local cells. Matrix and vector values for rows on different machines are sent to the owner using point-to-point communication. The linear system is subsequently solved in parallel.

3.5   Summary of Finite Element Algorithms

The building blocks of a distributed finite element calculation are described above. In an actual implementation, additional technical details must be addressed. To perform adaptive mesh refinement, an error estimate is needed to decide which cells to refine, solutions must be transferred between meshes, and hanging nodes must be handled. Hanging nodes come from degrees of freedom on interfaces between two cells on different refinement levels. For finite element systems with several components, like velocity, pressure, and temperature with different elements as used in Section 4, the global indices must be sorted by vector components. See [3] for details.

3.6   Communication Patterns

Because of the complex nature and algorithmic diversity of the operations outlined above it is difficult to analyze the MPI communication patterns appearing in the code.

The distribution of the mesh given by `p4est` has an interesting property: the number of *neighbors* for each processor (number of owners of the ghost cells) is bounded by a small number independent of the total problem size and the number of processors[2]. Most communication necessary in our code is therefore in the form of point-to-point messages between processors and a relatively small number of their neighbors, which results in optimal scaling.

Further efficiency is gained by *hiding* the latency of communication where possible: MPI communication uses non-blocking transfers and computations proceed while waiting for completion instead of leaving processors idle.

The amount of MPI communication within `deal.II` (outside that handled by `p4est`) consists of the parts described in section 3 and was created with the massive parallel implementation in mind. Exchanging degrees of freedom on the ghost cells is consequently done with neighbors only and is effectively hidden using non-blocking transfers. Other algorithms using communication, like error estimation and solution transfer, behave similarly. All other communication is done inside the linear algebra package.

As in all good finite element codes, the majority of compute time in our applications is spent in the linear solvers. For massively parallel applications, either PETSc or Trilinos provides this functionality. Limiting factors are then the speed of scalar product evaluations and matrix-vector products. The former

---

[2] The number of neighbors in all experiments is always smaller than fifty, but much lower for typical meshes. This seems to be a property of the space filling curve.

is a global reduction operation, it only consists of scalar data. On the other hand, matrix-vector products require this sort of communication but do not require the global reduction step. Furthermore, matrix-vector products are often not the most difficult bottleneck because communication can be hidden behind expensive local parts of the product. In summary, good scalability in the linear algebra is achieved if a low latency network like InfiniBand is available, as it is on most current high performance clusters.

4   Numerical Results

4.1   Scalability Test

We start by testing a Poisson equation in 2 and 3 dimensions with adaptive and global refinement, respectively. In Figure 1 we show the weak scalability from 8 to 1000 processors with about 500,000 degrees of freedom per processor on the three dimensional unit cube. We measure computation times for different parts of the program and average memory consumption on each machine. All parts but the solver (BiCGStab preconditioned with an algebraic multigrid) scale linearly. Figure 2 shows individual iterations in a fully adaptive refinement loop for a two dimensional Poisson equation on 1024 processors (left). The problem size increases over several refinement cycles from 1.5 million to 1.5 billion degrees of freedom. The right panel shows strong scalability starting from 256 and going to 4096 processors for the same cycle with a fixed problem size within the adaptive iteration loop. We have excellent scalability with respect to problem size and the number of processors. The memory consumption is nearly constant even when the problem size increases by over a factor of one hundred.

4.2   Results for a Mantle Convection Problem

Our second test case is a more complicated problem modeling thermal convection in the Earth's mantle. Details and motivation for the discretization and solver choices are given in [13] and [9].

In the Earth's mantle, fluid flow is strongly dominated by viscous stresses and is driven (among other factors) by temperature differences in the material, while

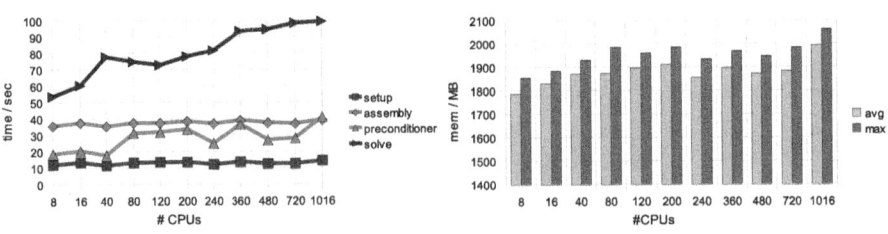

Fig. 1. 3D Poisson Problem, regular refinement, 500,000 degrees of freedom per processor. Left: Weak scaling up to 1016 processors. Right: Average peak memory for the same data.

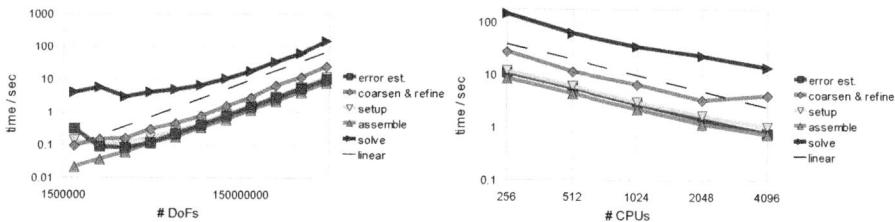

Fig. 2. 2D Poisson Problem, fully adaptive. Left: Weak scaling on 1024 processors. Right: Strong scaling, problem size of around 400 million DoFs.

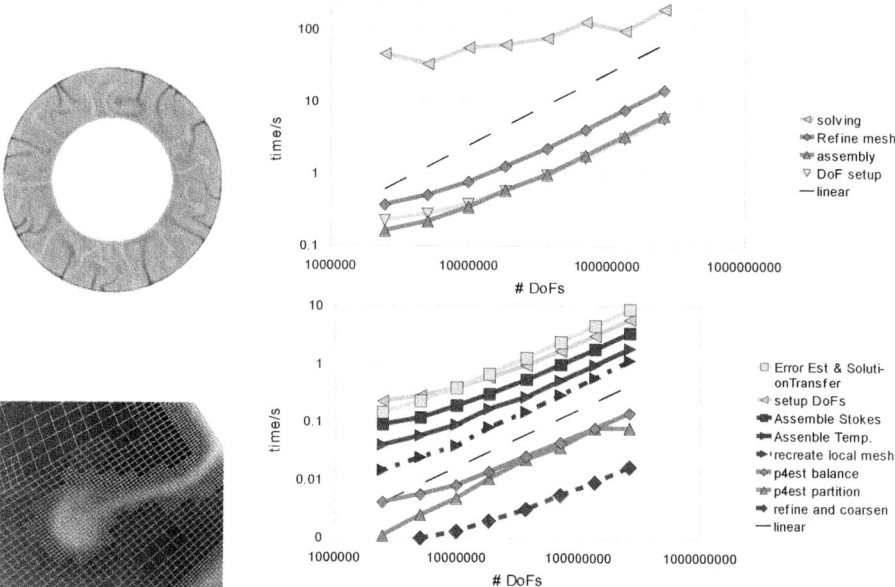

Fig. 3. 2D mantle convection. Left: snapshot of the temperature for a fixed time step and a zoom. Right: Solution times on 512 processors with overview (top) and detailed functions (bottom).

inertia is negligible at realistic velocities of a few centimeters per year. Thus the buoyancy-driven flow can be described by the Boussinesq approximation:

$$- \nabla \cdot (2\eta \varepsilon(\mathbf{u})) + \nabla p = -\rho \, \beta \, T \, \mathbf{g},$$
$$\nabla \cdot \mathbf{u} = 0, \qquad (1)$$
$$\frac{\partial T}{\partial t} + \mathbf{u} \cdot \nabla T - \nabla \cdot \kappa \nabla T = \gamma.$$

Here, $\mathbf{u}, p, T$ denote the three unknowns in the Earth's mantle: velocity, pressure, and temperature. The first two equations form a Stokes system for velocity and pressure with a forcing term stemming from the buoyancy through the temperature T. The third is an advection-diffusion equation. Let η be the viscosity of the fluid and κ the diffusivity coefficient for the temperature (both assumed to be constant here for simplicity), $\varepsilon(\mathbf{u}) = \frac{1}{2}(\nabla\mathbf{u} + (\nabla\mathbf{u})^T)$ denotes the symmetrized gradient, ρ is the density, β is the thermal expansion coefficient, g is the gravity vector, and γ describes the external heat sources. Fig. 3, left, shows a snapshot from the evolution of the turbulent mixing within the Earth's mantle.

In figure 3, right, we present timing of seven adaptive refinement steps for a single fixed time step. We observe good scalability; the solver itself scales better than linearly due to the relatively fine mesh for the two dimensional solution and because the solution on the coarser meshes is reused as a starting guess.

5   Conclusions

We present a general framework for massively parallel finite element simulation. The results are convincing, showing that even complex problems with more than a billion unknowns can be solved on a large cluster of machines. The developments in `deal.II` outlined here make the maximum solvable problem size two orders of magnitude larger than previously possible.

There are several reasons for the good scalability results. Most importantly the workload is distributed evenly, because every processor has roughly the same number of locally active cells. In addition the algorithms described in section 3 introduce no significant overhead in parallel. This includes the total memory usage. As described in section 3.6 most of the communication is restricted to the neighbors.

Acknowledgments. Timo Heister is partly supported by the German Research Foundation (DFG) through GK 1023. Martin Kronbichler is supported by the Graduate School in Mathematics and Computation (FMB). Wolfgang Bangerth was partially supported by Award No. KUS-C1-016-04 made by King Abdullah University of Science and Technology (KAUST), by a grant from the NSF-funded Computational Infrastructure in Geodynamics initiative through Award No. EAR-0426271, and by an Alfred P. Sloan Research Fellowship.

The computations were done on the Hurr[3] cluster of the Institute for Applied Mathematics and Computational Science (IAMCS) at Texas A&M University. Hurr is supported by Award No. KUS-C1-016-04 made by King Abdullah University of Science and Technology (KAUST).

References

1. Balay, S., Buschelman, K., Eijkhout, V., Gropp, W.D., Kaushik, D., Knepley, M.G., Curfman McInnes, L., Smith, B.F., Zhang, H.: PETSc users manual. Technical Report ANL-95/11 - Revision 3.0.0, Argonne National Laboratory (2008)

---

[3] 128 nodes with 2 quad-core AMD Shanghai CPUs at 2.5 GHz, 32GB RAM, DDR Infiniband, running Linux, OpenMPI, gcc 4.

2. Balay, S., Buschelman, K., Gropp, W.D., Kaushik, D., Knepley, M.G., Curfman McInnes, L., Smith, B.F., Zhang, H.: PETSc Web page (2009), http://www.mcs.anl.gov/petsc
3. Bangerth, W., Burstedde, C., Heister, T., Kronbichler, M.: Algorithms and Data Structures for Massively Parallel Generic Finite Element Codes (in preparation)
4. Bangerth, W., Hartmann, R., Kanschat, G.: deal.II Differential Equations Analysis Library, Technical Reference, http://www.dealii.org
5. Bangerth, W., Hartmann, R., Kanschat, G.: deal.II — a General Purpose Object Oriented Finite Element Library. ACM Transactions on Mathematical Software 33(4), 27 (2007)
6. Bruaset, A.M., Langtangen, H.P.: A comprehensive set of tools for solving partial differential equations; DiffPack. In: Dæhlen, M., Tveito, A. (eds.) Numerical Methods and Software Tools in Industrial Mathematics, pp. 61–90. Birkhäuser, Boston (1997)
7. Burstedde, C., Burtscher, M., Ghattas, O., Stadler, G., Tu, T., Wilcox, L.C.: Alps: A framework for parallel adaptive pde solution. Journal of Physics: Conference Series 180(1), 012009(2009)
8. Burstedde, C., Wilcox, L.C., Ghattas, O.: p4est: Scalable algorithms for parallel adaptive mesh refinement on forests of octrees. Submitted to SIAM Journal on Scientific Computing (2010)
9. Heister, T., Kronbichler, M., Bangerth, W.: Generic finite element programming for massively parallel flow simulations. In: Eccomas 2010 Proceedings (submitted, 2010)
10. Heroux, M.A., Bartlett, R.A., Howle, V.E., Hoekstra, R.J., Hu, J.J., Kolda, T.G., Lehoucq, R.B., Long, K.R., Pawlowski, R.P., Phipps, E.T., Salinger, A.G., Thornquist, H.K., Tuminaro, R.S., Willenbring, J.M., Williams, A., Stanley, K.S.: An overview of the Trilinos project. ACM Trans. Math. Softw. 31, 397–423 (2005)
11. Heroux, M.A., et al: Trilinos Web page (2009), http://trilinos.sandia.gov
12. Kirk, B., Peterson, J.W., Stogner, R.H., Carey, G.F.: libMesh: A C++ Library for Parallel Adaptive Mesh Refinement/Coarsening Simulations. Engineering with Computers 22(3-4), 237–254 (2006)
13. Kronbichler, M., Bangerth, W.: Advanced numerical techniques for simulating mantle convection (in preparation)
14. Langtangen, H.P.: Computational Partial Differential Equations: Numerical Methods and Diffpack Programming. Texts in Computational Science and Engineering. Springer, Heidelberg (2003)
15. Logg, A.: Automating the finite element method. Arch. Comput. Methods Eng. 14(2), 93–138 (2007)
16. Patzák, B., Bittnar, Z.: Design of object oriented finite element code. Advances in Engineering Software 32(10–11), 759–767 (2001)
17. Renard, Y., Pommier, J.: Getfem++. Technical report, INSA Toulouse (2006), http://www-gmm.insa-toulouse.fr/getfem/

Parallel Zero-Copy Algorithms for Fast Fourier Transform and Conjugate Gradient Using MPI Datatypes

Torsten Hoefler and Steven Gottlieb*

National Center for Supercomputing Applications
University of Illinois at Urbana-Champaign, Urbana, IL, USA
{htor,sgottlie}@illinois.edu

Abstract. Many parallel applications need to communicate non-contiguous data. Most applications manually copy (pack/unpack) data before communications even though MPI allows a *zero-copy* specification. In this work, we study two complex use-cases: (1) Fast Fourier Transformation where we express a local memory transpose as part of the datatype, and (2) a conjugate gradient solver with a checkerboard layout that requires multiple nested datatypes. We demonstrate significant speedups up to a factor of 3.8 and 18%, respectively, in both cases. Our work can be used as a template to utilize datatypes for application developers. For MPI implementers, we show two practically relevant access patterns that deserve special optimization.

1 Introduction

The Message Passing Interface (MPI) offers a mechanism called derived datatypes (DDT) to specify arbitrary memory layouts for sending and receiving messages. This mighty mechanism allows the integration of communication into the parallel algorithm and data layout and thus is likely to become an important part of application development and optimization. Not only do DDTs save implementation effort by providing an abstract and versatile interface to specify arbitrary data layouts, but they also provide a portable high-performance abstraction for data accesses. It is easy to show that datatypes are complete in that any permutation from a layout on the sender to a layout on the receiver can be expressed (different DDTs at sender and receiver are allowed as long as the *type maps* [1] match).

Zero-copy refers to a mechanism to improve application performance by avoiding copies in the messaging middleware. Several low-level communication APIs, such as InfiniBand [2] or DCMF [3] allow direct copies from a user-buffer on the sender to a user-buffer at the receiver. We extend this definition into the application space and argue that the specification of derived datatypes is **necessary** to enable *zero-copy algorithms*, i.e., no explicit buffer pack/unpack, for parallel applications. It has been shown that non-contiguous data can be transferred without additional copies using InfiniBand [4].

---

* On leave from Indiana University, Bloomington, IN, USA.

R. Keller et al. (Eds.): EuroMPI 2010, LNCS 6305, pp. 132–141, 2010.
© Springer-Verlag Berlin Heidelberg 2010

Many applications require sending data from non-contiguous locations, so we would expect that many MPI applications use datatypes to specify their communications. However, on the contrary, implementations of the DDT mechanism in MPI have been suboptimal so that manual packing and unpacking of data often yielded higher performance. In the last years, implementations have much improved [4,5,6,7,8] but the *folklore* about low performance still remains. Indeed, the number of success stories is low and limited to application benchmarks with relatively simple datatype layouts [9].

In this work, we demonstrate two complex use-cases for DDTs in parallel applications. The first example shows how to express the local transpose operations in a parallel Fast Fourier Transformation (FFT). The second example shows a complex 4-d stencil code with checkerboard layout.

2   Fast Fourier Transformations

Fast Fourier Transforms (FFT) have numerous applications in science and engineering and are among the most important algorithms today. One-dimensional (1-d) FFTs accept an array of N complex numbers as input and produce an array of size N as output. FFTs can also be done in place with negligible additional buffering. Such 1-d FFTs can be expressed as several multi-dimensional FFTs and application of so called *twiddle factors* [10, §12]. Such a decomposition is often used to parallelize FFTs because applying the twiddle factors is a purely local operation. Naturally multi-dimensional FFTs are also very important in practice, for example, 2-d FFTs for image analysis and manipulation and 3-d FFTs for real-space domains. Such n-d FFTs can be computed by performing 1-d FFTs in all n dimensions.

2.1   A Typical Parallel FFT Implementation

We discuss a typical parallel implementation of a $N_x \times N_y$ 2-d FFT with MPI. We assume that the array is stored in x-major order and distributed along the x dimension such that each process has N_x/P y-pencils. Figure 1 illustrates the whole procedure for a 4×4 FFT on two processes (0 and 1). Each process

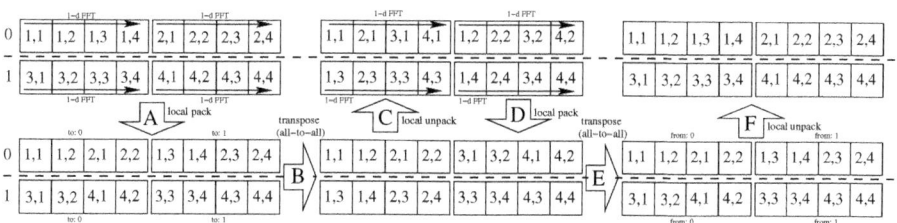

Fig. 1. Parallel two-dimensional FFT on two processes. The steps are explained below

holds two 4-element y-pencils in its local memory. The elements are shown with (x, y) indices in contiguous memory locations (left→right) in the figure. The two processes are drawn vertically and separated by a dashed line. The steps needed to transform the array and return it in the original layout are:

1. perform N_x/P 1-d FFTs in y-dimension (N_y elements each)
2. pack the array into a sendbuffer for the all-to-all (A)
3. perform global all-to-all (B)
4. unpack the array to be contiguous in x-dimension (each process has now N_y/P x-pencils) (C)
5. perform N_y/P 1-d FFTs in x-dimension (N_x elements each)
6. pack the array into a sendbuffer for the all-to-all (D)
7. perform global all-to-all (E)
8. unpack the array to its original layout (F)

Thus, in order to transform the two-dimensional data, it is rearranged six times. Each rearrangement is effectively a copy operation of the whole data. However, four rearrangements (pack and unpack) are related to the global transpose operation. Since MPI datatypes are complete, we can fold all pack and unpack operations into the communication and thus avoid the explicit copy for packing the data.

2.2   Constructing the Datatypes

We assume that the basic element is a complex number. A datatype for complex numbers can simply be created with MPI_Type_contiguous with two double elements.

The send-datatype can be constructed with MPI_Type_vector because each y-pencil is logically cut into P pieces that need to be redistributed to P processes. Thus, the blocklength is $\frac{N_y}{P}$. Each process typically holds $\frac{N_x}{P}$ pencils, thus, there are a total of $\frac{N_x}{P}$ such blocks. The stride between the blocks is one complete y-pencil of length N_y. The basic vector datatype is shown in Figure 2(a). Sending a single element of this datatype would transmit $\{(1,1),(1,2),(2,1),(2,2)\}$. The problem is now that the *comb*-shaped datatypes that need to be sent are interleaved as shown in Figure 2(b). Thus, one can't just sent two of those datatypes (as this would gather two contiguous combs instead of two interleaved combs).

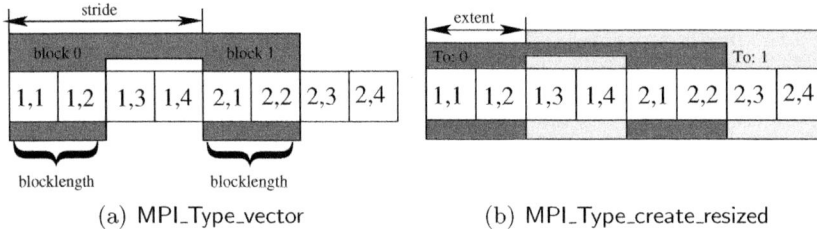

(a) MPI_Type_vector   (b) MPI_Type_create_resized

Fig. 2. Visualization of the send datatype creation

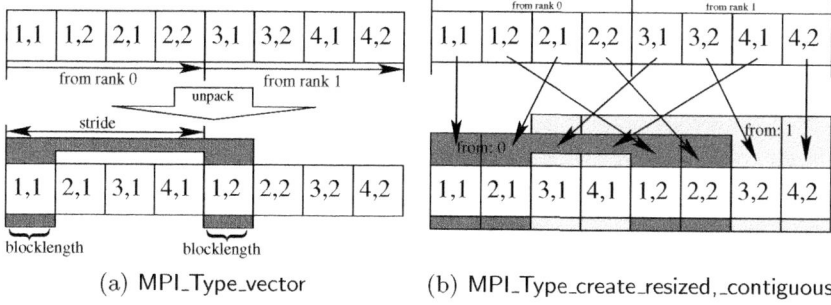

(a) MPI_Type_vector   (b) MPI_Type_create_resized,_contiguous

Fig. 3. Visualization of the receive datatype creation

MPI allows the user to change the *extent* of a datatype in order to allow such interleaved accesses. In our example we use MPI_Type_create_resized to change the extent to $\frac{N_y}{P}$ times the base-size as shown in Figure 2(b). The resulting datatype can be used as input to MPI_Alltoall by sending count=1 to each process.

Performing the unpack on the receiver is slightly more complex because the data arrives in non-transposed form from the sender. Thus, the receiver does not only need to unpack the data but also transpose each block locally. This can also be expressed in a single derived datatype. The top of Figure 3(a) shows how the data-stream arrives at the receiver (process 0) and the bottom the desired layout after unpack. Like in the sender-case, we create a MPI_Type_vector datatype. However, the blocklength is now one element because we need to transpose the array locally. We have $\frac{N_x}{P}$ blocks with a stride of N_x between them. The newly created comb-shaped type captures one incoming y-contribution of one process. To capture all, we need to create a contiguous datatype with $\frac{N_y}{P}$ elements. We have to change the extent to 1 with MPI_Type_create_resized as for the send datatype. Figure 3(b) shows the final datatypes for our example. Those types can be used as the receive type in MPI_Alltoall with count=1 per process (note that the send- and receive-types in MPI_Alltoall do not have to be identical as long as the type-map matches).

By using both created datatypes, we can effectively eliminate steps A, C, D, and F in Figure 1 which leads to a zero-copy FFT. An optimized MPI implementation would stream the data items directly from the send buffers into the receive buffers and apply the correct permutation (local transpose). This should lead to significant performance improvements over the state of the art because it avoids four explicit copies of the whole 2-d array. Higher-dimensional FFTs can be treated with similar principles.

2.3 Experimental Evaluation

We used two systems for our performance evaluation, *Odin* at Indiana University and *Jaguar* at the Oak Ridge National Laboratory. Odin consists of 128 compute nodes with dual-CPU dual-core Opteron 1354 2.1 GHz CPUs running Linux 2.6.18 and are connected with SDR InfiniBand (OFED 1.3.1). We used

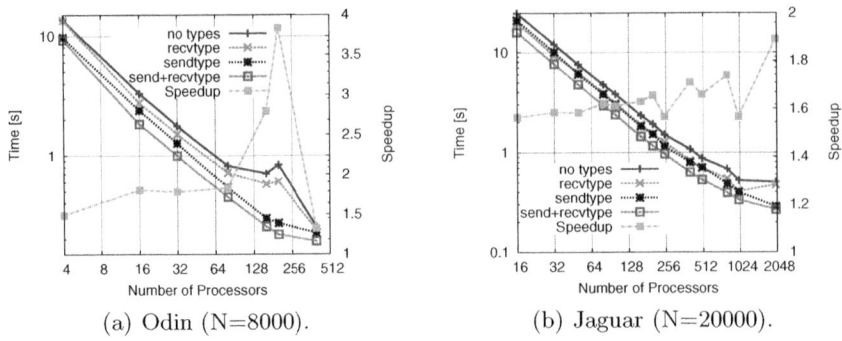

(a) Odin (N=8000).   (b) Jaguar (N=20000).

Fig. 4. Strong-scaling of a 2-d FFT with and without zero-copy (MPI Datatypes)

Open MPI 1.4.1 (openib BTL) and g++ 4.1.2 for our evaluation. Jaguar (XT-4) comprises 150152 2.1 GHz Opteron cores in quad-core nodes connected with a Torus network (SeaStar). Jaguar runs Compute Node Linux 2.1 and the Cray Message Passing Toolkit 3. All software was compiled with -O3 -mtune=opteron on both systems.

In all experiments, we ran one warmup round (using the same buffers as for the actual run). We repeated each run three times (in the same allocation) and found a maximum deviation of 4%. We report the smallest measured time for the complete parallel 2-d FFT of the three runs. The overhead to create the derived datatypes is included in the measurements that use derived datatypes.

Figure 4 shows the results for a strong-scaling $N \times N$ 2-d FFT on Odin and Jaguar. Using derived datatypes improves the performance of parallel FFTs on both systems by more then a factor of 1.5. The improvement typically grows with the number of processes as local FFTs get smaller. The anomaly at P=200 on Odin (Figure 4(a)) is reproducible. Datatypes also improved parallel scaling on Jaguar as shown in Figure 4(b) where the traditional FFT stopped scaling at 1024 processes and the version using derived datatypes scaled up to 2048 processes.

The performance of derived datatypes is system dependent and might as well not result in any speedup if the implementation performs a complete local pack/unpack. We found only one system, BlueGene/P, where the datatype implementation is slowing down the FFT significantly (up to 40%). The simple representation of the constructed vector datatype should not introduce significant overhead. This might point at an optimization opportunity or performance problem of the MPI implementation on BlueGene/P.

3   MIMD Lattice Computation Collaboration Application

The MIMD Lattice Computation (MILC) Collaboration studies Quantum Chromodynamics (QCD) the theory of the strong interaction [11]. Their suite of applications, known as the MILC code is publicly available for the study of lattice

QCD. This group regularly gets one of the largest allocations of computer time at NSF supercomputer centers. One application from the code suite, su3_rmd, is part of the SPEC CPU2006 and SPEC MPI benchmarks. It is also used to evaluate the performance of the Blue Waters computer to be built by IBM.

Lattice QCD approximates space-time as a finite regular hypercubic grid of points in four dimensions. The physical quarks are represented by 3-component complex objects at each point of the grid. The variables that describe the gluons, the carriers of the strong force are represented by 3×3 unitary matrices residing on each 'link' joining points of the grid. Currently, grids as large as $64^3 \times 192$ are in use. Much of the floating point work is involved in multiplying the 3×3 matrices together or applying the matrix to a 3-component vector. Routines for these basic operations are often optimized by assembly code or compiler intrinsics.

The code is easily parallelized by domain decomposition. Once that is done, the program must be able to communicate with neighboring processes that contain off-node neighbors of the points in its local domain. The MILC code abstracts all the communication into a small set of routines: start_gather, wait_gather, and cleanup_gather. These routines are all contained in a single file specific to the message passing library available on the target computer.

The MILC code allows very general assignments of grid points to the processes. At startup, a list of local grid points that need off-node neighbors for their computation is created for each direction $\pm x$, $\pm y$, $\pm z$, $\pm t$. There is one list corresponding to each other process that contains any needed neighbors for a particular direction. There are also similar lists for all the local grid points whose values will need to be sent to other processes. At the time a gather is called, the lists containing data that must be sent to other processes are processed and for each grid point in a list the value of the data to be gathered is copied (packed) into a buffer. The buffer is then sent to the neighboring process. The index list is used to allow for arbitrary decompositions of the grid; however, in practice, the most common data layout is just to break up the domain into hyperrectangular subdomains with checker boarding as described below. It is for this case that we have implemented derived datatypes to avoid copying the data to a buffer before sending it to the destination process. The receive portion does not require datatypes because the computation uses indirect addressing for all grid points. The index list of local grid points with remote data dependencies is set (once during initialization) to point to the correct element in the receive buffer.

3.1   Data Layout and Datatype Construction

The code consists of several computation phases that perform different tasks. There are compilation flags that allow timing and printing performance information for each phase. In this work, we will concentrate on the conjugate gradient (CG) solver since that routine takes the vast majority of the time in production runs. Checkerboarding, or even-odd decomposition is used in the iterative solver. A grid point is even (odd) if the sum of its coordinates is even (odd). Thus, the grid points are stored in memory so that all even sites are stored before the odd sites. If the coordinates of a point are denoted (x, y, z, t), the data is stored so that x is incremented first, then y is incremented, then z and finally t. That means that the edge of the domain in t is (almost) contiguously stored. If the local domain is of size $L_x \times L_y \times L_z \times L_t$, there are $L_x \times L_y \times L_z/2$ even sites stored contiguously and the same number of odd sites stored contiguously. Note that our current implementation of datatypes requires that each of the local dimensions is even. During the CG solver, we are usually only transferring one checkerboard at a time. (In other phases of the code, we operate on all grid points, so we also define datatypes for even-and-odd gathers. These are defined with MPI_Type_hvector in the code example. The blocks of even and odd sites are identical patterns separated by the number of even sites on each process. This is converted to bytes by multiplying by the size of the object.) If we need to fetch values from the z-direction, however, the points are not all stored contiguously. For each value of t, there are $L_x \times L_y/2$ contiguous sites in each checkerboard. The datatype defined for the gathers in the z-direction consists of L_t repetitions of such contiguous data. For the gathers in the y-direction, there are $L_z \times L_t$ regions of $L_x/2$ contiguous sites. Listing 1 shows parts of the datatype layout routine which is called during initialization.

```
/* the basic elements */
MPI_Type_contiguous(6, MPI_FLOAT, &su3_vect_dt);
MPI_Type_contiguous(12, MPI_FLOAT, &half_wilson_vector_mpi_t);
MPI_Type_contiguous(18, MPI_FLOAT, &su3_matrix_mpi_t);

/* 48 field types, 3 for su3_vector, half_wilson_vector, and su3_matrix,
   2 for even and even and odd, 8 for directions */
MPI_Datatype neigh_dt_ddt[3][2][8];

/* t-direction, even points */
MPI_Type_contiguous(Lx · Ly · Lz/2, su3_vect_dt, &neigh_dt_ddt[0][0][3]);
/* t-direction, even and odd points */
MPI_type_hvector(2,1,sizeof(su3_vector)*even_sites, neigh_dt_ddt[0][0][3],
   &neigh_dt_ddt[0][1][3]);

/* z-direction, even points */
MPI_Type_vector(Lt, Lx · Ly/2, Lx · Ly · Lz/2, su3_vect_dt,
   &neigh_dt_ddt[0][0][2]);
/* z-direction, even and odd points */
MPI_type_hvector(2,1,sizeof(su3_vector)*even_sites, neigh_dt_ddt[0][0][2],
   &neigh_dt_ddt[0][1][2]);
...
```

Listing 1. Datatype Example for the Up Direction and su3_vector. MILC uses 48 different data layouts for sending.

Three other issues are simplified in the code example. We do not show code for negative directions or for gathers of matrices and pairs of vectors. We show the basic definitions for half_wilson_vector_mpi_t and su3_matrix_mpi_t, but not the corresponding definitions of field_neigh_dt[{1,2}][ ][ ]. Further, for the CG routine, we also need to gather from sites three grid points away in each direction. These require contiguous blocks three times as long and merely require changing some factors of $1/2$ to $3/2$.

3.2   Experimental Evaluation

We now present performance results comparing the version the datatype version with the original pack/unpack version. We chose a weak scaling problem of size $L_x = L_y = L_z = L_t = 4$ per process which is similar to the Petascale benchmark problem that will be used to verify the Blue Waters machine on $> 3 \cdot 10^5$ cores. We ran each benchmark multiple times and report the average performance of all CG phases.

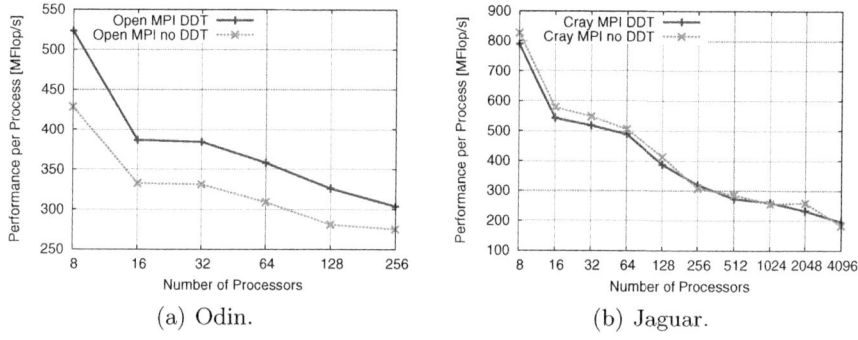

(a) Odin.   (b) Jaguar.

Fig. 5. Weak-scaling MILC run with a 4^4 lattice per process

Figure 5 shows the performance in MFlop/s of runs on Odin and Jaguar. The CG solver requires global sums in addition to the nearest neighbor gathers. These sums are the biggest impediment to scaling since the global sum time is expected to increase as the logarithm of the number of processes. For a fixed local grid size, *i.e.*, weak scaling, the time for the global sum will eventually dominate the time for the work that must be done on each process. This is reflected in the decreasing performance is the number of processors is increased beyond 16. The sharp dropoff between 8 and 16 is due to the fact the one additional direction has off-node neighbors. Most other parts of the code do not require global sums. We see a speedup up to 18% by using derived datatypes on Odin while we see no benefit, indeed an average performance penalty of 3% on Jaguar.

The performance degradation on Jaguar is surprising because the data access of the MPI_Type_vector definition of the used datatype can be easily expressed as two loops [5,7] while the original MILC packing routine traverses an array of indices which adds more pressure to the memory subsystem. This points at possible optimization opportunities in Cray's MPI because the simple structure of the datatype should, even in a simple implementation, not introduce significant overheads.

4   Conclusions

We demonstrated two applications that can take significant advantage of using MPI's derived datatype mechanism for communication. Such techniques essentially enable parallel zero-copy algorithms and even allows one to express additional local transformations (as demonstrated for FFT). Performance results of FFT and a CG solver show improvements up to a factor of 3.8 and 18% respectively. However, we also found performance degradation, which indicate optimization opportunities in the MPI libraries on BlueGene/P and Jaguar systems, in some cases.

We expect that our results will influence two groups: (1) application developers are encouraged to use MPI datatypes to simplify and optimize their code, and (2) MPI implementers should use the presented algorithms as examples for practically relevant access patterns that might benefit from extra optimizations. The source code of both applications is publicly available and can be used for evaluating datatype implementations.

Acknowledgments. The authors want to thank Bill Gropp (UIUC), Jeongnim Kim (UIUC), Greg Bauer (UIUC), and the anonymous reviewers for helpful comments. This work is supported by the Blue Waters sustained-petascale computing project, which is supported by the National Science Foundation (award number OCI 07-25070) and the state of Illinois.

References

1. MPI Forum: MPI: A Message-Passing Interface Standard. Version 2.2 (2009) http://www.mpi-forum.org/docs/mpi-2.2/mpi22-report.pdf
2. The InfiniBand Trade Association: Infiniband Architecture Specification , Release 1.2. InfiniBand Trade Association vol.1(2003)
3. Kumar, S., et al.: The deep computing messaging framework: generalized scalable message passing on the blue gene/p supercomputer. In: ICS 2008: Proceedings of the 22nd annual international conference on Supercomputing, pp. 94–103. ACM, New York (2008)
4. Santhanaraman, G., Wu, J., Huang, W., Panda, D.K.: Designing zero-copy message passing interface derived datatype communication over infiniband: Alternative approaches and performance evaluation. Int. J. High Perform. Comput. Appl. 19, 129–142 (2005)
5. Träff, J.L., Hempel, R., Ritzdorf, H., Zimmermann, F.: Flattening on the fly: Efficient handling of mpi derived datatypes. In: Proceedings of the 6th European PVM/MPI Users' Group Meeting on Recent Advances in Parallel Virtual Machine and Message Passing Interface, London, UK, pp. 109–116. Springer, Heidelberg (1999)

6. Gabriel, E., Resch, M., Rühle, R.: Implementing and benchmarking derived datatypes in metacomputing. In: HPCN Europe 2001: Proc. of the 9th Intl. Conference on High-Performance Computing and Networking, London, UK, pp. 493–502. Springer, Heidelberg (2001)

7. Gropp, W., Lusk, E., Swider, D.: Improving the performance of mpi derived datatypes. In: Proceedings of the Third MPI Developer's and User's Conference, pp. 25–30. MPI Software Technology Press (1999)

8. Byna, S., Gropp, W., Sun, X.H., Thakur, R.: Improving the performance of mpi derived datatypes by optimizing memory-access cost. In: IEEE International Conference on Cluster Computing, p. 412 (2003)

9. Lu, Q., Wu, J., Panda, D., Sadayappan, P.: Applying MPI Derived Datatypes to the NAS Benchmarks: A Case Study. In: Proc. of the Intl. Conf. on Par. Proc. Workshops (2004)

10. Press, W.H., Teukolsky, S.A., Vetterling, W.T., Flannery, B.P.: Numerical recipes in C: the art of scientific computing, 2nd edn. Cambridge University Press, Cambridge (1992)

11. Bernard, C., Ogilvie, M.C., DeGrand, T.A., DeTar, C.E., Gottlieb, S.A., Krasnitz, A., Sugar, R., Toussaint, D.: Studying Quarks and Gluons On Mimd Parallel Computers. International Journal of High Performance Computing Applications 5, 61–70 (1991)

Parallel Chaining Algorithms

Mohamed Abouelhoda and Hisham Mohamed

Center for informatics sciences, Nile University,
Smart Village, Giza, Egypt
mabouelhoda@nileuniversity.edu.eg,
hisham.mohamed@nileu.edu.eg

Abstract. Given a set of weighted hyper-rectangles in a k-dimensional space, the chaining problem is to identify a set of colinear and non-overlapping hyper-rectangles of total maximal weight. This problem is used in a number of applications in bioinformatics, string processing, and VLSI design. In this paper, we present parallel versions of the chaining algorithm for bioinformatics applications, running on multi-core and computer cluster architectures. Furthermore, we present experimental results of our implementations on both architectures.

Keywords: Chaining Algorithms, Bioinformatics, Sparse Dynamic Programming.

1 Introduction

Given a set of hyper-rectangles in a k-dimensional space, each associated with a certain weight, the chaining problem is to determine a chain of colinear and non-overlapping hyper-rectangles such that the total weight of the included hyper-rectangles is maximum. Figure 1 shows an example of some blocks in 2D space and an example of an optimal chain with highest score. These blocks can also be represented in parallel coordinates, where the coordinates are drawn as parallel lines and the intervals making up a block in each coordinate are connected by lines, see Figure 1(right).

The chaining algorithms, which solve the chaining problem, are per se of theoretical interest, because they provide a polynomial time solution to the maximum clique problem for trapezoid graphs. In practice, these algorithms have been used to solve the channel routing problem in VLSI design [6], and the longest common increasing subsequence problem [2] in string processing. In the bioinformatics domain, the chaining algorithms have received much more attention due to their use in many applications in computational comparative genomics. For example, they are used to speed up the alignment of multiple genomes [7,4,11], cDNA/EST mapping [10], and the comparison of restriction maps [8].

In all these bioinformatics applications, the blocks represent regions of similarity in the given genomes and an optimal chain approximates optimal alignment of the given genomic sequences. In Figure 1, the x_1 and x_2 correspond to two genomes, and the seven blocks correspond to regions of similarity between them. The chain of the blocks b_1, b_3, and b_7 are the alignment anchors,

R. Keller et al. (Eds.): EuroMPI 2010, LNCS 6305, pp. 142–151, 2010.

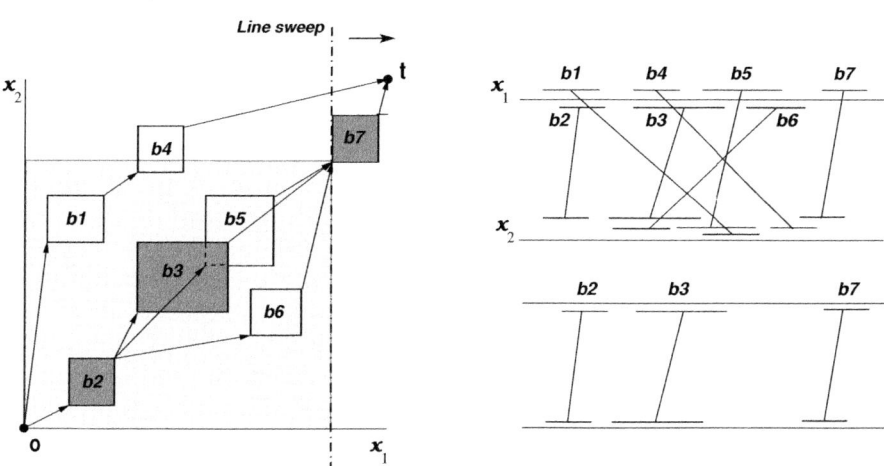

Fig. 1. Left: Representation of a set of 2D blocks. The origin (0) and terminus (t) are unit area blocks. An edge connects two blocks, if they are colinear and non-overlapping (not all edges are drawn). The blocks b_2 and b_3 are colinear and non-overlapping. The blocks b_1 and b_3 are non-overlapping but not colinear. The chain of b_2, b_3, and b_7 represents an optimal chain. The line sweep at the start point of b_7 searches for the block with maximum score (which is b_3) in the highlighted region. Right: Parallel coordinate representation of the blocks (upper part) and optimal chain (lower part). The intervals making up a block are represented by thick lines on the coordinates and a thin line connects the one on x_1 to that on x_2.

and the regions between the blocks correspond either to mismatches or gaps (insertions/deletions).

Sequential chaining algorithms were first introduced in [5] under the name *sparse dynamic programming* to speed up the alignment of *two* sequences. This technique was then generalized in [12,9,1] to multiple genomic sequences under the name of *chaining*, which became more common in bioinformatics. For $k = 2$, the chaining problem can be solved in $O(m \log m)$ time and $O(m)$ space, where m is the number of blocks. (It takes $O(m \log \log m)$ time if the blocks are given sorted w.r.t. one of the coordinates.) For $k > 2$, the problem is solved in sub-quadratic time based on range maximum queries (RMQ), which is a substantial improvement over the quadratic graph based algorithm.

The exponentially increasing amount of biological data requires much faster algorithms to compare more and longer genomes. In this paper, we introduce parallel chaining algorithms for the multicore and computer cluster architectures. To the best of our knowledge, our work is the first that addresses this issue from theoretical as well as practical point of view. This is not a straightforward task because the chaining algorithms are based on the line sweep paradigm and range maximum queries, which requires careful parallelization strategies to yield practical results.

This paper is organized as follows: In Section 2, we review the sequential chaining algorithms. In Section 3, we introduce our parallel chaining algorithms.

Section 4 contains implementation details and experimental results. Conclusions are in Section 5.

2   Review of Sequential Chaining Algorithms

We use the term *block*, denoted by b, to refer to a hyper-rectangle in a k dimensional space ($\mathbb{N}^k$). A block can be defined by the pair $(beg(b), end(b))$, where $beg(b) = (beg(b).x_1, .., beg(b).x_k)$ and $end(b) = (end(b).x_1, .., end(b).x_k)$ are its two extreme corner points $(beg(b).x_i < end(b).x_i, \forall i \in [1..k])$. Equivalently, it can be defined by the k-tuple $([beg(b).x_1..end(b).x_1], \dots, [beg(b).x_k..end(b).x_k])$, where each component is an interval on the x_i axis, $1 \leq i \leq k$. With each block, we associate an arbitrary weight $b.weight$, which can be, e.g., its area or perimeter. For ease of presentation, we consider the point $0 = (0, \dots, 0)$ (the *origin*) and $t = (t_1, \dots, t_k)$ (the *terminus*) as blocks with unit side-length and weight zero. We will also assume that each block b has $beg(b).x_i > 0$ and each block b' has $end(b').x_i < t_i$, for all $i \in [1..k]$.

Definition 1. *The relation $\ll$ on the set of blocks is defined as follows: $b' \ll b$ iff $end(b').x_i < beg(b).x_i, \forall i \in [1..k]$. If $b' \ll b$, we say that b' precedes b. The blocks b' and b are* colinear *and* non-overlapping *if either b' precedes b or b precedes b'.*

In other words, the blocks b and b' are colinear with $b' \ll b$, if b' starts and ends in the region $([0..beg(b).x_1-1], \dots, [0..beg(b).x_k-1])$; Figure 1 shows an example.

Definition 2. *The score of a chain of blocks $b_1 \ll b_2 \ll .. \ll b_t$ is $\Sigma_{i=1}^{i=t} b_i.weight$. Given a set of blocks, an optimal chain is the highest scoring chain among all possible chains that start with 0 and ends with t.*

The chaining problem can be solved using the recurrence

$$b.score = b.weight + \max\{b'.score | b' \ll b\}, \tag{1}$$

where $b.score$ denotes the maximum score of all chains ending with the block b.

The geometric based solution for this recurrence is based on the line sweep paradigm and *range maximum queries* (RMQ) to find the block $b' \ll b$ that maximizes the score at b. This solution works as follows:

The start and end points of the blocks are sorted w.r.t. their x_1 coordinate and are processed in this order to simulate a line (hyper-plane in $\mathbb{N}^k$) that sweeps the points w.r.t. their x_1 coordinate. All end points are initialized as *inactive*. If an end point of block b is scanned by the sweeping line, it becomes *active* (and so does the respective block) with score $b.score$. (For short, we might speak of the score of a point to mean the score of the respective block.) While scanning the start point of a block b, we search for the block b' that maximizes Recurrence 1 among the active blocks by means of RMQs.

Because all the points (representing the blocks) are given in advance, we use a semi-dynamic data structure D supporting RMQ with activation to manipulate the points. In this data structure, we store the $(k-1)$-dimensional end point

$(end(b).x_2, .., end(b).x_k)$ of each block b and attach the score of a block to it. That is, we ignore the first dimension and reduce the RMQ dimension by one. This is correct, because the points are scanned w.r.t. x_1 coordinate and any previously scanned and activated block b' should satisfy $end(b').x_1 < beg(b).x_1$. Figure 1 shows an example where a start point is scanned and searches for an end point with maximum score.

The complexity of this algorithm depends on the complexity of the RMQ supported by D. If D is implemented as a range tree, then it takes $O(m \log^{k-2} m \log \log m)$ time and $O(m \log^{k-2} m)$ space. If it is implemented as a kd-tree, then it takes $O(m^{2-\frac{1}{k-1}})$ time and $O(m)$ space. (Activation of a point takes $O(\log^{k-2} m \log \log m)$ and $O(\log m)$ time using the range tree and kd-tree, respectively) However, the kd-tree is superior to the range tree in practice due to its linear space and the programming tricks used for the semi-dynamic points [3]. For more details on this algorithm, we refer the reader to [1].

3   Parallelization of Chaining Algorithms

Our parallelization strategy works on different levels of increasing complexity: 1) *space decomposition*, 2) *anti-chain decomposition*, and 3) *wavefront decomposition*.

3.1   Space Decomposition

As expected, profiling results of the sequential chaining program revealed that the RMQ's take most of the chaining time. Hence, we present a space division strategy of 2 levels to reduce this time.

Bucket division: The set of the end points are divided into equal size buckets $BUK_1, .., BUK_{P_b}$, such that the points in BUK_i occur before the points in BUK_{i+1} w.r.t. x_2. Note that each bucket BUK_i is defined over the region $([0..\infty], [L_i.x_2..L_{i+1}.x_2], .., [0, \infty])$, where L_i is a line (hyper-plane) separating the points in BUK_i from BUK_{i+1}. Figure 2 (left) shows an example of the space divided into three buckets.

Hierarchical division: The d-dimensional points are first divided into two equal sets w.r.t. to dimension y_1. The y_1 component of the median point represents a line that splits the point set into two subsets in the space. The point in each subset is then divided w.r.t. dimension y_2, then the resulting subsets are further divided into dimension y_3, and so on. After division w.r.t. y_d, the subsets are divided again with respect to y_1, and so on. After y_d divisions, the space is divided into disjoint hyper-rectangles, each containing equal number of points. The well-informed reader can easily figure out that this division strategy is the one used in the kd-tree.

Putting all together, the set of end points are first divided into P_{buk} buckets. Then each bucket is further split hierarchically into P_h sub-buckets. On each sub-bucket, we construct the RMQ data structure. In fact, this division can be regarded as parallelizing the search over kd-tree, where the tree is split into multiple subtrees. Algorithm 1 describes a parallel version of the chaining algorithm based on this idea.

Algorithm 1

1. Sort all start and end points of the m blocks in ascending order w.r.t. their x_1 coordinate and store them in the array points.
2. For each block b, create the point $(end(b).x_2, .., end(b).x_k)$ and store it in a list $Temp$
3. Create $P_d = P_{buk} \times P_h$ sub-buckets over the set of points in $Temp$
4. Create a data structure D_i for each sub-bucket i; the point in each D_i is initialized as inactive
5. **for** $1 \leq i \leq 2m$
6.   **if** (points[i] is a start point)
7.   determine the block b with $beg(f).x_1 = $ points[i]
8.   **for** $1 \leq j \leq P_d$
9. $q_j := RMQ_{D_j}([0..beg(b).x_2-1], .., [0..beg(b).x_k-1])$
10.   determine the block b_j corresponding to q_j
11. $maxscore=\max\{b_1.score, b_2.score, .., b_j.score\}$
12. $b.score = b.weight + maxscore$
13.   connect the block b_j whose $score = maxscore$ to b
14.   **else**   * it is an end point *\
15.   determine the block b with $end(b).x_1 = $ points[i] and the D_j containing $end(b).x_1$
16.   activate $end(b)$ in D_j with score $b.score$

In Algorithm 1, the **for** loop in lines 8-10 can run in parallel over the different data structures $D_1, .., D_j$. Thus, the time complexity is $O(m \log^{k-2}(\frac{m}{P_d}) \log \log(\frac{m}{P_d}) + mP_d)$ time using the range tree and $O(m(\frac{m}{P_d})^{1-\frac{1}{k}} + mP_d)$ time using the kd-tree, assuming $P_d = P_h \times P_{buk}$ processors.

In this algorithm, the point set is still scanned in sequential order which reduces the amount of parallelization. To improve this, we introduce two additional strategies for parallelization of the outer **for** loop in Algorithm 1.

3.2   Anti-chain Decomposition

Definition 3. *A point p is said to be* dominating *q, if $q.x_i < p.x_i$, $\forall i \in [1..k]$. Two points p and q are called* anti-dominant *if neither p dominates q nor q dominates p.*

Lemma 1. *Two blocks b_1 and b_2 with $beg(b_1).x_1 < beg(b_2).x_1$ cannot belong to the same chain (hence, called antichains) if 1) the points $beg(b_1)$ and $beg(b_2)$ are anti-dominant or 2) the points $beg(b_1)$ and $end(b_2)$ are anti-dominant.*

Proof. It follows from the non-overlapping and collinearity relation of Definition 1.

Corollary 1. *Consecutive start points* points[j], .., points[$j+r$] *in the list* points *of Algorithm 1 belong to antichain blocks. Furthermore, consecutive end points* points[j'], .., points[$j'+r'$] *in the same list are antichain blocks.*

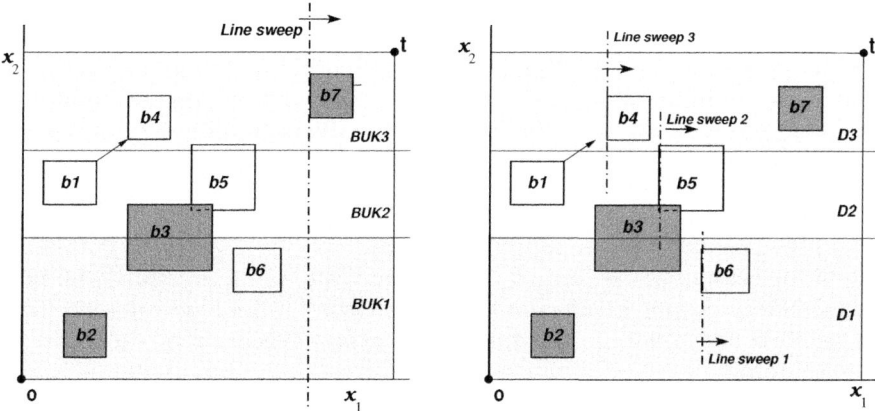

Fig. 2. Left: Space is divided into three buckets highlighted in different colors. A single sweep line scans all the points. Right: Multiple sweep lines work on different strips. (strips overlaid on the buckets). Lower strips lead the higher ones.

Based on this corollary, we can process multiple blocks in parallel either in searching for the best block or for activation. (For activation, the end points should be in different buckets to avoid race conditions.) For example, the blocks b_1 and b_2 in Figure 2 are antichains that can be processed in parallel. Algorithm 2 shows how the antichain properties can be implemented.

Algorithm 2

1. Sort all start and end points of the m blocks in ascending order w.r.t. their x_1 coordinate and store them in the array points.

2. For each block b, create the point $(end(b).x_2, .., end(b).x_k)$ and store it in a list Temp.

3. Create $P_d = P_{buk} \times P_h$ sub-buckets over the set of points in Temp,

4. Create a data structure D_i for each sub-bucket i; the point in each D_i is initialized as inactive

5. **while** $i \leq 2m$ *(i is initialized with zero)*

6. *Lookahead to determine consecutive start points $S_{list} = $ points$[i]$, points$[i+1]$, .., points$[r]$*

7. *$i = r + 1$*

8. **foreach** *point in S_{list}*

9. *determine the respective block b and attach it to the highest scoring block as in Algorithm 1*

10. *Lookahead to determine consecutive end points $E_{list} = $ points$[i]$, points$[i+1]$, .., points$[r']$*

11. *$i = r' + 1$*

12. **foreach** *point in E_{list}*

activate the respective block b in the respective data structure D_j as in Algorithm 1

The *Lookahead* function takes total $O(m)$ time and can be implemented in advance to determine the locations of consecutive start/end points; i.e., determine where a start point is followed by an end point or vice versa in the array points. Let m_s denote the number of these locations, and let m_p denote the maximum number of consecutive points that can be processed in parallel. Then the time complexity of the algorithm becomes $O(m_s \frac{m_p}{P_a}(T_D))$, where T_D is the time for RMQ search given in Algorithm 1, and P_a is the number of used processors.

The number of antichain blocks is very high in genome comparison due to repeats. (E.g., about 50% of the human genome is composed of repeats.) A region that is repeated two times in one genome and matches one region in the other genome makes 2 antichain blocks arranged on a line parallel to one axis. In Figure 1, the blocks b_1 and b_2 represent repeats in the genome corresponding to x_2.

3.3   Wavefront Decomposition

In our algorithms so far, a single sweep line visits all the start and end points of the blocks existing in different space divisions. Now we briefly discuss the use of multiple sweep lines in different space divisions, hence the name *wavefront decomposition*.

We create P_s sets (called *strips*), each containing an almost equal number of blocks. These strips are constructed with respect to the x_2 coordinate of the blocks' start points. Roughly, we can assume that the strip borders are overlaid on that of the buckets defined in Subsection 3.1. We say that a start point of a block b is *computable*, if all the blocks $b' \ll b$ are processed. We then assign each strip to a processor that runs the line-sweep procedure described in Algorithm 2. But each processor stops the line sweep at a start point if it is not yet computable and waits until the status changes. (The computability can be checked by checking status array declaring the position of the line sweep in each strip.) The effect of this waiting can be visualized in Figure 2 (Right), where the line sweep of strip i is leading that of strips $j > i$, assuming bottom up numbering of the strips. Algorithm 3 shows how to implement this idea.

Algorithm 3

1. Sort all start and end points of the m blocks in ascending order w.r.t. their x_1 coordinate and
*   store them in the array* points.
2. Divide the array points *into P_s sub-arrays* points$_1$, points$_2$, .., points$_{P_s}$ *of equal size.*
3. For each block b, create the point $(end(b).x_2, .., end(b).x_k)$, and store it in a list $Temp$
4. Create $P_d = P_{buk} \times P_h$ sub-buckets over the set of points in $Temp$
5. Create a data structure D_i for each sub-bucket i; the point in each D_i is initialized as inactive
6. **foreach** *sub-array* points$_i$
7.   **while** $i \leq$ |points$_i$| *(i is initialized with zero)*
10.   *Lookahead to determine consecutive start points $S_{list} =$* points$[i]$, points$[i+1]$, .., points$[r]$
11. $i = r + 1$
12.   **foreach** *point in S_{list}*
13.   *determine the respective block b*
14.   **wait until** *b is computable and do the following*
15.   *attach it to the highest scoring block as in Algorithm 1*
16.   *Lookahead to determine consecutive end points $E_{list} =$* points$[i]$, points$[i+1]$, .., points$[r']$
17. $i = r' + 1$
18.   **foreach** *point in E_{list}*
29.   *determine the respective block b*
20.   **wait until** *b is computable and do the following*
21.   *activate the respective block b in the respective data structure D_j as in Algorithm 1*

The **for** loop at line 6 can run in parallel to simulate the multiple sweep lines in the different strips, as discussed above. The complexity of this algorithm depends on the number of strips. Assuming P_s processors are assigned to handle the strips, the complexity becomes $O(\frac{m_s}{P_s} \frac{m_p}{P_a} (T_D))$, where T_D is the time for parallel RMQ search.

4   Implementation and Experimental Results

4.1   Implementation

The algorithms presented in this paper were implemented to run on computer cluster using MPI and to run on multicore computer using OpenMP. We used the kd-tree for answering RMQ's, because it is superior to the range tree in its linear space consumption and query time. Theoretically, the total number of processors in our algorithm is $P_d + P_a + P_s$, where P_d is the processors for parallel RMQ (space division), P_a is for anti-chain parallelization, and P_s is for wavefront decomposition. But practically, we have a limited number of processors. Therefore, some of the tasks have to be scheduled over the available processors.

Implementation on computer cluster: In the experiments presented here, we used the following simple strategy to assign jobs to the available cluster nodes. We set $P_d = P_{cluster}$, where $P_{cluster}$ is the number of cluster nodes. We created $P_s = P_{cluster}/2$ strips. The data to be transfered among the nodes includes, status of computability among blocks in different strips, regions of the RMQs, points of highest scores, and points to be activated. For higher dimensions, such amount of data increases the communication time and dramatically affects the overall running time. To overcome this, we used a strategy where we broadcast the blocks to all the nodes. Hence, it is enough that the nodes exchange only references to the points rather than the points themselves This also applies for the region of an RMQ, because it is defined by a start point, and this saves the transfer of status information. The cost of broadcasting is $O(T_{com}(m \log P_{cluster}))$, where T_{com} is the communication time and $P_{cluster}$ is the number of cluster nodes. For handling antichain parallelization, the processing of the consecutive start/end points is queued and dispatched to the cluster nodes. The experimental results given below is based on this implementation.

Implementation on multicore architecture: In the experiments presented here, we created n_b buckets for parallel RMQ such that $n_b = P_{core}$, where P_{core} is the number of available cores. In OpenMP, we created n_s strips and assigned them to different threads. The RMQ's are scheduled (pooled) over all available cores using nested threads. Note that for a block in strip $i \in [1..n_s]$, it is enough to launch only i RMQ's in the strips/buckets $[1..i]$, because the higher buckets are already out of the query range. This means that the threads handling the lower strips need to launch less number of RMQ's, and these finish before those associated with higher strips. This might leave some cores non-busy. But the antichain blocks are processed by means of other threads as well; hence all the cores are busy.

Table 1. Times and speed up of chaining on multicore machine and a computer cluster. The column titled "dataset" contains subsets of the genomes mentioned in the text, the number of genomes involved is given in the prefix of the dataset name. The column titled "blocks" contains the number of generated blocks. The last dataset " 5 Bacteria v2" is generated with less number of blocks. The column titled "seq." contains the sequential time in minutes. The numbers in the sub-tables "Cores" and "Nodes" contain the speed up for multicore and computer cluster architectures, respectively.

Dataset	blocks	seq.	Cores				Nodes				
			2	4	6	8	2	4	6	8	10
5 Bacteria	1058749	15m	1.6	2.7	2.8	4.1	1.4	2.1	2.3	2.9	3.5
6 Bacteria	2035732	18m	1.4	2.4	3	4.6	1.4	2.1	2.5	3	3.7
9 Bacteria	2381401	51m	1.6	3.2	5.3	6.5	1.4	3.0	4.4	5.7	8.5
5 Bacteria v2	715381	2m	1.3	1.4	1.5	1.7	1.1	1.3	1.4	1.5	1.5

4.2   Experimental Results

Our experiments ran on an 8 cores DELL PowerEdge machine, each with 2.5 GHz CPU and 512KB cache. The machine has 64GB RAM. Our computer cluster is composed of 10 compute nodes of the same specifications. The switch connecting the nodes is 1 Gigabit Ethernet.

We used biological sequences composed of 9 bacterial genomes, and the program Multimat [7] was used to compute regions of similarity which make up the blocks (block side-length at least 13): The genomes are of the bacteria *M. Tuberculosis* H37Rv (4.26MB), *M. KMS* (5.54MB), *C. Muridarum* (1.03MB), *M. Tuberculosis* CDC1551 (4.25MB), *M. Avium* (4.7 MB), *M. MCS* (5.51MB), *M. Smegmatis* MC2 (6.75MB), *M. Ulcerans* Agy99 (5.44MB), and *M. Tuberculosis* F11 (4.27MB).

Table 1 shows instances of the chaining problem with different number of genomes and different number of blocks. It also shows the sequential running times and the speed up (sequential time/parallel time) for these data sets and over increasing number of cores and cluster nodes.

In general, the computer cluster implementation is slower than that of multicore due to the communication time. One possible interpretation is that the more decomposition of the *kd*-tree, the faster the search time, which leads to finer granularity and domination of the communication time. We also note that the higher the number of genomes (i.e., the higher the RMQ dimension), the more speed up and better scalability. This can be explained by the fact that the search time increases with the RMQ dimension, which leads to coarser granularity.

5   Conclusions

In this paper, we presented parallel chaining algorithms using a three-levels strategy. The parallelization of these algorithms is challenging, because the algorithm is based on the line sweep paradigm and range maximum queries.

We performed limited experiments to verify our approach using MPI and OpenMP for computer cluster and multicore architectures, respectively. The experimental results showed practical improvement over the sequential algorithms. In future work, we will conduct more experiments using larger machines with more cores and nodes to address the assignment of processors to the different parallelization levels we introduced, i.e., we will address how many processors are assigned to handle parallel RMQ's, how many handle antichain parallelization, and how many handle strips, optimizing the usage of available cores and nodes. We will also address the scalability of the chaining algorithms in more details, and investigate the usage of MPI on a single multicore machine, and compare the results to that of OpenMP.

Variations of the chaining algorithm discussed in [1] can be easily parallelized in the same way with the same time complexity, because these variations just depend on different weighting and scoring of blocks. In an extended version of our paper, we will address these variations in more details.

Acknowledgments. We thank the anonymous reviewers for the valuable comments that helped us improve this paper. We also thank Mohamed Zahran for profile analysis. This work was supported by a grant from ITIDA, Nile University, and Microsoft CMIC.

References

1. Abouelhoda, M.I., Ohlebusch, E.: Chaining algorithms and applications in comparative genomics. J. Discrete Algorithms 3(2-4), 321–341 (2005)
2. Baker, B., Giancarlo, R.: Longest common subsequence from fragments via sparse dynamic programming. In: Bilardi, G., Pietracaprina, A., Italiano, G.F., Pucci, G. (eds.) ESA 1998. LNCS, vol. 1461, pp. 79–90. Springer, Heidelberg (1998)
3. Bently, J.L.: K-d trees for semidynamic point sets. In: Proc. of 6th Annual ACM Symposium on Computational Geometry, pp. 187–197 (1990)
4. Deogen, J.S., Yang, J., Ma, F.: EMAGEN: An efficient approach to multiple genome alignment. In: Proc. of Asia-Pacific Bioinf. Conf., pp. 113–122 (2004)
5. Eppstein, D., Galil, Z., Giancarlo, R., Italiano, G.F.: Sparse dynamic programming. J. Assoc. Comput. Mach. 39, 519–567 (1992)
6. Felsner, S., Müller, R., Wernisch, L.: Trapezoid graphs and generalizations, geometry and algorithms. Discrete Applied Mathematics 74(1), 13–32 (1997)
7. Höhl, M., Kurtz, S., Ohlebusch, E.: Efficient multiple genome alignment. Bioinformatics 18(Suppl. 1), 312–320 (2002)
8. Myers, E.W., Huang, X.: An $O(n^2 \log n)$ restriction map comparison and search algorithm. Bulletin of Mathematical Biology 54(4), 599–618 (1992)
9. Myers, E.W., Miller, W.: Chaining multiple-alignment fragments in sub-quadratic time. In: Proc. of SODA, pp. 38–47 (1995)
10. Shibuya, S., Kurochkin, I.: Match chaining algorithms for cDNA mapping. In: Benson, G., Page, R.D.M. (eds.) WABI 2003. LNCS (LNBI), vol. 2812, pp. 462–475. Springer, Heidelberg (2003)
11. Treangen, T., Messeguer, X.: M-GCAT: Interactively and efficiently constructing large-scale multiple genome comparison frameworks. BMC Bioinformatics 7, 433 (2006)
12. Zhang, Z., Raghavachari, B., Hardison, R.C., et al.: Chaining multiple-alignment blocks. J. Computional Biology 1, 51–64 (1994)

Precise Dynamic Analysis for Slack Elasticity: Adding Buffering without Adding Bugs*

Sarvani Vakkalanka, Anh Vo, Ganesh Gopalakrishnan, and Robert M. Kirby

School of Computing, Univ. of Utah, Salt Lake City, UT 84112, USA

Abstract. Increasing the amount of buffering for MPI sends is an effective way to improve the performance of MPI programs. However, for programs containing non-deterministic operations, this can result in *new* deadlocks or other safety assertion violations. Previous work did not provide any characterization of the space of *slack elastic* programs: those for which buffering can be safely added. In this paper, we offer a precise characterization of slack elasticity based on our formulation of MPI's *happens before* relation. We show how to efficiently locate *potential culprit sends* in such programs: MPI sends for which adding buffering can increase overall program non-determinism and cause new bugs. We present a procedure to minimally enumerate potential culprit sends and efficiently check for slack elasticity. Our results demonstrate that our new algorithm called POE$_{MSE}$ which is incorporated into our dynamic verifier ISP can efficiently run this new analysis on large MPI programs.

1 Introduction

A common myth is that if an MPI program does not deadlock under zero buffering for the sends, it will not deadlock under increased buffering. This myth has been expressed in many places [3,4,6]. Previous work [10,16] shows that while *deterministic* message passing programs enjoy this property, non-deterministic ones do not: they can exhibit new deadlocks or new non-deadlock safety violations when buffering is added. Those programs unaffected by increased buffering are called *slack elastic* [10]. Therefore, any developer wanting to improve MPI program performance by increasing runtime buffering needs to first make sure that the program is slack elastic.

Importance of detecting slack inelasticity: It is expected that MPI will continue to be an API of choice for at least another decade for programming large-scale scientific simulations. Large-scale MPI implementations must be formally verified to avoid insidious errors suddenly cropping up during field operation. Of all the formal verification methods applicable to MPI, dynamic methods have shown the most promise in terms of being able to instrument and analyze large programs. Examples range from scalable semi-formal approaches such as [5] that analyze large MPI programs for specific properties such as deadlocks, and formal *dynamic partial order* reduction methods such as in our ISP tool [18,19,20,21].

---

* Supported in part by Microsoft, NSF CNS-0935858, CCF-0903408.

R. Keller et al. (Eds.): EuroMPI 2010, LNCS 6305, pp. 152–159, 2010.

A relatively neglected aspect of previous MPI program formal analysis methods is how buffering affects correctness. A dynamic verifier for MPI must not only verify a given MPI program on a given platform – it must also ideally verify it *across all future* platforms. In other words, these tools must conduct *verification for safe portability*. Users of an MPI program must be allowed to switch to an MPI runtime that employs aggressive buffer allocation strategies or port the program to a larger machine and increase the `MP_EAGER_LIMIT` environment variable value without worrying about new bugs; for instance:

• Some MPI runtimes [7] perform credit-based buffer allocation where a program can perceive time-varying eager limits. This can create situations where the eager limit varies even for different instances of the same MPI send call.

• Future MPI programs may be generated by program transformation systems [2], thus creating 'unusual' MPI call patterns. Therefore, simple syntactic rules that guard against slack inelastic behavior are insufficient.

• Misunderstanding about buffering are still prevalent. Here are examples: *If you set* `MP_EAGER_LIMIT` *to zero, then this will test the validity of your MPI calls* [4]. A similar statement is made in [6]. The paper on the Intel Message Checker [3] shows this as the recommended approach to detect deadlocks.

Past work [10,16] did not offer a precise characterization of slack elastic programs nor a practical analysis method to detect its violation. Many important MPI programs (*e.g.*, MPI-BLAST [12], and ADLB [9]) exploit non-deterministic receives and probes to opportunistically detect the completion of a previous task to launch new work. Therefore, avoiding non-determinism is not a practical option, as it can lead to code complexity and loss of performance. We offer the first precise characterization and an efficient dynamic analysis algorithm for slack inelasticity.

• We describe a new dynamic analysis algorithm called POE_{MSE} that can efficiently enumerate what we term *potential culprit sends*. These are MPI sends occurring in an MPI program to which adding buffering can increase the overall program non-determinism. This increased non-determinism potentially results in unexplored behaviors and bugs.

• We search for potential culprit sends over the graphs defined by MPI's *happens before*. We also contribute this happens-before relation (a significant extension of the *intra happens-before* [21]) as a new result. As we shall show, happens-before incorporates MPI's message non-overtaking and wait semantics.

• The new version of ISP containing POE_{MSE} is shown to perform efficiently on non-trivially sized MPI programs. Given the limited amount of space, we opt to present our results intuitively, with some background § 2, present the POE_{MSE} algorithm § 3, and results § 4. Formal details are presented in [17].

2   Motivating Examples

We begin with the familiar 'head to head' sends example Figure 1(a). To save space and leave out incidental details of MPI programming, we adopt the following abbreviations (explained through examples). We prefer to illustrate our ideas using non-blocking send/receive operations just to be able to discuss the effects of buffering on MPI_Waits. We use $S_{0,0}(1)$ to denote a non-blocking send (MPI_Isend) targeting process 1 ($P1$). The subscript $0,0$ says that this is send number 0 of $P0$ (counting from zero). Without loss of generality, we assume all these communication requests carry the same tag, happen on the same communicator, and communicate some arbitrary data. Also, any arbitrary C statement including conditionals and loops may be placed between these MPI calls. We use $W_{0,1}(h_{0,0})$ to denote the MPI_Wait corresponding to $S_{0,0}(1)$. Similarly, $R_{0,2}(0)$ stands for an MPI_Irecv sourcing $P0$, and $W_{0,3}(h_{0,2})$ is its corresponding MPI_Wait. It is clear that this example will deadlock under zero buffering but not under infinite buffering. This is a well-known MPI example underlying almost all "zero buffer" deadlock detection tests mentioned earlier.

Deadlock Detection by Zeroing Buffering is Inconclusive: The MPI program in Figure 1(b) will deadlock when either $S_{0,0}$ or $S_{1,0}$ or both are buffered. There is no deadlock when both $S_{0,0}$ and $S_{1,0}$ have buffering.

P_0	P_1
$S_{0,0}(1)$	$S_{1,0}(0)$
$W_{0,1}(h_{0,0})$	$W_{1,1}(h_{1,0})$
$R_{0,2}(0)$	$R_{1,2}(0)$
$W_{0,3}(h_{0,2})$	$W_{1,3}(h_{1,2})$

(a) Zeroed buffering finds deadlocks)

P_0	P_1	P_2
$S_{0,0}(1)$	$S_{1,0}(2)$	$R_{2,0}(*)$
$W_{0,1}(h_{0,0})$	$W_{1,1}(h_{1,0})$	$W_{2,1}(h_{2,0})$
$S_{0,2}(2)$	$R_{1,2}(0)$	$R_{2,2}(0)$
$W_{0,3}(h_{0,2})$	$W_{1,3}(h_{1,2})$	$W_{2,3}(h_{2,2})$

(b) Zeroed buffering misses deadlocks

Fig. 1. Where zeroing buffering helps, and where it does not

If both $S_{0,0}(1)$ and $S_{1,0}(2)$ have zero buffering, their corresponding MPI_Wait operations $W_{0,1}(h_{0,0})$ and $W_{1,1}(h_{1,0})$ remain blocked until (at least) the corresponding receives are issued. Of the two sends, only $S_{1,0}(2)$ can proceed by matching $R_{2,0}(*)$ (standing for a wildcard receive, or a receive with argument MPI_ANY_SOURCE). This causes the waits $W_{1,1}(h_{1,0})$ to unblock, allowing $R_{1,2}(0)$ to be posted. This allows $S_{0,0}(1)$ to complete and hence $W_{0,1}(h_{0,0})$ will return, allowing $S_{0,2}(2)$ to issue. Since $W_{2,1}(h_{2,0})$ has unblocked after $R_{2,0}(*)$ matches, $R_{2,2}(0)$ can be issued, and then the whole program finishes without deadlocks. Now, if $S_{0,0}(1)$ were to be buffered, the following execution can happen: First, $S_{0,0}(1)$ can be issued. The corresponding wait, *i.e.* $W_{0,1}(h_{0,0})$ can return regardless of whether P_2 has even posted its $R_{2,0}(*)$. This can now result in $S_{0,2}(2)$ to be issued, and this send competes with $S_{1,0}(2)$ to match with $R_{2,0}(*)$. Suppose $S_{0,2}(2)$ is the winner of the competition, *i.e.* it matches $R_{2,0}(*)$. This now leads to a deadlock with respect to $R_{2,2}(0)$ because the only process able to satisfy this request is P_0, but unfortunately this process has no more sends.

Neither Zero Nor Infinite Buffering Helps: In the example of Figure 1(b), even if all sends are given infinite buffering, we will run into the deadlock with respect to $R_{2,2}(0)$ described earlier. This may give us the impression that either zero buffering or infinite buffering will catch all deadlocks. This is not so! The example in Figure 2(a) will not deadlock when none of the sends are buffered or

P_0	P_1	P_2
$S_{0,0}(1)$	$S_{1,0}(2)$	$R_{2,0}(*)$
$W_{0,1}(h_{0,0})$	$W_{1,1}(h_{1,0})$	$W_{2,1}(h_{2,0})$
$S_{0,2}(2)$	$R_{1,2}(0)$	$R_{2,2}(0)$
$W_{0,3}(h_{0,2})$	$W_{1,3}(h_{1,2})$	$W_{2,3}(h_{2,2})$
$S_{0,4}(2)$	$R_{1,4}(2)$	$S_{2,4}(1)$
$W_{0,5}(h_{0,4})$	$W_{1,5}(h_{1,4})$	$W_{2,5}(h_{2,4})$
$R_{0,6}(1)$	$S_{1,6}(0)$	$R_{2,6}(*)$
$W_{0,7}(h_{0,6})$	$W_{1,7}(h_{1,6})$	$W_{2,7}(h_{2,6})$

(a) Specific buffering causes deadlock

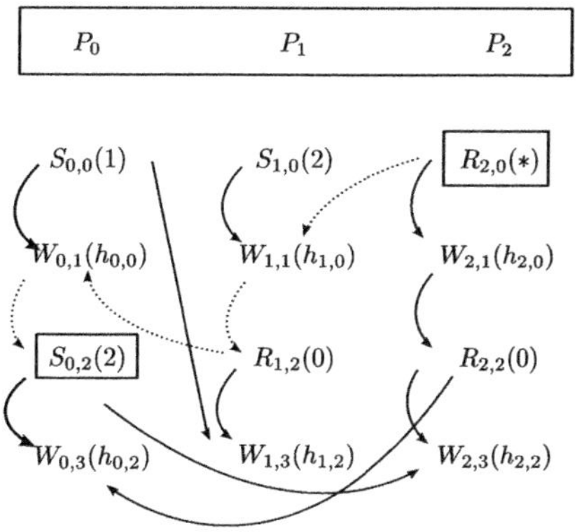

(b) Slack Inelasticity modeled as Happens-before path breaking

Fig. 2. More Buffering-related Deadlocks

all the sends are buffered. However, it will deadlock only when (i) $S_{0,0}$ is buffered and (ii) both $S_{1,0}$ and $S_{2,4}$ are not buffered, as detailed in § 3.

There is one common simple explanation for all the behaviors seen so far. To present that, we now introduce, through examples, the *happens before* (HB) relation underlying MPI that we have discovered. We adapt Lamport's happens-before [8] – widely used in programming to study concurrency and partial order semantics – for MPI (for full explanations, please see [17,21]). We will now use Figure 2(b), which essentially redraws Figure 1(b) with the HB edges added, to illustrate the precedences MPI happens-before. There is a HB edge between: (i) Every non-blocking send/receive and its wait. This can be seen in Figure 2(b) as the solid arrows from, say $S_{0,0}(1)$ to $W_{0,1}(h_{0,0})$. (ii) MPI waits and their successive instructions. We show this as, for example, $W_{2,1}$ to $R_{2,2}$. Notice that $W_{1,1}$ to $R_{1,2}$ and $W_{0,1}$ to $S_{0,2}$ are also HB ordered; these are shown dotted because they happen to fall on a HB path (*HB-path*) which we highlight via dotted edges. (iii) Collective operations (including barriers) and their successive MPI instructions, two non-blocking sends targeting the same destination (MPI non-overtaking), and two receives sourcing from the same source (exhaustive list is available [17]).

So far we introduced *intra* HB; the rule for inter HB is roughly as follows: if two instructions can match, their successors are in the *inter* HB relation with respect to each other. For example, since $S_{0,0}$ and $R_{1,2}$ can match in the execution of this program under zero-buffering, we have an inter HB edge from $S_{0,0}$ to $W_{1,3}$ and another from $R_{1,2}$ to $W_{0,1}$. (These happen to be shown dotted because they lie on the HB-path.)

Now, here is what buffering of $S_{0,0}$ does

• It allows $W_{0,1}$ to return early (since the message is buffered, this wait can return early – or it becomes a no-op). Now consider the dotted path from $R_{2,0}(*)$ to $S_{0,2}(2)$. Before buffering was added to $S_{0,0}$, this path had all its edges. When buffering was added to $S_{0,0}$ thus turning $W_{0,1}$ in effect to a no-op, we broke this HB-path.

- We show that when a send (such as $S_{0,2}(2)$) and a receive (such as $R_{2,0}(*)$) do not have an HB-path separating them, they become concurrent [17]. We saw this earlier because as soon as we buffered $S_{0,0}$, $S_{0,2}(2)$ became a *competing* match to $R_{2,0}(*)$. The **potential culprit send** is $S_{0,0}$ because it was by buffering this that we made $S_{0,2}(2)$ and $R_{2,0}(*)$ match, thus increasing the overall non-determinism in the program.

- Therefore the POE$_{\text{MSE}}$ algorithm is (detailed in § 3 but briefly now): (i) execute under zero buffering; (ii) build the HB-paths separating potential competing sends such as $S_{0,2}$ with wildcard receives such as $R_{2,0}(*)$; (iii) locate potential culprit sends such as $S_{0,0}$ (if any) that can break HB-paths thus making these new sends match a wildcard receive, increasing non-determinism; (iv) if the increased non-determinism leads to bugs, the potential culprit sends are actual culprit sends.

In this example buffering $S_{0,0}$ indeed lead to a deadlock, and so we located an actual culprit send. Hence, this program is *not slack elastic!*

3   The POE$_{\text{MSE}}$ Algorithm

The POE$_{\text{MSE}}$ algorithm is an extension of the POE algorithm [18,19,20,21]. We present POE$_{\text{MSE}}$ at a high level using the examples in Figure 1(a) first, Figure 1(b) next, and then that in Figure 2(a) last. All formal details are in [17].

We run the program as per our POE algorithm under zero buffering for all sends. Running the example in Figure 1(a) instantly reveals the deadlock, allowing the user to fix it. For Figure 1(b), POE will simply run through the code without finding errors in the first pass. Being a stateless dynamic verifier, ISP only keeps a stack history of the current execution. The happens-before graph is built as in Figure 2(b). Next, ISP unwinds the stack history. For each wildcard receive encountered, it will find out which sends *could have matched* should other sends (potential culprit sends) have buffering. In our case, it will find that $S_{0,2}(2)$ could have matched $R_{2,0}(*)$ if we were to break the HB-path by buffering $S_{0,0}$ as said before.

Coming to Figure 2(a), when POE$_{\text{MSE}}$ initially executes forward, it runs the whole program under zero buffering. It would find that $R_{2,0}(*)$ matched $S_{1,0}(2)$, and $R_{2,6}(*)$ matched $S_{2,4}(2)$. Then stack unwinding proceeds as follows:

- When $R_{2,6}(*)$ is popped from the stack, POE$_{\text{MSE}}$ will try to force another sender to match the wildcard receive $R_{2,6}(*)$. It does not find any culprit sends that can be buffered to make it happen. This is clear because if another send were to match $R_{2,6}(*)$, it has to come from $P1$ (MPI's non-overtaking). However, there is no such sender.

- Thus, POE$_{\text{MSE}}$ will unwind more, finally popping $R_{2,0}(*)$. At this point, POE$_{\text{MSE}}$ will find that the culprit send of $S_{0,0}(1)$ indeed works, because buffering this send immediately turns $W_{0,1}$ into a no-op, breaking the HB-path $S_{0,0}(1) \rightarrow W_{0,1} \rightarrow S_{0,2}(2) \rightarrow R_{2,0}(*)$ at $W_{0,1} \rightarrow S_{0,2}(2)$.

Table 1. ParMETIS$_b$ is ParMETIS$_*$ w. slack; ParMETIS$_*$ is ParMETIS modified to use wildcards

Number of interleavings (notice the extra necessary interleavings of POE_{MSE})	POE_{MSE}	POE
sendbuff.c	5	1
sendbuff-1a.c	2 (deadlock caught)	1
sendbuff2.c	1	1
sendbuff3.c	6	1
sendbuff4.c	3	1
Figure 2(a)	4 (dl caught)	2 (dl missed)
ParMETIS$_b$	2	1
Overhead of POE_{MSE} on ParMETIS / ParMETIS$_*$ (runtime in seconds (x) denotes x interleavings)	POE_{MSE}	POE
ParMETIS (4procs)	20.9 (1)	20.5 (1)
ParMETIS (8procs)	93.4 (1)	92.6 (1)
ParMETIS$_*$	18.2 (2)	18.7(2)

- Now POE$_{MSE}$ will replay forward from this stack frame, initially giving no buffering at all to subsequent sends. This will cause a head-to-head deadlock between $S_{1,0}(2)$ trying to send to P_2 and $S_{2,4}(1)$ trying to send to P_1.
- If we were to buffer $S_{1,0}(2)$, this head-to-head deadlock will disappear. This is why we may need to buffer some sends (culprit sends) and not buffer other sends (head-to-head deadlock inducing sends).

In [17], we explain all these details, provides actual pseudo-codes and mathematical proofs. Here are additional facts:

- In general, we may find *multiple potential culprit sends*. More than one might need to be buffered to break an HB-path. However, it is also important to break HB-paths in a minimal fashion –*i.e.*, giving a "flood of buffering" is not a good idea because we can mask later head-to-head deadlocks. Thus POE$_{MSE}$ always allocates buffering for potential culprit sends only.
- It is only for deadlock checking that we need these precautions. All non-deadlock safety violations can be checked with infinite buffering, which is simulated as follows: we `malloc` as much buffer space as the message the MPI send is trying to send, and copy away the data into it, and nullify `MPI_Wait`. Thus, ISP when running POE$_{MSE}$ really converts MPI waits into no-ops.

4   Results and Conclusions

We first study variants of the examples in Figure 1, Figures 1(b) and 2(a) (called `sendbuff`). These examples explore POE$_{MSE}$'s capabilities to detect the different matchings as well as deadlocks. For each of the sendbuff variants, POE$_{MSE}$ correctly discovers the minimal number of send operations to be buffered in order to enable other sends to match with wildcard receives. We also reproduced our example in Figure 2(b) as sendbuff-1a.c, where our algorithm indeed caught the deadlock at the second interleaving, where $S_{0,2}(2)$ is matched with $R_{2,0}(*)$.

Next we study large realistic examples that show that POE_{MSE} adds virtually no overheads. We used ParMETIS [14,15], a hypergraph partition library (14K LOC of MPI/C), as a benchmark for measuring the overhead of POE_{MSE} (shown in Table 1 as ParMETIS (xprocs) where x is the number of processes that we ran the benchmarks with. $ParMETIS_*$ is a modified version where we rewrote a small part of the algorithm using wildcard receives. In most of our benchmarks where no additional interleavings are needed, the overhead is less than 3%, even in the presence of wildcard receives, where the new algorithm has to run extra steps to make sure we have covered all possible matchings in the presence of buffering.

Finally, we study large examples with slack inelastic patterns inserted into them. This is reflected in Table 1 as $ParMETIS\_b$ where we rewrote the algorithm of ParMETIS again, this time not only to introduce wildcard receives, but also to allow the possibility of a different order of matching that can only be discovered by allowing some certain sends to be buffered. Our experiment shows that POE_{MSE} successfully discovered the alternative matching during the second interleaving. POE_{MSE} has handled all the large examples previously handled by POE with only negligible overhead in practice.

The POE_{MSE} algorithm in our actual ISP tool handles over 60 widely used MPI functions, and hence is practical for many large MPI programs.

4.1   Concluding Remarks

In addition to MPI, MCAPI [11], CUDA [1], and OpenCL [13] include the same kind of non-blocking calls and waits discussed here. Programmers worry about the cost (the amount of memory) tied up by buffering. Memory costs and memory power consumption already outweigh those figures for CPUs. Thus one likes to allocate buffer space wisely.

Non-deterministic reception is an essential construct for optimization. Unfortunately non-deterministic reception and buffering immediately leads to slack variant behaviors. The code patterns that cause slack inelastic behaviors are not that complex – meaning, they can be easily introduced.

We propose the POE_{MSE} algorithm that detects such behaviors based on an MPI-specific happens-before relation. It works by first locating all minimal sets of non-blocking message send operations that must be buffered, so as to enable other message send operations to match wildcard receives; and subsequently running the dynamic analysis over all such minimal send sets. The overhead of these steps is negligible in practice. A promising avenue of future research is to detect slack inelastic behaviors during the design of new APIs.

Acknowledgments. We thank Grzegorz Szubzda and Subodh Sharma for their valuable contributions, and Bronis R. de Supinski for his encouraging remarks.

References

1. Compute Unified Device Architecture (CUDA), http://www.nvidia.com/object/cuda_get.html
2. Danalis, A., Pollock, L., Swany, M., Cavazos, J.: Mpi-aware compiler optimizations for improving communication-computation overlap. In: ICS 2009 (2009)
3. DeSouza, J., Kuhn, B., de Supinski, B.R., Samofalov, V., Zheltov, S., Bratanov, S.: Automated, scalable debugging of mpi programs with Intel Message Checker. In: SE-HPCS 2005 (2005)
4. http://www.hpcx.ac.uk/support/FAQ/eager.html
5. Hilbrich, T., de Supinski, B.R., Schulz, M., Müller, M.S.: A graph based approach for MPI deadlock detection. In: ICS 2009, pp. 296–305 (2009)
6. http://www.cs.unb.ca/acrl/training/general/ibm_parallel_programming/pgm3.PDF
7. PE MPI buffer management for eager protocol, http://publib.boulder.ibm.com/infocenter/clresctr/vxrx/index.jsp?topic=/com.ibm.cluster.pe431.mpiprog.doc/am106_buff.html
8. Lamport, L.: Time, clocks, and the ordering of events in a distributed system. Commun. ACM 21(7), 558–565 (1978)
9. Lusk, R., Pieper, S., Butler, R., Chan, A.: Asynchronous dynamic load balancing, http://unedf.org/content/talks/Lusk-ADLB.pdf
10. Manohar, R., Martin, A.J.: Slack elasticity in concurrent computing. In: Jeuring, J. (ed.) MPC 1998. LNCS, vol. 1422, pp. 272–285. Springer, Heidelberg (1998)
11. http://www.multicore-association.org
12. http://www.mpiblast.org
13. OpenCL: Open Computing Language, http://www.khronos.org/opencl
14. ParMETIS - Parallel graph partitioning and fill-reducing matrix ordering. http://glaros.dtc.umn.edu/gkhome/metis/parmetis/overview
15. Schloegel, K., Karypis, G., Kumar, V.: Parallel static and dynamic multi-constraint graph partitioning. Concurrency and Computation: Practice and Experience 14, 219–240 (2002)
16. Siegel, S.F.: Efficient verification of halting properties for MPI programs with wildcard receives. In: Cousot, R. (ed.) VMCAI 2005. LNCS, vol. 3385, pp. 413–429. Springer, Heidelberg (2005)
17. Vakkalanka, S.: Efficient Dynamic Verification Algorithms for MPI Applications. PhD Dissertation (2010), http://www.cs.utah.edu/Theses
18. Vakkalanka, S., DeLisi, M., Gopalakrishnan, G., Kirby, R.M.: Scheduling considerations for building dynamic verification tools for MPI. In: Parallel and Distributed Systems - Testing and Debugging (PADTAD-VI), Seattle, WA (July 2008)
19. Vakkalanka, S., DeLisi, M., Gopalakrishnan, G., Kirby, R.M., Thakur, R., Gropp, W.: Implementing efficient dynamic formal verification methods for MPI programs. In: Lastovetsky, A., Kechadi, T., Dongarra, J. (eds.) EuroPVM/MPI 2008. LNCS, vol. 5205, pp. 248–256. Springer, Heidelberg (2008)
20. Vakkalanka, S., Gopalakrishnan, G., Kirby, R.M.: Dynamic Verification of MPI Programs with Reductions in Presence of Split Operations and Relaxed Orderings. In: Gupta, A., Malik, S. (eds.) CAV 2008. LNCS, vol. 5123, pp. 66–79. Springer, Heidelberg (2008)
21. Vakkalanka, S., Vo, A., Gopalakrishnan, G., Kirby, R.M.: Reduced execution semantics of mpi: From theory to practice. In: Cavalcanti, A., Dams, D.R. (eds.) FM 2009. LNCS, vol. 5850, pp. 724–740. Springer, Heidelberg (2009)

Implementing MPI on Windows: Comparison with Common Approaches on Unix[*]

Jayesh Krishna[1], Pavan Balaji[1], Ewing Lusk[1],
Rajeev Thakur[1], and Fabian Tillier[2]

[1] Argonne National Laboratory, Argonne, IL 60439
[2] Microsoft Corporation, Redmond, WA 98052

Abstract. Commercial HPC applications are often run on clusters that use the Microsoft Windows operating system and need an MPI implementation that runs efficiently in the Windows environment. The MPI developer community, however, is more familiar with the issues involved in implementing MPI in a Unix environment. In this paper, we discuss some of the differences in implementing MPI on Windows and Unix, particularly with respect to issues such as asynchronous progress, process management, shared-memory access, and threads. We describe how we implement MPICH2 on Windows and exploit these Windows-specific features while still maintaining large parts of the code common with the Unix version. We also present performance results comparing the performance of MPICH2 on Unix and Windows on the same hardware. For zero-byte MPI messages, we measured excellent shared-memory latencies of 240 and 275 nanoseconds on Unix and Windows, respectively.

1   Introduction

Historically, Unix (in its various flavors) has been the commonly used operating system (OS) for high-performance computing (HPC) systems of all sizes, from clusters to the largest supercomputers. In the past few years, however, Microsoft Windows has steadily increased its presence as an operating system for running HPC clusters, particularly in the commercial arena. Commercial applications in areas such as computational fluid dynamics, structural analysis, materials, industrial design and simulation, seismic modeling, and finance run on Windows clusters. Windows has also made inroads at the very high end of the spectrum. For example, the Dawning 5000A cluster at the Shanghai Supercomputer Center with 30,720 cores and running Windows HPC Server 2008 achieved more than 200 TF/s on LINPACK and ranked 10th in the November 2008 edition of the Top500 list [16].

Since the vast majority of HPC applications use MPI as the programming model, the use of Windows for HPC clusters requires an efficient MPI implementation. Given the historical prevalence of Unix in the HPC world, however,

---

[*] This work was supported in part by a grant from Microsoft Corp. and in part by the Office of Advanced Scientific Computing Research, Office of Science, U.S. Department of Energy, under Contract DE-AC02-06CH11357.

R. Keller et al. (Eds.): EuroMPI 2010, LNCS 6305, pp. 160–169, 2010.

the MPI developer community tends to have more expertise in implementing and tuning MPI on Unix platforms. In this paper, we discuss some of the issues involved in implementing MPI on Windows and how they differ from commonly used approaches for Unix. We particularly focus on issues such as asynchronous progress, process management, shared-memory access, and threads.

The MPICH implementations of MPI (both the older MPICH-1 and the current MPICH2 implementations) have supported both Unix and Windows for many years. Here we describe how we implement MPICH2 to support both Unix and Windows efficiently, taking advantage of the special features of Windows while still maintaining a largely common code base. We also present performance results on the Abe cluster at the National Center for Supercomputing Applications (NCSA, University of Illinois), where we ran MPICH2 with Unix and Windows on the same hardware.

The rest of this paper is organized as follows. In Section 2, we provide an overview of MPICH2 and its internal architecture. In Section 3, we discuss some of the differences in implementing MPI on Windows and Unix. Performance results are presented in Section 4. We discuss related work in Section 5 and conclude in Section 6 with a brief look at future work.

2 Background

MPICH2 [8] is a high-performance and widely portable implementation of MPI. It supports the latest official version of the MPI standard, MPI 2.2 [7]. MPICH2 has a modular architecture that is designed to make it easy to plug in new network devices, process managers, and other tools (see Figure 1). This design enables anyone to port MPICH2 easily and efficiently to new platforms.

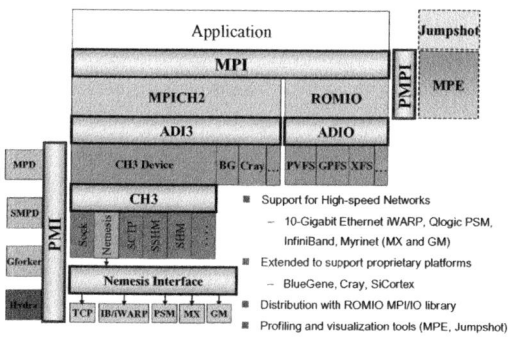

Fig. 1. MPICH2 architecture

A key feature of MPICH2 is a scalable, multinetwork communication subsystem called Nemesis [3]. Nemesis offers very low-latency and high-bandwidth communication by using efficient shared memory operations, lock-free algorithms, and optimized memory-copy routines. As a result, MPICH2 achieves a very low shared-memory latency of around 240–275 ns. We have developed an efficient implementation of Nemesis for Windows.

An MPI library typically requires thread services (e.g., thread creation, mutex locks), shared-memory services (for intranode communication), internode communication services (e.g., TCP/IP sockets), and OS process-management services. The APIs for these features can differ among operating systems and platforms. For portability, MPICH2 uses an internal abstraction layer for these services, which can be implemented selectively on different OS platforms. MPICH2 also includes a portable library for atomic operations, called OPA (Open Portable Atomics) [14], which provides OS-independent atomic primitives, such as fetch-and-increment and compare-and-swap. In addition, we have developed a runtime system that can launch MPI jobs on clusters with any flavor of Unix or Windows.

As a result, MPICH2 can run efficiently on both Windows and Unix operating systems while maintaining a largely common code base.

3   Implementing MPI on Windows versus Unix

From the MPI perspective, Windows and Unix are just different OS flavors, providing similar operating system services. However, a high-performance implementation of MPI on the two OS flavors can differ significantly. These differences can make building a widely portable, high-performance library a huge challenge. In this section, we discuss some of the functionality differences between the two operating systems and the corresponding challenges and benefits.

3.1   Asynchronous Progress

Windows supports an asynchronous model of communication, in which the user initiates an operation and the operating system ensures progress on the operation and notifies the user when the operation is completed. In Nemesis on windows we provide asynchronous internode communication by using an I/O completion object, exposed by the OS as a *completion port*, with TCP/IP sockets. To initiate communication, Nemesis posts a request to the kernel for the operation and waits for a completion event. When the request is completed, the kernel queues a completion packet on the completion port associated with the I/O completion object.

MPI implementations on Unix systems typically use *nonblocking* progress to implement internode communication using TCP/IP sockets. In this case, the library polls for the status of a socket and processes the requested/pending operation. Nonblocking progress differs from asynchronous progress, in which the OS performs the requested operation on the user's behalf. Nonblocking progress is typically implemented on Unix systems by using the POSIX `poll` system call.

Nonblocking progress is generally inefficient compared with asynchronous progress because of deficiencies in `poll`. It also requires the MPI implementation to do more work than with asynchronous progress where some work is offloaded to the operating system. The `poll` system call requires the set of socket descriptors to be polled to be contiguous in memory. This restriction increases bookkeeping and reduces scalability of libraries that allow for dynamic connections or that optimize memory allocated for the socket descriptors by dynamically expanding it. When `poll` returns, indicating the occurrence of an event, the user must search through the entire set of descriptors to find the one with the event. Some operating systems provide event-notification mechanisms similar to completion ports on Windows, for example, `epoll` in Linux and `kqueue` in BSD. However, these mechanisms are not widely portable across the various flavors of Unix. In Nemesis, we use an asynchronous internode communication module for Windows and a nonblocking internode communication module for Unix systems.

Another MPI feature that can take advantage of asynchronous services is generalized requests, which allow users to define their own nonblocking operations that are represented by MPI request objects. MPI specifies that the user is responsible for causing progress on generalized requests. On Unix, the user may be required to use an external thread. Windows allows users to register callback functions with asynchronous OS calls; this mechanism allows a user library to use generalized requests without needing an external thread to cause progress on the operations.

3.2   Process Management

Process management in MPI typically involves providing a mechanism to launch MPI processes and setting the appropriate runtime environment for the processes to be able to connect to each other.

Launching MPI Jobs. On Unix systems that do not have an external job launcher, MPI process managers typically use `fork` to launch processes locally and network protocols such as SSH to launch processes remotely. These network-protocol agents assume the existence of a standalone daemon process on each node that can interact with the remote protocol agent. Since Windows does not natively provide a network-protocol mechanism similar to SSH, we need a distributed process-management framework with standalone manager daemons on each Windows node.

When launching MPI jobs, the remote MPI processes must be launched with the user's credentials. When using a network protocol such as SSH, the protocol provides this service. On Windows, however, the process-manager daemon must do this job. In MPICH2, we have implemented a process manager, called SMPD, that provides process-management functionalities to MPI jobs on both Windows and Unix systems. On Windows, the standalone SMPD daemon impersonates the user launching the job by using the user's credentials. Where available, SMPD can also use technologies such as Active Directory and the job scheduler in Windows HPC Server 2008 to manage user credentials and launch MPI jobs.

Managing MPI Processes. Once the MPI processes are launched, the process manager is responsible for managing them. It must provide information to the processes so that they can connect to each other; handle `stdin`, `stdout`, and `stderr`; and handle termination and shutdown. SMPD provides these features using a communication protocol that is independent of the data model used by the individual nodes of a cluster. This allows users to run MPI jobs on a heterogeneous cluster containing both Unix and Windows nodes.

3.3   Intranode Communication

MPI implementations typically use some form of shared-memory communication for communicating between MPI processes running on a single node. Nemesis uses lock-free shared-memory queues for improving scalability and reducing overhead for intranode communication [3]. The use of these queues reduces the intranode communication latency for small messages and is particularly effective when the communicating processes share CPU data caches. When they do not, however, the performance often degrades.

An MPI implementation can also use OS services that allow users to transfer data directly between the memories of two processes. This approach can improve performance for large-message transfers among processes that do not share a cache. A variety of standard and nonstandard methods for doing so are available on Unix [2]. Windows provides an OS service for directly accessing the address space of a specified process, provided the process has appropriate security privileges. For small messages, however, we observed that this service has more overhead than the lock-free shared-memory queues in Nemesis. Therefore, we use the remote-copy method only for large messages in Nemesis on Windows.

3.4   Threads

The MPI standard clearly defines the interaction between user threads and MPI in an MPI program [5]. The user can request a particular level of thread support from the MPI implementation, and the implementation can indicate the level of thread support provided. On both Windows and Unix, MPICH2 supports the MPI_THREAD_MULTIPLE level, which allows any user thread to make MPI calls at any time. This feature requires some thread-locking mechanisms in the implementation in order to make it thread safe.

Unix platforms typically use a POSIX threads (Pthreads) library, whereas Windows has its own version of threads. MPICH2 uses an OS-independent thread-abstraction layer that enables it to use different threads libraries and thread-locking mechanisms portably. The default version of MPICH2 uses a global lock to control access to an MPI function from multiple threads. A thread calling an MPI function tries to obtain this global lock and then releases the lock after completing the call or before blocking on an OS request. When a lock is released, a thread waiting for the global lock gets access to the lock and performs progress on its MPI communication. We are also developing a more efficient version of MPICH2 that supports finer-grained locks [1].

4   Experimental Evaluation and Analysis

In this section, we evaluate the different strategies discussed in the paper and compare the results. We ran all tests on the Abe cluster at NCSA, which has both Unix and Windows nodes. Each node consists of 2 quad-core Intel 64 (Clovertown) 2.33 GHz processors with a 2x4 MB L2 cache, 4x32 K L1 cache, and 8 GB

RAM. The Unix nodes ran Linux 2.6.18 and used Intel C/C++ 10.1 compilers. The Windows nodes were installed with Windows Server 2008 HPC Edition SP2 and the Visual Studio 2008 compilers. On both Unix and Windows, we compiled MPICH2 with aggressive optimization and disabled error checking of user arguments. The interconnection network used was gigabit Ethernet.

4.1 Asynchronous Progress

We compared asynchronous and nonblocking progress by calculating the amount of overlap of communication and computation with the two strategies. To measure the performance of nonblocking progress on Windows, we implemented a version of Nemesis that uses the `select` system call since it is more portable across various versions of Windows than `poll`. The default version of Nemesis on Windows uses asynchronous progress with I/O completion ports.

We measured communication latency and bandwidth by using an MPI version of the popular Net-PIPE benchmark [11], called NetMPI, which performs a nonblocking receive, a blocking send, and an `MPI_Wait` multiple times in a loop. We modified the benchmark to perform several nonblocking sends and receives at a time and used `MPI_Testall` to test for their completion without blocking. Between calls to `MPI_Testall`, we performed some computation for ≈250 ns. The less time spent by the MPI implementation in `MPI_Testall`, the more time available to the user to perform computation.

Figure 2 shows the breakdown of the time spent within the MPI library

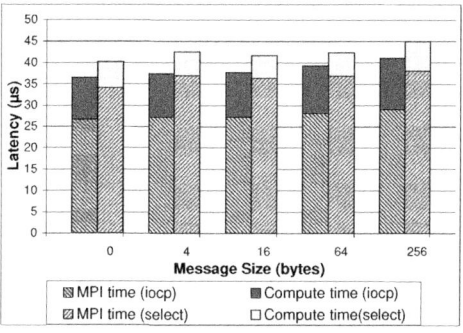

Fig. 2. Time spent in the MPI library (MPI time) and time available for user computation (Compute time) when using asynchronous progress (iocp) versus non-blocking progress (select) on Windows for internode communication

and the time available for the user's computation when sending a message by using asynchronous progress versus nonblocking progress. As expected, asynchronous progress results in less time being spent within the MPI library and more time available for user computation. The reason is that with asynchronous progress the MPI library delegates the reading/writing of data from/to TCP/IP sockets to the OS, whereas with nonblocking progress the library polls for events and then performs the read/write. When I/O is delegated to the OS, the library has little work to do and quickly returns to the application.

4.2 Intranode Communication

In this experiment, we compared the intranode communication performance using lock-free shared-memory queues and direct remote-memory access for large messages. We also compare the intranode performance of MPICH2 on Windows and Unix for small and large messages. We used the Ohio State University (OSU) microbenchmarks [15] to measure latency and bandwidth in all cases.

We have implemented a read remote virtual memory (RRVM) module in Nemesis that performs remote memory access for large messages (≥16 KB) on Windows. Figure 3 shows the intranode communication bandwidth when using lock-free shared-memory queues versus RRVM on Windows. We considered two cases, one where the communicating processes shared a 4 MB L2 cache and another case where the processes were launched on cores that do not share a data cache. We observe that the shared-memory queues perform better than the RRVM module for some message sizes when the processes share a data cache. However the RRVM module delivers a better overall performance because it performs significantly better when the processes don't share a cache.

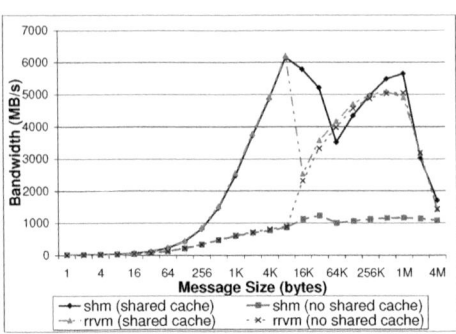

Fig. 3. Intranode communication bandwidth using shared-memory queues (shm) versus direct copy (rrvm) on Windows

The jagged graph in the shared-cache case is because at 16 KB, the double-buffering scheme used for communication runs out of L1 cache. The performance again improves from 128 KB because Nemesis switches to a different protocol that allows pipelining of messages. The bandwidth then drops at 2 MB because the double-buffering scheme runs out of L2 cache. We will investigate whether tuning some parameters in Nemesis can help smoothen the curve.

Figure 4 shows the intranode communication latency for small messages and bandwidth for large messages on Windows and Unix. The latency results on the two operating systems are excellent (240 ns on Unix and 275 ns on Windows for zero-byte MPI messages) and comparable (only ≈35 ns apart for small messages). For small messages, we use lock-free shared-memory queues for intranode communication on both systems. Therefore, we observe a performance degradation on both Unix and Windows for small messages when the communicating processes do not share a cache. For large messages, the bandwidth on Unix degrades substantially when the processes do not share a cache, whereas on Windows the performance is good in both cases because of the use of direct copy.

4.3   Internode Communication

We also studied MPI internode communication performance using TCP on both Windows and Unix. We measured the latency and bandwidth for internode communication by using the OSU microbenchmarks. Figure 5 shows the results.

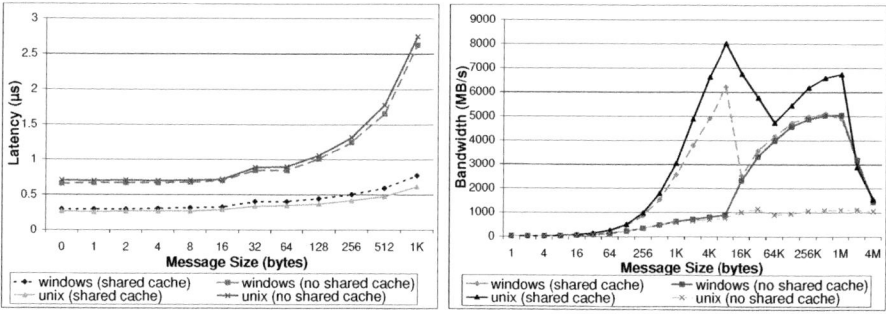

Fig. 4. Intranode communication latency and bandwidth on Windows and Unix

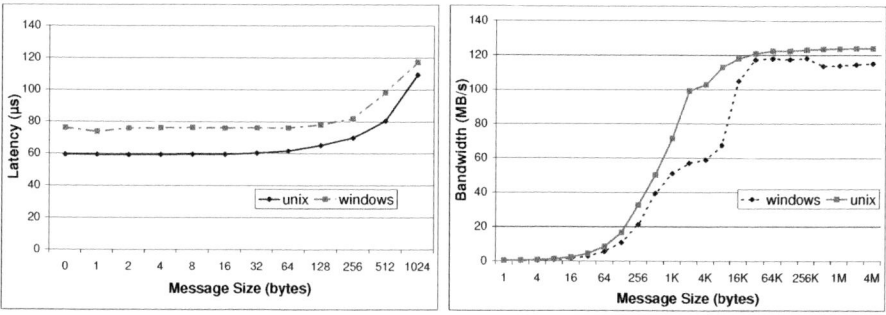

Fig. 5. Internode MPI communication latency and bandwidth on Windows and Unix using Nemesis over TCP/IP

We observe that, in these experiments, MPICH2 on Unix performs better than MPICH2 on Windows, with respect to both latency and bandwidth. We are investigating the cause of the difference, but we expect that further tuning and optimization of the Nemesis TCP module for Windows will eliminate the performance gap. We note that these microbenchmarks measure only point-to-point communication performance, whereas overall application performance depends also on the scalability of the underlying communication subsystem and the ability to overlap communication with computation. We expect the Windows version to have good overall application performance because the asynchronous model with completion ports is scalable and supports overlap. To verify this, we plan to conduct experiments with applications as described in Section 6.

4.4   Cost of Supporting Thread Safety

The MPI library incurs some overhead to support multiple user threads. To measure this overhead, we modified the OSU micro-benchmark to use `MPI_Init_thread` instead of `MPI_Init` and measured the latency for intranode communication with `MPI_THREAD_SINGLE` and `MPI_THREAD_MULTIPLE`. Note that, in both cases, the program has only one thread, but in one case support for multi-threading is disabled and in the other it is enabled, requiring the implementation to acquire thread locks in case multiple threads make MPI calls.

Figure 6 shows the overhead as the percentage increase in latency over the `MPI_THREAD_SINGLE` case when multithreading is enabled. We observe that the overhead is significantly lower on Windows than on Unix. The Unix version uses Pthread mutex locks for thread safety; the Windows version uses intraprocess locks (critical sections). We note that, on Windows, we initially used interprocess thread locks (mutexes), but their performance was much worse. By switching to intraprocess locks (critical sections), the performance improved significantly. Intraprocess locks are sufficient for Nemesis because it uses lock-free shared-memory queues for interprocess communication.

5   Related Work

Although MPI implementations have traditionally been developed on Unix, several MPI implementations are now available on Windows. Microsoft and Intel have developed MPI implementations for Windows [6,10], which are both derived from MPICH2. DeinoMPI [4] is another implementation of MPI, derived from MPICH2, for Windows. In addition, Open MPI has recently added support for Windows [13]; and MPI.NET [9] is an implementation that provides C# bindings for Microsoft's .NET environment.

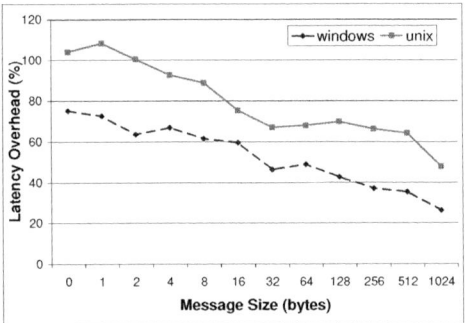

Fig. 6. Overhead in intranode MPI communication latency on Windows and Unix when multithreading is enabled

6   Conclusions and Future Work

We have discussed several issues in implementing MPI on Windows and compared them with approaches on Unix. We have also discussed how we implemented MPICH2 to exploit OS-specific features while still maintaining a largely common code base. Performance results with both Windows and Unix on the same hardware demonstrate that the performance of MPICH2 on both operating systems is comparable. We observed some difference in internode communication performance, which we plan to investigate and optimize on Windows. MPICH2 takes advantage of the asynchronous communication features in Windows, which enable applications to overlap communication with computation.

Windows HPC Server 2008 introduced a new low-latency RDMA network API, called Network Direct [12], that enables applications and libraries to use the advanced capabilities of modern high-speed networks, such as InfiniBand. We plan to implement a Nemesis module for Network Direct and study the performance of MPICH2 with high-speed networks on Windows. We also plan to evaluate application-level performance with MPICH2 on Windows, including commercial MPI applications at large scale.

References

1. Balaji, P., Buntinas, D., Goodell, D., Gropp, W., Thakur, R.: Fine-grained multithreading support for hybrid threaded MPI programming. International Journal of High Performance Computing Applications 24(1), 49–57 (2010)
2. Buntinas, D., Goglin, B., Goodell, D., Mercier, G., Moreaud, S.: Cache-efficient, intranode, large-message MPI communication with MPICH2-Nemesis. In: Proc. of the 2009 International Conference on Parallel Processing, pp. 462–469 (2009)
3. Buntinas, D., Mercier, G., Gropp, W.: Design and evaluation of Nemesis, a scalable, low-latency, message-passing communication subsystem. In: Proc. of 6th IEEE/ACM Int'l Symp. on Cluster Computing and the Grid (CCGrid) (May 2006)
4. Deino MPI, http://mpi.deino.net/
5. Gropp, W., Thakur, R.: Thread safety in an MPI implementation: Requirements and analysis. Parallel Computing 33(9), 595–604 (2007)
6. Intel MPI, http://software.intel.com/en-us/intel-mpi-library/
7. Message Passing Interface Forum: MPI: A Message-Passing Interface Standard, Version 2.2 (September 2009), http://www.mpi-forum.org
8. MPICH2 – A high-performance portable implementation of MPI, http://www.mcs.anl.gov/mpi/mpich2
9. MPI.NET: A high performance MPI library for.NET applications, http://osl.iu.edu/research/mpi.net/
10. Microsoft MPI, http://msdn.microsoft.com/en-us/library/bb524831(VS.85).aspx
11. NetPIPE: A network protocol independent performance evaluator, http://www.scl.ameslab.gov/netpipe/
12. Network Direct: A low latency RDMA network API for Windows. http://msdn.microsoft.com/en-us/library/cc9043(v=VS.85).aspx
13. Open MPI, http://www.open-mpi.org
14. Open Portable Atomics library, https://trac.mcs.anl.gov/projects/openpa/wiki
15. OSU Micro-Benchmarks (OMB), http://mvapich.cse.ohio-state.edu/benchmarks/
16. Top500 list (November 2008), http://www.top500.org/lists/2008/11

Compact and Efficient Implementation of the MPI Group Operations

Jesper Larsson Träff

Department of Scientific Computing, University of Vienna
Nordbergstrasse 15C, A-1090 Vienna, Austria
traff@par.univie.ac.at

Abstract. We describe a more compact representation of MPI *process groups* based on strided, partial sequences that can support all group and communicator creation operations in time proportional to the size of the argument groups. The worst case lookup time (to determine the global processor id corresponding to a local process rank) is logarithmic, but often better (constant), and can be traded against maximum possible compaction. Many commonly used MPI process groups can be represented in *constant space* with *constant lookup time*, for instance the process group of MPI_COMM_WORLD, and all consecutive subgroups of this group, but also many, many others). The representation never uses more than one word per process, but often much less, and is in this sense strictly better than the trivial, often used representation by means of a simple mapping array. The data structure and operations have all been implemented, and experiments show very worthwhile space savings for classes of process groups that are believed to be typical of MPI applications.

1 Introduction

Among the fundamental abstractions of the *Message-Passing Interface* (MPI) [10,7] are the notions of *process group* and *communicator*. A process group is a set of consecutively ranked processes that are mapped to a set of distinct processors, such that no two processes map to the same processor. A communicator has an associated process group (two, for the case of inter-communicators), and processes within this group can communicate without interference from communication on other communicators. Process groups are *local* objects that are created and maintained by the individual processors. In MPI a process group is simply a(n ordered) list of the different processors $[m_0, m_1, \ldots, m_{p-1}]$ of the group. Each m_i is a *global processor id* (an index or a pointer, for instance), and p is called the *size* of the group. The index i, $0 \leq i < p$, of processor m_i in a process group is called the *(group) local rank* of the processor.

For the efficient implementation of MPI a fast mapping $\pi : \{0, \ldots p - 1\} \rightarrow \{0, \ldots, m - 1\}$ from local ranks to global ids for any given process group is needed. MPI implementations typically represent this mapping by an array associated with the process group that maps directly from local to global rank.

R. Keller et al. (Eds.): EuroMPI 2010, LNCS 6305, pp. 170–178, 2010.

This implementation which requires p words of storage for a group of size p is not scalable in space, and will run into problems on very large systems with limited memory per processor (relative to the number of processors) [1]. It is therefore relevant to investigate more space efficient group representations that still allow fast lookup and can support the MPI group manipulation operations as well. The problem was considered in [3], which described limited solutions for restricted cases of group operations for the OpenMPI implementation [4]. Very recently, much more thorough, scalable communicator and group implementations were presented in [6], which proposes and evaluates both local, more compact group representations as well as means for sharing group representations among processes. For the local representations a framework is given making it possible to use several different representations with various tradeoffs between compactness and lookup time. The paper also investigates alternative definitions of an MPI process group facility, that can be more amenable to compact and shared implementations. The present paper in contrast develops one possible, local representation based on strided sequences in detail.

We describe a simple representation of groups based on finding and compacting strided (arithmetic) partial sequences in the processor lists. We aim to show that this in many cases give good compaction (down to constant space for many common application groups), that is only in some cases offset by non-constant, logarithmic in the worst case, lookup time. The contention is that many MPI applications creates subgroups with a regular, partially strided structure, and the representation are well suited to such cases. Application studies are needed to determine to what extent this claim is well founded. Other compaction schemes may have properties that make them good in other situations.

The representation can support explicit, eager algorithms for all of the MPI group constructor operations, taking only m bits of extra work space for all (except one) of these operations. The solution is a general representation of arrays, and does not exploit the special (injective) properties of the process to processor mapping. Better solutions by exploiting such properties or the specific group operations that need to be supported in MPI are definitely possible. Succinct representations of permutations and other objects have been studied in [8,9,2]. In [5] a scheme for representation of functions with constant time lookup and compression close to the entropy of the sequence of function values is given. This looks as theoretically desired for MPI, however, it is not clear what the actual compression will be for the relative regular maps that may arise in MPI applications, also not whether construction times are acceptable for supporting the group constructors; implementation results are not available. We note that since the number of possible process rank to processor id maps is $m!/(m-p)!$, the number of bits required to distinguish between these maps is $\Theta(\log(m!/(m-p)!)) = \Theta(m \log m)$.

2 Process Group Operations

The MPI standard defines a number of local operations on process groups for constructing new groups out of old ones [7, Section 6.3]. A group consisting of all started processors (in some externally defined order) is initially associated with the communicator MPI_COMM_WORLD, and all other groups are built from this (bar new processors that can be added by dynamic process management).

We outline a more or less standard, eager, explicit implementation of the group operations that in all cases except one takes only m extra bits of work space. Each group operation allocates and deallocates an m-bit array.

Let G, G_1, G_2 be process groups, and assume they are represented in such a way that it is possible to scan through the processors in the prescribed order in linear time in the size of the group. The new group is constructed by listing its processors in order. From this a (compact) representation of the group needs to be constructed eagerly, preferably without having to explicitly expand (uncompact) any the processor lists.

- MPI_Group_size(G) and MPI_Group_rank(G) return the size of the group and the local rank of the calling processor in the group (if belonging, MPI_PROC_-NULL otherwise). For the latter each group constructor needs to record the new local rank of the calling processor.
- MPI_Group_union(G_1, G_2) is the group (list) consisting of the processors of G_1 (in the order of G_1) followed by the processors in G_2 that are *not* in G_1. The union is computed by scanning through the processors of G_2 setting bit m_j in the work array, next scanning through G_1 resetting bit m_i, then listing the processors of G_1 in order followed by the processors of G_2 in order for which the corresponding bit m_j is still set.
- MPI_Group_intersection(G_1, G_2) is the group (list) consisting of the processors of G_1 (in the order of G_1) that are also in G_2. This is computed by scanning through the processors of G_1 resetting bit m_i in the work array, next scanning through G_2 setting bit m_j, then listing the processors of G_1 in order for which the corresponding bit m_i is set.
- MPI_Group_difference(G_1, G_2) is the group (list) consisting of the processors of G_1 that are *not* in G_2. This is computed by scanning through the processors of G_1 setting bit m_i in the work array, next scanning through G_2 resetting bit m_j, then listing the processors of G_1 in order for which the corresponding bit m_i is still set.
- MPI_Group_incl(G, L) is the group (list) consisting of the processors with local rank in the list L (without repetitions) in that order. The group is computed by simply listing the processors in G in the order $[m_{L[0]}, m_{L[1]}, \ldots, m_{L[n-1]}]$ where n is the length of L.
- MPI_Group_excl(G, L) is the group (list) consisting of the processors of G that are *not* in L in the order of G. This is computed by first scanning through the processors of G setting bit m_i, then scanning through L resetting each bit $m_{L[i]}$, then listing the processors of G in order for which the corresponding bit m_i is still set.
- MPI_Group_range_incl(G, n, R) is the group formed by inclusion of the ranks given implicitly by the n ranges R. Each range is a triple (f, s, l) of first (local) rank f, stride s and last rank l which defines a list of local ranks $f, f+s, f+2s, \ldots f + \lfloor \frac{l-f}{s} \rfloor s$.

- MPI_Group_range_excl$G, n, R)$ is the group formed by exclusion of the ranks given implicitly by the n ranges R.
- MPI_Group_translate_ranks(G_1, L, G_2) is the list the local ranks in G_2 of the processors with local ranks in L of G_1 in the order of L (if existing, otherwise MPI_PROC_NULL). For translation an m-sized integer array is needed. First all global ranks of G_1 with local rank in L are marked as untranslated MPI_PROC_NULL in the array. Scanning through G_2 all m_j are marked with their local rank j in the array. Scanning through L the local rank in G_2 (or MPI_PROC_NULL) can be found at index $m_{L[i]}$ of G_1 in the array.
- MPI_Group_compare(G_1, G_2) compares the two groups for identity (same processors in the same order) or similarity (same processors, different order). Identity is checked by scanning both groups in parallel and checking whether the processors occur in the same order. For similarity, scan through G_1 and set the corresponding bits m_i, then scan through G_2 and reset the bits m_j. Scanning again through G_1 if all bits m_i are now reset, G_1 is a subgroup of G_2. If similarly G_2 is a subgroup of G_1, the two groups are similar.

In each of the operations (except for MPI_Group_translate_ranks the m-bit work array is allocated (m-integer work array for the translation operation), but no potentially expensive initialization is needed. This is taken care of by the scans through the groups, so that the time complexity is determined only by the size of the group arguments. For a compact representation of groups with processors $[m_0, \ldots, m_{p-1}]$ with the property that the next element of the sequence m_{i+1} can be produced after m_i in constant time without having to explicitly expand the whole list of processors (which would take time and p words of space) the MPI group operations can be implemented with m *bits* of work space, and time in most cases proportional to the size of the group arguments. Only for some of the operations taking a list L of group ranks (e.g. MPI_Group_incl) the lookup time per group member needs to be factored in. The representation developed in the next section has this property.

The m-bit work array obviously can be saved if the inverse mapping π_G^{-1} is also available for each of the groups. To keep the work array small, processors m_i are best represented by indices (into an array of processor descriptions of size m) and not pointers to arbitrary memory locations. We finally note that no sorted lists of processor ids are needed.

3 Compact Representation of Mappings

We now give a general, fairly straightforward method for compact representation of arrays with certain regularities with worst case logarithmic lookup time (by binary search). For arrays without the considered regularities lookup time will be constant, and the representation will not take more than p words (plus small constant extra space). In other words, the space consumption will never be worse than the trivial implementation by a mapping array, and considerably better for arrays of strided ranges without compromising lookup time too much. For many somewhat regular arrays the representation will have both fast lookup time and take considerably less than linear, often constant space. The compaction is based on identifying *runs* (non-standard use of the term) in the sequence of processors of the form $m + di$ for $i = 0, \ldots, l - 1$. Theoretically better results can surely be achieved by considering and exploiting the fact that groups are represented by injective mappings, which is not done here.

Consider a group with processor indices $[m_0, m_1, \ldots, m_{p-1}]$. The sequence of processor indices is partitioned into r not necessarily maximal *runs* of the form $m_i + jd_i$ for $0 \leq j < l_i$ where l_i is the length of the run. The number of such runs is at most $\lceil p/2 \rceil$. We can assume w.l.o.g. that the length of each run is either 2 or greater than 5 (break runs shorter than 6 into runs of length 2; it does not matter that this may change the sequence of runs). The ith such run is represented by the pair $[m_i, d_i]$ and its length l_i. The sequence of processor indices in a group $[m_0, \ldots, m_1, \ldots, m_{r-1}, \ldots, m_{p-1}]$ can thus be represented by the sequence of runs

$$[m_0, d_0] \ [m_1, d_1,] \ \ldots \ [m_{r-1}, d_{r-1}] \tag{1}$$
$$k_0 \qquad k_1 \qquad \ldots k_{r-1} \qquad k_r$$

and the sequence of corresponding run lengths. In (1) k_i is the group local rank of the first processor in the run starting with m_i. Instead of storing the run lengths explicitly, we use the sequence of first local ranks $[k_0, k_1, \ldots k_{r-1}, k_r]$ with $k_r = p$ being the start index of a virtual last run. The length of run i is then simply $l_i = k_{i+1} - k_i$. The global rank (processor index) of local rank i can now be computed by first finding (by binary search) the run j with $k_j \leq i < k_{j+1}$, and then computing $m_j + (i - k_j)d_j$. This gives a compaction down to three times the number of identified runs (m_i, d_i, k_i for run i), with lookup time that is logarithmic in the number of runs. In the worst case of $\lceil p/2 \rceil$ runs this is $\lceil 3/2p \rceil$, and thus worse than the trivial array representation. Also in this case, lookup time is $O(\log p)$.

To improve both compaction (to linear in the worst case) and lookup time we need a more compact representation of the lookup sequence $[k_0, k_1, \ldots, k_r]$. We achieve this by representing supersequences of runs of the *same length* by only the index r_i of the first run in the original sequence (1), and as before the group local rank k_i of the first processor in the first run of the supersequence. Lookup is again accomplished by binary search for an index j with $k_j \leq i < k_{j+1}$. The run index is then found as

$$s = r_j + \lfloor (i - k_j)/l_i \rfloor$$

and the corresponding processor as

$$m_s + d_s \lfloor (i - k_j) \bmod l_i \rfloor$$

where $l_j = \frac{k_{j+1} - k_j}{r_{j+1} - r_j}$ is the length of the runs in the supersequence of runs from r_j to r_{j+1}. This is illustrated in (2).

$$\overset{\text{run length } \frac{k_1-k_0}{r_1-r_0}}{\overbrace{[m_0, d_0]}} \quad [m_1, d_1,] \ldots \quad \overset{\text{run length } \frac{k_2-k_1}{r_2-r_1}}{\overbrace{[m_2, d_2]}} \quad \ldots [m_{r-1}, d_{r-1}] \tag{2}$$

$$k_0 \qquad\qquad\qquad\qquad k_1 \qquad\qquad\qquad \ldots k_{g-1} \qquad k_g$$
$$r_0 \qquad\qquad\qquad\qquad r_1 \qquad\qquad\qquad \ldots r_{g-1} \qquad r_g$$

Because either all runs have length two or are longer than 5, the space needed for both the encoding of the runs (two words) and the search structure (two words per supersequence of same length runs) the total space consumed is at most p words. A sequence longer than 5 saves enough space to store also the search indices for the previous supersequence. The search times is likewise improved to logarithmic in the number of supersequences.

For the worst case of $\lceil p/2 \rceil$ runs of length two, there is only one supersequence. Thus the lookup-time for "random" arrays where all runs are of length two is constant, since no binary search is needed. Likewise, the lookup time of arrays that can be represented by a few supersequences or runs will be constant. An important case is MPI_COMM_WORLD which is in most MPI implementations represented by a single $(m + i)$ run.

In the worst case where short and longer runs are alternating without repetition lookup time will still be $O(\log p)$. At the cost of less compression, runs can be restricted to being either of length two or very long, e.g. $O(p/\log p)$, which improves lookup time correspondingly, e.g. $O(\log \log p)$. It is important that the division into runs can be done greedily online as the m_i's are becoming known, but this is trivially possible. A first run is started by setting $m = m_0$ and $d = m_1 - m_0$. Each time a new m_i is given, extend the length of the previous run by one if $m_i = m_{i-1} + d$, otherwise close the run and start a new one. If the run being closed is shorter than 5 (but longer than two), split the previous run into runs of length two. It is likewise important (for the MPI group construction operations) to note that the sequences of m_i's can be listed in linear time. For this the index of the current run and its length need to be maintained while computing the m_i processor ids one after another. This is necessary and sufficient for the implementation of the group operations as explained in Section 2.

The sequences of first processor ids m_i and corresponding factors d_i for the runs are stored as two separate arrays. These can again be (recursively) compressed by the above technique, exploiting regularities in either sequence, and is important for achieving compact representation of some common MPI groups (see next section).

4   Experimental Evaluation

The group operations, a lookup operation, and iterator constructs to scan through all processors of a group in sequence have been implemented as outlined in the previous sections. Here we evaluate only the compression scheme by counting the space consumption and the number of operations involved in looking up all processors for a sample of groups constructed using the MPI group operations. The aim is to show that many commonly encountered MPI groups can indeed be represented in constant space and with constant time lookup, that no groups use more than the claimed p words (plus a small constant), and that lookup times are even in the worst case tolerable.

For each individual lookup for some local group rank i we count the number of needed dereferencings (when an address has been dereferenced it is assumed cached, and not counted again for that operation). We perform a lookup for all ranks, and list the total number of such dereferencing operations and the average number per lookup operation. In the experiment we set p equal to a million processors [1].

Table 1. Lookup operation counts for various process groups

Group	Space	1-Level space	Lookup operation count Total	Average
MPI_COMM_WORLD	16	6	2 000 000	2.00
Reverse MPI_COMM_WORLD	16	6	2 000 000	2.00
SMP local roots union	16	20	2 000 000	2.00
SMP reverse nodes	16	250004	2 000 000	2.00
Worst case alternating	750016	1000004	17 951 472	17.95
Linear increasing	4260	5656	10 560 038	10.56
Exponentially increasing	78	82	5 288 652	5.29
Random	1000018	1000004	2 000 000	2.00

The groups considered are the initial MPI_COMM_WORLD group, these processors in reverse order (created by MPI_Group_incl), a group of all processors in some random order, groups where the distances in processor id between rank i and rank $i + 1$ increase linearly and exponentially, respectively, a worst case group where short and long runs alternate, and groups created for SMP clusters. For the latter it is assumed that processors are divided into processor nodes of size 8. First, a group is formed by the union of first all processors with local rank 0 on their SMP node, then local rank 1, followed by local rank 2, and so forth. This is the local roots union group. Second, a group is formed by the processors on the last SMP node, the next to last SMP node, down to the processors on the first SMP node. This is the reverse nodes group. All groups have size 1,000,000.

The results are shown in Table 4. We list the space consumption in words with the full recursive compaction scheme, and with the scheme where only compaction of the original processor sequence is performed (no compaction of the m_i and the d_i sequence, referred to as "1-level space" in the table), the total number of dereferencing operations for accessing all processor ids, and the average number of such operations per lookup. The table shows that space consumption cannot be any worse than the trivial implementation by means of an array, at the cost of two instead of one array dereferencing operation. For the simple, very regular groups considered here, space can, especially by applying the scheme recursively, be reduced to constant (16 words for the SMP reverse nodes group, down from 250004), and with lookup time that might only for the worst case alternating group be problematic. This, as explained, can be improved (at the cost of up to p words in total) by accepting only runs of a certain minimum length (e.g. $p/\log p$).

These synthetic results must of course be complemented by results with real applications with nontrivial use of group and communicator operations, but to asses the compaction that can be achieved in practice, and to estimate how much the sometimes increased lookup times affect overall application time.

5   Compact Binary Search

There are techniques for compressing the ordered array used for binary search, while still allowing efficient search in logarithmic time with little extra overhead. We sketch such an improvement here that reduces the space by a factor of two. We assume the ordered array R of process ranks stores w-bit integers.

Let $u < w$ be some number of prefix bits. Create a new array R' with the integers as in R (in the same order), but with the u-bit prefix for each entry chopped off. Refer to the sequences of numbers with the same prefix as the *blocks*. The space needed for R' is then $p(w - u)$ bits. We assume that all blocks are non-empty; the case where some blocks are empty can be catered for, but is not described here.

Construct a lookup table of size 2^u ($\log_2 p$) bit pointers that for each prefix points to the beginning of its block. To search for a rank i we first find the block for the u-bit prefix of i by table lookup, and then do binary search for the $(w - u)$-bit suffix of i in this block. The return index for i is computed by adding the block index (from the table lookup) to the found index within the block. The total space for this variant is $p(w - u) + 2^u(\log_2 p)$ bits.

Assume now for the MPI case that $p, m \leq 2^w$ are (say) at most 2^{24}. Instead of storing the process ranks in an array of 32-bit numbers, this can now be done with an array of 16-bit numbers plus the lookup table of $2^8 = 256$ pointers, which saves almost half the space with little extra overhead for the lookup. The idea can be generalized, but is only sketched here, and has not (yet) been implemented.

6   Concluding Remarks

We described a simple method to represent arrays in a more compact (space efficient) fashion at the cost of a non-constant worst-case dereferencing time. The method supports linear time iteration through the array, and is therefore immediately suited to eager implementation of the MPI group operations. We gave such an implementation, and showed that many common process groups can indeed be represented in constant space and with constant lookup time. The methods presented here may therefore be a first step toward space efficient, and thus more scalable representation of MPI objects like process groups as well as other, similar mappings used for representation of hierarchical communication systems. We hope that this will be of relevance to MPI implementations and stimulate further research in this direction. Some highly relevant theoretical results were mentioned, that may have practical merit to the specific MPI related problems dealt with here. Practically, it would be interesting to see how this improvement affects the memory footprint of MPI libraries on applications creating a non-trivial amount of groups and communicators. A further interesting issue for MPI and other parallel processing interfaces is whether process to processor mappings need always be represented explicitly, and whether interfaces and operations can be defined that allow faster (implicit) group operations with much less space. This issue was also taken up in [6].

References

1. Balaji, P., Buntinas, D., Goodell, D., Gropp, W., Kumar, S., Lusk, E., Thakur, R., Träff, J.L.: MPI on a million processors. In: Ropo, M., Westerholm, J., Dongarra, J. (eds.) Recent Advances in Parallel Virtual Machine and Message Passing Interface. LNCS, vol. 5759, pp. 20–30. Springer, Heidelberg (2009)
2. Barbay, J., Navarro, G.: Compressed representations of permutations, and applications. In: 26th International Symposium on Theoretical Aspects of Computer Science (STACS). Dagstuhl Seminar Proceedings, vol. 9001, pp. 111–122 (2009)
3. Chaarawi, M., Gabriel, E.: Evaluating sparse data storage techniques for MPI groups and communicators. In: Bubak, M., van Albada, G.D., Dongarra, J., Sloot, P.M.A. (eds.) ICCS 2008, Part I. LNCS, vol. 5101, pp. 297–306. Springer, Heidelberg (2008)
4. Gabriel, E., Fagg, G.E., Bosilca, G., Angskun, T., Dongarra, J.J., Squyres, J.M., Sahay, V., Kambadur, P., Barrett, B., Lumsdaine, A., Castain, R.H., Daniel, D.J., Graham, R.L., Woodall, T.S.: Open MPI: Goals, concept, and design of a next generation MPI implementation. In: Kranzlmüller, D., Kacsuk, P., Dongarra, J. (eds.) EuroPVM/MPI 2004. LNCS, vol. 3241, pp. 97–104. Springer, Heidelberg (2004)
5. Hreinsson, J.B., Krøyer, M., Pagh, R.: Storing a compressed function with constant time access. In: Fiat, A., Sanders, P. (eds.) ESA 2009. LNCS, vol. 5757, pp. 730–741. Springer, Heidelberg (2009)
6. Kamal, H., Mirtaheri, S.M., Wagner, A.: Scalability of communicators and groups in MPI. In: ACM International Symposium on High Performance Distributed Computing, HPDC (2010)
7. MPI Forum. MPI: A Message-Passing Interface Standard. Version 2.2 (September 4, 2009), www.mpi-forum.org
8. Munro, J.I., Raman, R., Raman, V., Rao, S.S.: Succinct representations of permutations. In: Baeten, J.C.M., Lenstra, J.K., Parrow, J., Woeginger, G.J. (eds.) ICALP 2003. LNCS, vol. 2719, pp. 345–356. Springer, Heidelberg (2003)
9. Munro, J.I., Rao, S.S.: Succinct representations of functions. In: Díaz, J., Karhumäki, J., Lepistö, A., Sannella, D. (eds.) ICALP 2004. LNCS, vol. 3142, pp. 1006–1015. Springer, Heidelberg (2004)
10. Snir, M., Otto, S.W., Huss-Lederman, S., Walker, D.W., Dongarra, J.: MPI: The Complete Reference. MIT Press, Cambridge (1996)

Characteristics of the Unexpected Message Queue of MPI Applications

Rainer Keller and Richard L. Graham

Oak Ridge National Laboratory
{keller,rlgraham}@ornl.gov*

Abstract. High Performance Computing systems are used on a regular basis to run a myriad of application codes, yet a surprising dearth of information exists with respect to communications characteristics. Even less information is available on the low-level communication libraries, such as the length of MPI Unexpected Message Queues (UMQs) and the length of time such messages spend in these queues. Such information is vital to developing appropriate strategies for handling such data at the library and system level. In this paper we present data on the communication characteristics of three applications GTC, LSMS, and S3D. We present data on the size of their UMQ, the time spend searching the UMQ and the length of time such messages spend in these queues. We find that for the particular inputs used, these applications have widely varying characteristics with regard to UMQ length and show patterns for specific applications which persist over various scales.

1 Introduction

The Message Passing Interface (MPI) [5] is the ubiquitous parallel programming paradigm for large-scale parallel applications. For current PetaFlop computers such as RoadRunner at LANL, Jugene in Juelich or the current leader in the Top500 list, JaguarPF at ORNL, MPI is the standard for communication. With ever-increasing machine sizes, the requirements with regard to scalability in terms of resource usage for the application as well as libraries change. Here, the requirements on the MPI library itself is no different; sparse resources should be used conservatively, communication algorithms should take advantage of hardware capabilities and network topologies and data-structures and algorithms adapt to different usages in applications. Therefore it would be helpful for MPI-developers, system designers and application developers alike to quantitatively and qualitatively understand the communication characteristics and specifically the characteristics of the most crucial data structures, the unexpected message queue (from hereon UMQ) of large-scale applications. This paper provides this data for applications running with input data sets important to the Oak Ridge Leadership Computing Facility (OLCF). To produce this data, a new tracing

---

* This work was supported by the U.S. Department of Energy and performed at ORNL, managed by UT-Battelle, LLC under Contract No. DE-AC05-00OR22725.

R. Keller et al. (Eds.): EuroMPI 2010, LNCS 6305, pp. 179–188, 2010.

library is developed based on the Peruse [8] specification to gather UMQ usage information from the Open MPI [6] library.

This paper is organized as follows: Sec. 2 gives an overview of the tracing library developed and the methodology used to gather the performance data. Section 3 introduces the applications used for this study and shows the MPI-internal data gathered for these applications. In Sec. 4 the related work is presented. Finally Sec. 5 concludes the paper.

2   Design and Implementation

The MPI standard does not impose any restrictions on the communication mechanisms or protocols of the implementation; however it requires the library to

- match messages between a given pair of processes in order, based on the message envelope, which consists of the (src, dest, tag, communicator)-tuple,
- progress and finish either one or both calls of a pair of MPI_Send/ MPI_Recv once they have matched, regardless of other communication in the system,
- complete a message transfer, in case the matching receive has been posted; which may require buffering of messages.

This internal buffering of application messages cannot be traced using the usual PMPI-based tools. Apart from the communication characteristics such as number of point-to-point calls, source-destination traffic, little is known of these internal characteristics, especially when running at scale.

Understanding the characteristics of the UMQ is of importance for application developers, MPI library developers, as well as system designers trying to provision resources for communications. Unexpected receives generally consume more resources on the receive side than expected receives [11]. The nature and order of magnitude of these resources depends on the MPI implementation, and the strategy each implementation chooses for handling such data. For example, header only data my be kept at the receiver so that it can be retrieved from the source once the match is made, or one may decide to store both header data and payload. Payloads may be stored in scarce network buffers, or may be copied to other buffers. In addition, unexpected messages go through the matching logic both when they arrive and a match is attempted against pre-posted receives, and potentially each time a new relevant receive is posted, until they are matched. From the user's perspective, understanding their application specific communication characteristics gives them the option to change their application to reduce library memory usage, and the matching costs. Application and system designers often optimize their designs for the common use case, so understanding the communication characteristics of real applications is invaluable. This helps in informing both matching and memory management strategies

In order to expose internals of the MPI implementation, the Peruse specification [8] has been defined and subsequently implemented in Open MPI [9]. Peruse offers a mechanism to query and select events in a portable way, and allows an MPI implementation to only provide events that do not limit the implementation or otherwise introduce excessive overhead. Furthermore it allows an application or tracing library to hook callbacks into these Peruse-specified events, which are invoked when the event is triggered in the MPI library. This offers a low-overhead way to trace the messages traverse through the MPI stack.

In this paper we introduce a new library based on the Peruse interface. The library is intended to be easy to use, have low overhead and scale with the application. These requirements were met in the following way: The library provides hooks for `MPI_Init` and `MPI_Finalize` for C and Fortran, such that the application programmer may just link to the library and be able to gather statistics. These are saved or output by the wrapper of the `MPI_Finalize` call. Furthermore it provides a small API in C and Fortran to customize the gathering and outputing of the data gathered, e. g. the data may be written multiple times during the execution, or the statistics may be reset to differentiate between parts of the code. The last requirement was met by taking care not to introduce bottlenecks, e. g. collective reduction functions are provided to reduce the information printed, and the whole UMQ data gathered may be written to disk using collective parallel MPI I/O. A tricky part proved to provide the proper `MPI_Info` hints to the MPI I/O implementation to aggregate the writes. Further details of the implementation are described in [10].

3   Application Measurement

In our study, we concentrate on three codes important to Oak Ridge National Laboratory. The following describes the applications and measurements done on them. If not otherwise noted, all libraries and applications have been compiled with the Portland Group Compiler PGI-9.0.4 with compiler options `-fastsse`, `-Mipa=fast,inline` and instruction scheduling for the AMD hex-core processor on JaguarPF `-tp=istanbul-64`. The developer trunk r22760 of Open MPI was used.

Gyrokinetic Toroidal (GTC). The Gyrokinetic Toroidal Code (GTC) is a 3-D particle-in-cell code [12] to simulate the transport in fusion plasmas such as the tokamak fusion reactor ITER. The code is one of the benchmark codes used within the compute centers NCCS and NERSC. Similar to the studies in [14], we weakly scale the input up to 8192 processes on the Cray XT5 at ORNL, JaguarPF. For this experiment however, we turned off the data diagnosis code, which writes the 3D domain to one file per process. Due to memory constraints at scale, all runs were done with only 4 processes per node.

Locally-Self-Consistent Multiple-Scattering (LSMS). The Locally Self-consistent Multiple-Scattering (LSMS) Code [13] simulates the interactions between electrons and atoms in magnetic materials. LSMS has been under development at ORNL since 1994 and was used to study large numbers of atoms, in ensembles of up to 10k atoms. Using the so-called Wang-Landau (WL-LSMS) scheme for work distribution, the code may scale up to the whole machine size of

JaguarPF; this work received the Gordon Bell Prize at SC 2009 [4]. WL-LSMS distributes chunks of electron configurations to groups of locally optimizing lsms-processes using non-blocking point-to-point communication. For this paper, the communication due to this work distribution however is not of interest, as it is very coarse grained and sends only a few bytes. Rather the communication requirements of the main kernel is of interest, which applies density functional theory to the relativistic wave equation for electron behavior. Therefore, to examine the communication behavior of the most time-consuming part, in this paper lsms_main has been employed without the Wang-Landau method.

The first test-case used consists of a standard benchmark for LSMS. Here a system of iron atoms in a body-centered cubic crystal structure in which the iron atoms are equally spaced, with a cut-off radius of 12.5 Bohr. The system is weakly scaled up to 4096 processes, doubling the amount of Fe atoms. To reduce the overall time, the number of evaluated energy points and number of iterations is reduced.

S3D. The S3D code is a highly scalable direct numerical solver (DNS) for the fully compressible Navier-Stokes equations with detailed chemistry [7], developed at Sandia National Laboratories. It has been scaled to the full machine size of JaguarPF. The input data-set used was 25^3 grid points per processor and weakly scaled. The chemical reaction was enabled and the 8th order spatial derivative was used. As for GTC the saving of intermediate files, such as restart files were turned off.

Maximum Length of UMQ. As the first comparison, Fig. 1 shows the maximum length of the UMQ. The left column shows the maximum length of the UMQ as the number of processes is scaled up, as well as the shortest maximum length of UMQ[1] with both lines stating the rank of the process above the line. The right column shows the length of the UMQ over the ranks for the largest test-case run with each application. As the vertical axis takes into account the value of the single-most process with the longest UMQ, which in this case is always process with rank zero, the axis seems to be coarse grained, but better shows the relation of distribution versus highest value. The reason process zero has in almost all cases, the longest UMQ is due to being the root in collective communication which uses the same underlying communication mechanisms and adds to the pressure on the UMQ resources.

One may see that different codes have different communication characteristics. For GTC the length of the UMQ is bounded to 24 even for the largest run (once the file output was turned off), while the length of LSMS' increases linearly with the number of processes run. This is due to the communication pattern of the distribution of the so-called τ-Matrix, a 32^2 matrix of double-complex data (overall 16 kB) to all of the neighbors within the cut-off radius: first the data to send is assembled, then sent using non-blocking MPI_Isend, while internal matrices are solved and saved, and finally the buffers used in the matrix multiplication are overwritten with the data received in a non-blocking MPI_Irecv.

---

[1] The lower bound of all processes' maximum length of UMQ in a run.

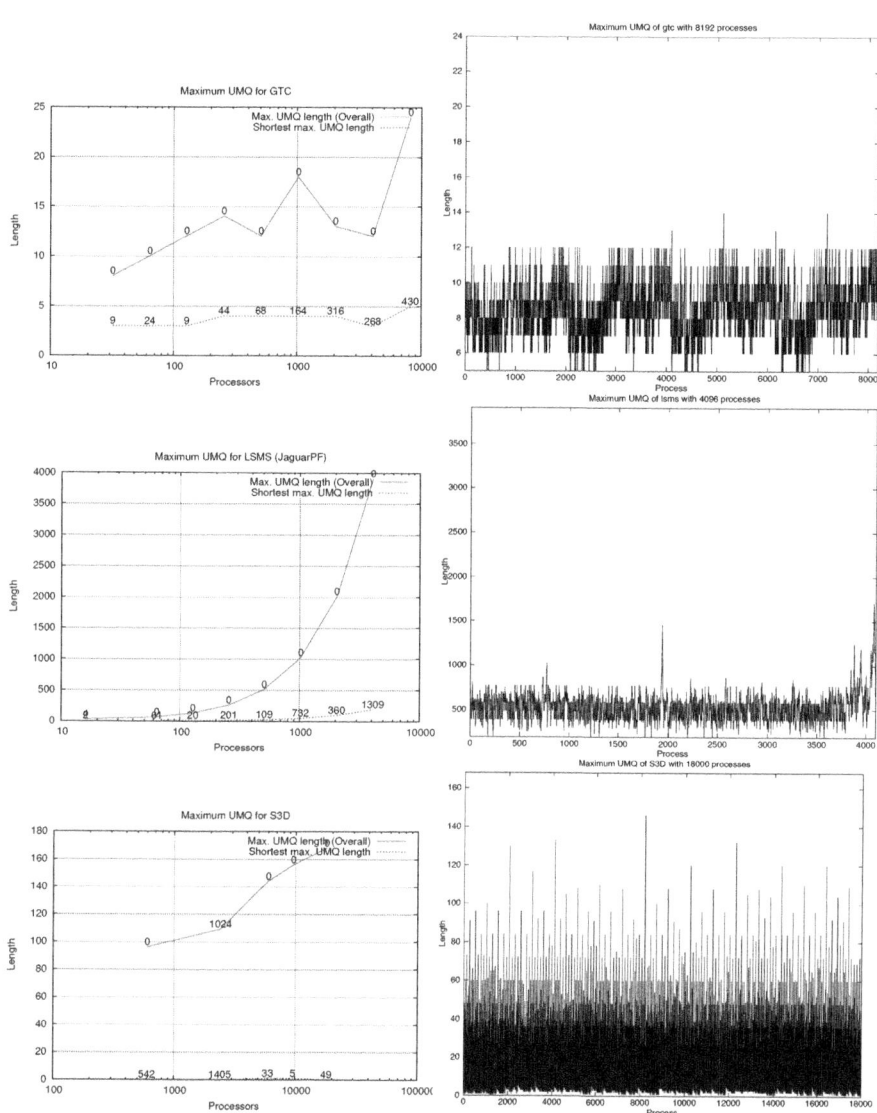

Fig. 1. Maximum length of UMQ and distribution of length at the largest case

Table 1. Statistical values of the length of the UMQ over all processes

Application	Size	Min	Median	Max	Mean	Std. Dev.
GTC	2048	4	7	13	7	1.34
GTC	4096	3	7	12	6	1.65
GTC	8192	5	8	24	8.2	1.52
LSMS	1024	49	121	1022	122	42.43
LSMS	2048	97	241	2010	249	73.52
LSMS	4096	192	504	3906	522	169.31
S3D	6000	1	4	144	8	12.79
S3D	9600	1	4	156	9	12.86
S3D	18000	0	4	168	9	12.86

Therefore, in the current version of the code, messages may aggregate on the UMQ, while the matrix multiplication is being performed. This part of the code is currently being rewritten.

Furthermore Figure 1 shows patterns in the length of the UMQ. These patterns are most visible for GTC with its shorter maximum UMQ length, but are also visible with S3D. For GTC, this pattern may be explained with the actual physical domain – a toroid. Due to page constraints all information cannot be shown here. The statistical parameters of the maximum length of UMQ for the largest test-case are given in Table 1. As one may see, the mean and median do not differ much for GTC, with a low standard deviation, while S3D has a varying maximum UMQ with large standard deviation, while LSMS has the longest maximum UMQ with large standard variation and process zero having an over 7 times longer UMQ than the mean UMQ length.

Search Time of UMQ. Another interesting statistical analysis is the time spent in each process searching the UMQ. This event is triggered, when a MPI_Recv is posted and the UMQ is not empty. Figure 2 shows the results again only for the largest test run for all three applications. This is measured by mpistat using the events PERUSE_COMM_SEARCH_UNEX_Q_BEGIN and PERUSE_COMM_SEARCH_UNEX_Q_END. Here the search time is aggregated for each process, i. e. only the total time spent searching the UMQ is plotted. Again, the results are very different for every application and the statistical data is presented in Table 2. As may be seen the characteristics on relation between median and mean are similar to the previous table.

While GTC with its short UMQ also exhibits a low total search time with an even distribution among the processes, LSMS with its rather long max. UMQ, shows a high total time spent searching the UMQ (max. of 4.6 s). However, it is very unevenly distributed. In fact, of the 4096 processes 3738 processes have a total search time smaller than the mean total search time (0.247 s), aka the remaining 358 of the processes search longer in the UMQ. In comparison to the overall statistics (see Table 2) this reduces the mean search time to 0.107 s and reduces the std. deviation to 0.0159.

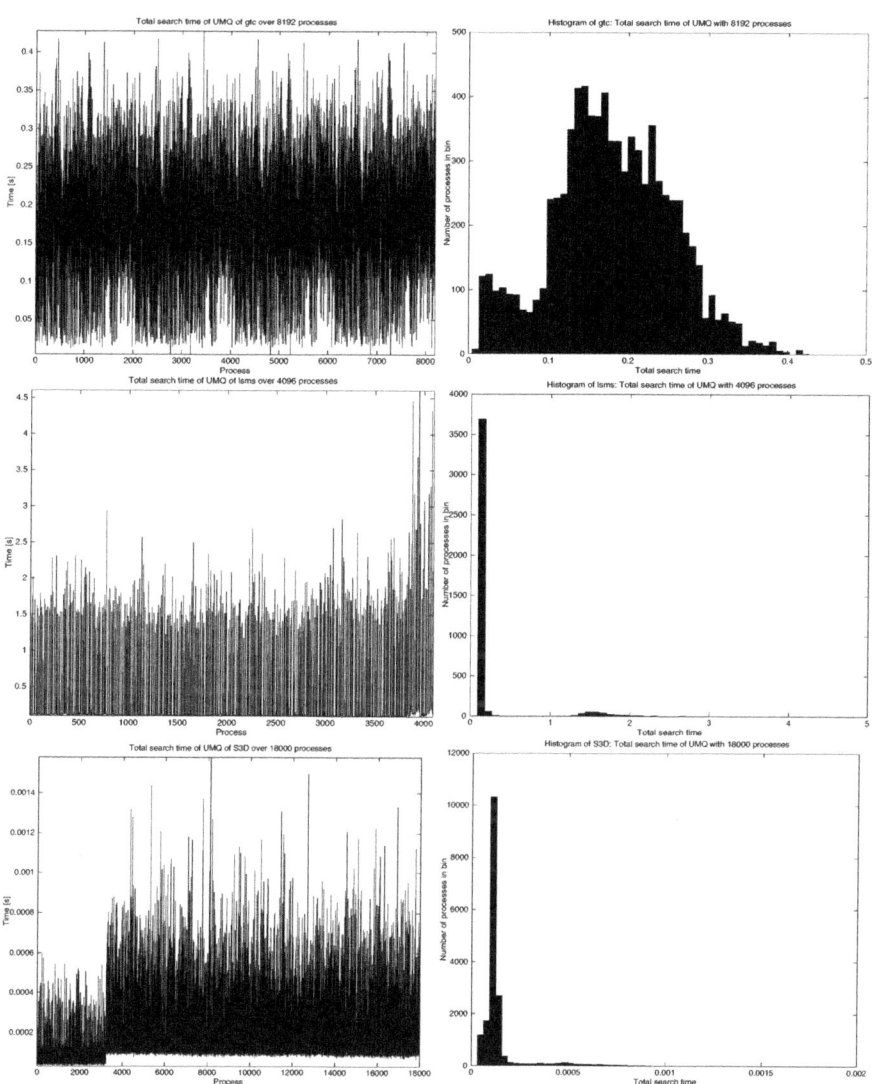

Fig. 2. Total time spent per process searching the UMQ at the largest case

Table 2. Statistical values of total search time in UMQ over all processes

Application	Size	Min	Median	Max	Mean	Std. Dev.
GTC	2048	0.002413	0.038883	0.143237	0.042010	0.019658
GTC	4096	0.002598	0.063949	0.221349	0.068989	0.028306
GTC	8192	0.004523	0.176509	0.427667	0.179208	0.074060
LSMS	1024	0.010401	0.017356	0.150949	0.025008	0.025214
LSMS	2048	0.030489	0.034856	0.591989	0.064125	0.097730
LSMS	4096	0.090373	0.104260	4.603960	0.246608	0.477655
S3D	6000	0.000017	0.000067	0.001229	0.000081	0.000079
S3D	9600	0.000022	0.000092	0.001323	0.000118	0.000112
S3D	18000	0.000032	0.000110	0.001577	0.000144	0.000132

Table 3. Statistical values of the time messages stayed in UMQ for process 0

Application	Size	Min	Median	Max	Mean	Std. Dev.
GTC	2048	0.000001	0.039768	0.471423	0.067976	0.078905
GTC	4096	0.000002	0.032429	0.841412	0.091685	0.123482
GTC	8192	0.000001	0.000030	2.648100	0.033560	0.110619
LSMS	1024	0.000003	1.408290	3.867880	1.547093	1.311374
LSMS	2048	0.000002	0.012800	9.030230	3.233932	3.763306
LSMS	4096	7.786120	10.165600	12.459800	10.026571	1.234413
S3D	6000	0.000022	0.022476	0.063193	0.019889	0.010769
S3D	9600	0.000035	0.019729	0.183491	0.076856	0.081038
S3D	18000	0.000027	0.064833	0.168463	0.065595	0.036252

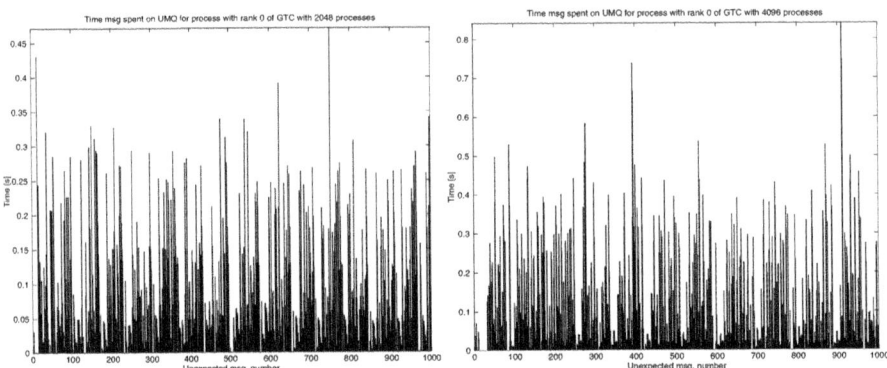

Fig. 3. Time messages spent on UMQ for first 1000 messages on process 0 of GTC for 2048 and 4096 processes

Another artifact that may be seen for S3D at all scales: Process ranks larger than a certain critical rank all have higher total search times, as may be seen in Fig. 2 at the bottom.

Time Messages Spent on UMQ. The time messages spent on the UMQ is again highly dependent on the application. In Fig. 3 we show the results for the GTC application with 2048 and 4096 processes for process with rank zero. Again a pattern is visible. Few messages stay very long on the UMQ (up to 0.471 s for the 2048 process and 0.841 s for the 4096 process case), while the majority of messages stay only shortly on the UMQ. This diagram however is over-exaggerating, as the statistics show in Table 3: the mean time is a lot lower than as might be expected from the graph.

4 Related Work

Much work has been published focusing on the scaling behavior of large-scale parallel applications, fewer papers exist which focus on the communication requirements of these applications. However, very few papers have focused on the internals of the MPI-libraries unexpected message queue handling. In [3] and [2], the authors research the collective and point-to-point communication requirements of the NAS parallel Benchmarks [1] and three applications (LAMMPS, CTH and ITS) for job sizes of up to 64 processes. Similar to our work, the method allows the analysis of the length of the UMQ, but does not state the length messages stayed in the UMQ. Additionally, the papers provide the posted receive queues, and the average search depth in the UMQ until successfully matching a message, which is currently not yet possible in our implementation. Both of these papers show similar application-specific characteristics, however the scale here is rather limited.

5 Conclusion

Little is known of the buffer requirements of relevant applications at scale. This paper presents a library, that allows the in-depth analysis of the unexpected message queue (UMQ) behavior even at scale. The library provides reduced output of the main numbers such as maximum length of the UMQ, total & longest time a message stayed on the UMQ and the overall as well as the longest time processes spent searching the UMQ. Additionally, the complete data gathered may be saved for each statistic. The paper shows, that applications have varying characteristics on the usage of the UMQ and that process with rank 0 has the longest UMQ. This tool helps find scalability bottlenecks, as shown with LSMS. Furthermore applications show certain patterns which may be used as fingerprint to detect application behavior in subsequent work.

Acknowledgments. The authors would like to thank Ramanan Sankaran, Markus Eisenbach and Joshua Ladd for providing the applications, test-cases and the valuable discussions.

References

1. Bailey, D.H., et al: The NAS parallel benchmark. Technical report, NAS Applied Research Branch (1994)
2. Brightwell, R., Goudy, S., Rodrigues, A., Underwood, K.D.: Implications of application usage characteristics for collective communication offload. International Journal of High-Performance Computing and Networking – Special Issue: Design and Performance Evaluation of Group Communication in Parallel and Distributed Systems (IJHPCN) 4(3/4), 104–116 (2006)
3. Brightwell, R., Underwood, K.D.: An analysis of NIC resource usage for offloading MPI. In: Parallel and Distributed Processing Symposium, International, vol. 9, p. 183a (2004)
4. Eisenbach, M., Zhou, C.-G., Nicholson, D.M.C., Brown, G., Larkin, J.M., Schulthess, T.C.: A scalable method for *ab initio* computation of free energies in nanoscale systems. In: SC, Portland, Oregon, USA, November 14-20. ACM, New York (2009)
5. Message Passing Interface Forum. MPI: A Message-Passing Interface Standard, Version 2.2 (September 2009)
6. Gabriel, E., et al.: Open MPI: Goals, concept, and design of a next generation MPI implementation. In: Kranzlmüller, D., Kacsuk, P., Dongarra, J. (eds.) EuroPVM/MPI 2004. LNCS, vol. 3241, pp. 97–104. Springer, Heidelberg (2004)
7. Hawkes, E.R., Sankaran, R., Sutherland, J.C., Chen, J.H.: Direct numerical simulation of turbulent combustion: fundamental insights towards predictive models. Journal of Physics 16, 65–79 (2005)
8. Jones, T., et al: MPI Peruse – an MPI extension for revealing unexposed implementation information. Internet (May 2006), http://www.mpi-peruse.org
9. Keller, R., Bosilca, G., Fagg, G., Resch, M.M., Dongarra, J.J.: Implementation and usage of the PERUSE-interface in open MPI. In: Mohr, B., Träff, J.L., Worringen, J., Dongarra, J. (eds.) PVM/MPI 2006. LNCS, vol. 4192, pp. 347–355. Springer, Heidelberg (2006)
10. Keller, R., Graham, R.L.: MPI queue characteristics of large-scale applications. In: Cray User Group (May 2010) (Submitted for publication)
11. Koop, M.J., Sridhar, J.K., Panda, D.K.: TupleQ: Fully-asynchronous and zero-copy MPI over InfiniBand. In: International Parallel and Distributed Processing Symposium (IPDPS), pp. 1–8 (May 2009)
12. Lin, Z., Hahm, T.S., Lee, W.-L.W., Tang, W.M., White, R.B.: Turbulent transport reduction by zonal flows: Massively parallel simulations. Science 281, 1835–1837 (1998)
13. Wang, Y., Stocks, G.M., Shelton, W.A., Nicholson, D.M.C., Temmerman, W.M., Szotek, Z.: Order-n multiple scattering approach to electronic structure calculations. Phys. Rev. Lett. 75(11), 2867–2870 (1995)
14. Wu, X., Taylor, V.: Using processor partitioning to evaluate the performance of MPI, OpenMP and hybrid parallel applications on dual- and quad-core Cray XT4 systems. In: Cray User Group Conference, May 4–7 (2009)

Dodging the Cost of Unavoidable Memory Copies in Message Logging Protocols

George Bosilca[1], Aurelien Bouteiller[1], Thomas Herault[1,2], Pierre Lemarinier[1], and Jack J. Dongarra[1,3]

[1] University of Tennessee, TN, USA
[2] Universite Paris-Sud, INRIA, France
[3] Oak Ridge National Laboratory, TN, USA
{bosilca,bouteiller,herault,lemarinier,dongarra}@eecs.utk.edu

Abstract. With the number of computing elements spiraling to hundred of thousands in modern HPC systems, failures are common events. Few applications are nevertheless fault tolerant; most are in need for a seamless recovery framework. Among the automatic fault tolerant techniques proposed for MPI, message logging is preferable for its scalable recovery. The major challenge for message logging protocols is the performance penalty on communications during failure-free periods, mostly coming from the payload copy introduced for each message. In this paper, we investigate different approaches for logging payload and compare their impact on network performance.

1 Introduction

A general trend in High Performance Computing (HPC), observed in the last decades, is to aggregate an increasing number of computing elements [1]. This trend is likely to continue as thermic issues prevent frequency increase to progress at Moore's law rate, leaving massive parallelism, with hundred of thousands of processing units, as the only solution to feed the insatiable demand for computing power. Unfortunately, with the explosion of the number of computing elements, the hazard of failures impacting a long-living simulation becomes a major concern. Multiple solutions, integrated to middleware like MPI [2], have been proposed to allow scientific codes to survive critical failures, i.e. permanent crash of a computing node. Non-automatic fault tolerant approaches, where the middleware puts the application in charge of repairing itself, have proven to be, at the same time, very efficient in term of performance, but extremely expensive in terms of software engineering time [3]. As a consequence, only a small number of targeted applications are able to benefit from the *capability* of modern leadership computing centers; the typical workload of HPC centers suggests that most scientists still have to scale down their jobs to avoid failures, outlining the need for a more versatile approach.

Automatic fault tolerant approaches, usually based on rollback-recovery, can be grouped in two categories: coordinated or uncoordinated checkpointing mechanisms. Coordinated checkpointing relies on a synchronization of the checkpointing wave, an often blocking protocol, and a rollback of every process, even in the

R. Keller et al. (Eds.): EuroMPI 2010, LNCS 6305, pp. 189–197, 2010.

event of a single failure, which leads to a significant overhead at large scale [4]. Uncoordinated checkpointing let individual processes checkpoint at any time. As a benefit, checkpoint interval can be tailored on a per-node basis, and the recovery procedure effectively sandboxes the impact of failures to the faulty resources, with limited non disruptive actions from the neighboring processes.

However, the stronger resiliency of uncoordinated checkpointing comes at the price of more complexity, to solve the problems posed by *orphan* and *in-transit* messages. Historically, research on message logging have mostly focused on handling the costly orphan messages, by introducing different protocols (optimistic, pessimistic, causal). Recent works have nevertheless tremendously decreased the importance of the protocol choice [5], leading the once negligible overhead incurred by in-transit messages to now dominate. The technique considered as the most efficient today to replay in-transit messages is called sender-based message logging: the sender keeps a copy of every outgoing message. Although sender-based logging requires only a local copy, done in memory, and could theoretically be overlapped by actual communication over the network, it has appeared experimentally to remain a significant overhead.

The bandwidth overhead of the sender-based copy is now standing alone in the path of ubiquitous automatic and efficient fault tolerant software. In this article, we consider and compare multiple approaches to reduce or overlap this cost to a non-measurable overhead in the Open MPI implementation of message logging: Open MPI-V [6]. The rest of the paper is organized as follow: in section 2, we discuss the other approaches that have been taken, then we present the Open MPI architecture, and the different approaches to create copy of messages locally in section 3, that we compare on different experimental platforms in section 4, to conclude in section 5.

2   Related Works

Most of the existing works on message logging have focused on reducing the number of events to be logged: [7], the bottleneck of disk I/O was the main challenge in Message Logging, and the proposed solution consisted in reducing the generality of the targeted application to accept only behaviors that can be tolerated without logging messages. Other works [8,9] reduced the kind of failures that can be tolerated to increase the asynchrony of the logging requirements, thus hoping to recover the I/O time with more computation. However, these approaches still require logging of messages, and the data can be passed back to the user application only when it has been copied completely.

To the best of our knowledge, no previous work has studied how the message payload should be logged by the sender, and how this level could be optimized. Many works have recently considered the more general issue of copying memory regions in multicore systems using specific hardware [10,11], or how the memory management can play a significant role in the communication performance [12,13]. However, the interactions between simultaneously transferring the data to the Network Interface Card and obtaining an additional copy in the application space has not been addressed.

3   Strategies for Sender-Based Copies

Open MPI [14] is an open source implementation of the MPI-2 standard. It includes a generic message logging framework, called the PML V, that can be used for debugging [15] and fault tolerance [5]. One of the fault tolerant methods of the PML V is the pessimist message logging protocol. In this protocol, two mechanisms are used: event logging and sender-based message logging. The event logging mechanism defuse the threat on recovery consistency posed by orphan messages, those who carry a dependency between the non deterministic future of the recovering processes and the past of the survivors. The outcome of every non deterministic event is stored on a stable remote server; upon recovery, this list is used to force the replay to stay in a globally consistent state. In this paper, we focus our efforts on improving the second mechanism, message payload copy, thus we do not modify the event logging method. The necessity of the sender-based message logging comes from in-transit messages, *i.e.* messages sent in the past of the survivors but not yet received by the recovering processes. Because only the failed processes are restarted, messages sent in the past from the survivors can not be regenerated. The sender-based message logging approach keeps a memory copy of every outgoing message on the sender, so that any in-transit message is either regenerated (because the sender also failed and therefore is replaying the execution as well), or is readily available.

There are mostly two parameters governing the payload logging: 1) the backend storage system, and 2) the copy strategy from the user memory to the backend storage system. We have designed three backend storages: a) a file that is mapped in memory, b) heap memory as backend, allocated using memory mapping of private anonymous memory, and c) a dummy backend storage, that does not implement message logging, but provides us a mean to measure the overhead due only to the copy itself. We have also designed three copy methods: a) a pack method, that copy the message in one go into the backend space, b) a convertor method, that chops the copy of the message according to the Open MPI pipeline, and c) a thread method, that creates an independent thread responsible of doing the copies. In the following, we describe with more details these strategies.

Backend Storage
Memory Mapped file. It should be noted that there is no necessity for the log to be persistent: if a process crashes, it will restart in its own history, and recreate the messages that have been logged after the last checkpoint (still, messages preceding the last checkpoint must be saved with the checkpoint image, because they are part of the state of the process). However, a file backend is natural, because the volume of message to be logged can be significant, and this should not reduce the amount of memory available for the application. Mapping the backend file into memory is the most convenient way of accessing it.

We designed this backend file as a growing storage space, on which we open a moving window using the mmap system call. When the window is too small to accept a new message (we use windows of 256 MB, unless some message exceed the size of the window), we wait that all messages are logged (depending on the copying method, described later), make the file grow if necessary, and move the window entirely to a free area of the file.

Heap Memory. If the amount of memory available on the machine is large enough to accept at the same time the application and the copy of the messages payload (up to garbage collection time), then the payload logging can be kept in memory. This second method uses anonymous private memory allocated with the mmap system call to create such a backend for our message logging system.

Dummy Storage. In order to measure independently the overhead introduced by the copy method itself, we also designed a Dummy Storage that does not really implement message logging: after a message is logged, the pointer to store the message payload is moved back to the beginning of the same memory area, reallocated if the size of the message is larger than the largest message seen until the call. When messages are sent often, the pages related to this area will most likely be present in the TLB, and for very short messages, it is even possible that the area itself remains in the CPU cache between two emissions. Though this storage cannot be considered as a backend storage for message logging, it helps us evaluate the overheads of the copy methods themselves, without considering other parameters like TLB misses and pages fault.

Copy Method
Pack. The Pack method consists in copying the payload of the message using the memcpy libc call, from the user space to the backend storage space, when the PML V intercepts the message emission for the first time. This interception can happen just after the message has been given to the network card for short messages, or just after the first bytes of the message have been given to the network card, and the network card cannot send more without blocking for longer messages.

Conv. When converting the user data to a serialized form usable by the network cards, the Open MPI data type engine can introduce a pipeline, to send multiple messages of a predetermined maximal size on the network cards, instead of sending a very large single message. Up to four messages can be given to the network card simultaneously, which will send one after the other. The data type engine tries to keep this pipeline as filled as possible, to ensure that the network card has always something to send. Using the Conv method, PML V intercepts each of these, and introduces the message payload copy at this time. This is what the Conv (short for convertor) method does: if the pipeline is enabled, each time a chunk of data is copied from the user data to the network card, the PML V copies the same amount of bytes from the user data to the backend storage. The size of the chunks in the pipeline is a parameter of this method.

Thread. The last copying method is based on a thread. A copying thread is created during the initialization. This thread waits on a queue for copies. When this queue is not empty, the thread pops the first element of the queue, and copies the whole user memory onto the backend storage, using the memcpy libc call. When a message emission is intercepted by the PML V, if the message is short, it is copied as for the Pack method. If the message is long enough and could not be sent to the network in one go, a copy request is created and pushed at the end of the request queue. When the application returns from the MPI call, it synchronizes with the copy thread, and waits that the related messages have been entirely logged before returning from the MPI call, to ensure message integrity. To ensure a fair comparison, at constant hardware resources, this thread is pinned on the same core as the MPI process that produces the message.

4   Experimental Evaluation

The Dancer cluster is a small 8 node cluster, each node based on a Intel Q9400 2.5Ghz quad core processor, with 4GB of memory. All nodes are connected using a dual Gigabit Ethernet links, and four feature an additional Myricom MX10G. Linux 2.6.31.2 (CAoS NSA 1.29) is deployed. The software is compiled using gcc 4.4 with -O3 flags, and uses the trunk of Open MPI (release 21423) modified to include the different logging techniques presented in Section 3. For every run, we forced Open MPI to use only the high-speed Myricom network of dancer using the MCA parameters: -mca btl mx,self. All latency and bandwidth measurements were obtained using the MPI version of the NetPIPE-3.7 benchmark [16]. NetPIPE evaluates latency and bandwidth by computing three times the average value on a varying number of iterations, and taking the best value of the three evaluations.

Figure 1 presents the reference latency (Fig. 1(a)) and bandwidth (Fig. 1(b)) of Open MPI on the specified network, and of the memory bus of the machines used. These figures are presented here as a absolute reference of peak performance achievable without message logging. A first observation is that the high memory bandwidth and low latency compared to the High-Speed network card should enable a logging in memory with little performance impact for messages

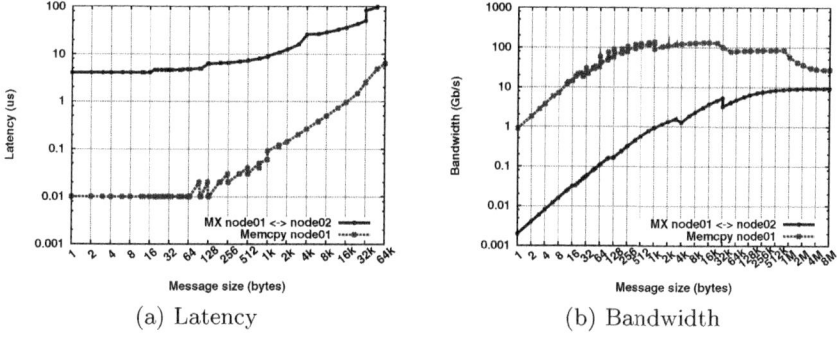

(a) Latency   (b) Bandwidth

Fig. 1. Reference MPI MX NetPIPE performance between two dancer nodes compared to memcopy

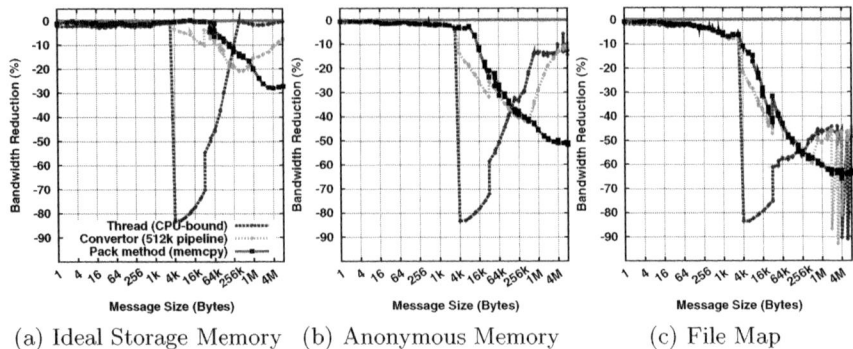

(a) Ideal Storage Memory   (b) Anonymous Memory   (c) File Map

Fig. 2. NetPIPE MX bandwidth between two dancer nodes, according to the storage method

of less than 1MB. For larger messages, the bandwidth of the memory bus will become a bottleneck for the logging, and unless the time taken to transfer the message on the network can be recovered by the logging mechanism, overheads are to be expected.

A few characteristics of the underlying network and the Open MPI implementation can moreover be observed from these two figures: one can clearly see the gaps in performance for messages of 4KB (default size of the MX frame), and 32KB (change of communication protocol from eager to rendez-vous in the Open MPI library). In the rest of the paper, all other measurement will be presented relative to the bandwidth performance of the high-speed network card, to highlight the overheads due to message logging.

Each of the first figures grouped under Figure2 consider a specific storage medium, and compare for a given medium the overheads of the different logging methods as function of the message size.

First, we consider Figure 2(a) that uses as a storage medium the "Ideal" Storage. As described in Section 3, the Ideal storage uses a single memory area to log all the messages (thus overriding existing log with new messages). The goal of this experiment is to demonstrate the overheads due to the copy itself (and when it happens) without other effects, like page faults, etc... One can see that the logging method has no significant impact up to (and excluding) messages of 4KB. At 4KB, the Thread method suffers a huge overhead that decreases the performance by 80%, while the other methods suffer a lower overhead.

A single MX frame is of 4KB (on this platform). Thus, for messages of 4KB of payload, or more, multiple MX frames are necessary to send the message (this is true for messages of 4KB of payload too, since the message header must also be sent). When the message fits in a single frame, the logging thread can be scheduled while the message circulates on the network and is handled by the receiving peer. When the message doesn't fit in a single MX frame, the Open MPI engine requires scheduling to ensure the lowest possible latency. Since both threads are bound on the same core, they compete for the core, and the relative performances decrease.

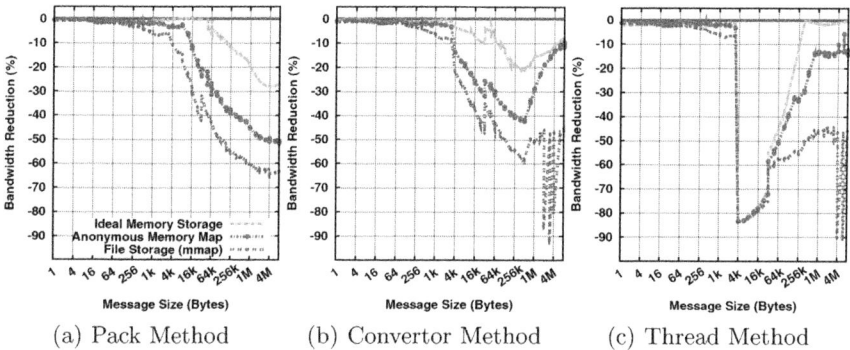

(a) Pack Method   (b) Convertor Method   (c) Thread Method

Fig. 3. NetPIPE MX bandwidth between two dancer nodes, according to the copy method

On one hand, when the number of frames needed for a single message is low, the MPI thread and the logging thread must alternate with a high frequency on the core (since the MPI call exits only when the message has been sent and logged). On the other hand, when the number of frames needed is high, the thread that is scheduled on the CPU can either log the whole message in one quantum, or use all available frames in the MX NIC to send as much data as possible in one go. Thus, when the number of frames increases, the relative overhead due to the logging thread decreases.

The pack method decreases almost linearly with the message size, since all copies are made sequentially after the send. Because the network is eventually saturated, the relative overhead reaches a plateau. The Convertor method uses a pipeline of 512KB. Thus, until messages are 512KB long, it behaves similarly as the Pack method. The difference is due to a slightly better cache re-use from the Pack method that send the message, then logs it, instead of first logging it during the pack operation, then sending it on the network. When the messages size is larger than the pipeline threshold, the Convertor method introduces some parallelism (although not as much as the Thread method), that is used to recover the communication time with logging time.

The other two figures (2(b) and 2(c)) demonstrate a similar behavior before 4K, although the overheads begin to be notifiable a little sooner for all methods when logging on a File. This is due to file system overheads (inodes and free blocks accounting), and memory management (TLB misses) when more pages are needed to log the messages. When the file system is effectively used (messages of 4MB and 8MB end up consuming all available buffer caches of the file system), a high variability in the relative overhead becomes observable (Figure 2(c)).

These phenomenon are more observable on the second group of three figures under Fig. 3. These figures consider each a specific logging method, and expose the impact of the medium on the overheads due to a logging method, as function of the message size. As can be seen, using a mmaped file as a storage space introduces the highest overhead, significantly higher than the overhead due to in-memory storage, even when the kernel buffers of the file system are large enough to hold this amount of data. This is due to accounting in the file system (free blocks lists, inodes status), forced synchronization of the journaling information, and a conservative policy for the copy of the data to the file system.

The difference of overhead between an anonymous memory map (in the heap of the process virtual memory), and the Ideal storage space is mainly due to TLB misses introducing additional page reclaims. This cost is unavoidable to effectively log the messages, but it is small for small messages, and amortized for very large messages. As a consequence, logging should happen in memory as long as the log can be kept small enough to fit there, and the system should resort to mmaped files only when necessary.

Figures 2(b) and 2(c) lead us to the conclusion that an hybrid approach, with different thresholds depending on the storage medium, and on the message size, should be taken: up to messages of 2KB, the method has little influence, however after this, the Pack or the convertor methods should be preferred up to messages of 128KB. For messages higher than 128KB, the use of an asynchronous thread, even if it must share the core with the application thread, is the preferred method of logging.

5   Conclusion

In this paper, we studied three techniques to log the payload of messages in a sender-based approach, in the Open MPI PML V framework that implement message logging fault-tolerance. Because the copying of the message payload must be achieved before the corresponding MPI emission is complete (either when the blocking send function exits, or when the corresponding wait operation exits), copying this payload is a critical efficiency bottleneck of any message logging approach.

One of the techniques proposed is to use an additional thread to process the copying asynchronously with the communication; a second uses the pipeline installed by the Open MPI communication engine to interlace transmissions towards the network, and copies in memory; the third simply copy the payload after it has been sent, and before the completion of the communication at the application level.

We also demonstrated that the medium used to store the payload has a significant impact on the performances of the payload logging process. We concluded that depending on the medium for storage, and the message size, different strategies should be chosen, advocating for a hybrid approach that will have to be tuned specifically for each hardware.

References

1. Meuer, W.H.: The top500 project: Looking back over 15 years of supercomputing experience. Informatik-Spektrum 31(3), 203–222 (2008)
2. The MPI Forum: MPI: a message passing interface. In: Supercomputing 1993: Proceedings of the 1993 ACM/IEEE conference on Supercomputing, pp. 878–883. ACM Press, New York (1993)

 3. Fagg, G.E., Gabriel, E., Bosilca, G., Angskun, T., Chen, Z., Pjesivac-Grbovic, J., London, K., Dongarra, J.J.: Extending the MPI specification for process fault tolerance on high performance computing systems. In: Proceedings of the International Supercomputer Conference (ICS) 2004, Primeur (2004)
 4. Lemarinier, P., Bouteiller, A., Herault, T., Krawezik, G., Cappello, F.: Improved message logging versus improved coordinated checkpointing for fault tolerant MPI. In: IEEE International Conference on Cluster Computing (Cluster 2004). IEEE CS Press, Los Alamitos (2004)
 5. Bouteiller, A., Ropars, T., Bosilca, G., Morin, C., Dongarra, J.: Reasons to be pessimist or optimist for failure recovery in high performance clusters. In: IEEE (ed.): Proceedings of the 2009 IEEE Cluster Conference, New Orleans, Louisiana, USA (2009)
 6. Bouteiller, A., Bosilca, G., Dongarra, J.: Redesigning the message logging model for high performance. In: Proceedings of the International Supercomputer Conference (ISC 2008), Dresden, Germany. Wiley, Chichester (2008) (to appear)
 7. Strom, R.E., Bacon, D.F., Yemini, S.: Volatile logging in n-fault-tolerant distributed systems. In: Society, I.C. (ed.) Proceedings of the Eighteenth International Symposium on Fault Tolerant Computing (1988)
 8. Strom, R.E., Yemini, S.: Optimistic recovery: an asynchronous approah to fault-tolerance in distributed systems. In: Proceedings of the 14th International Symposium on Fault-Tolerant Computing. IEEE Computer Society Press, Los Alamitos (1984)
 9. Manivannan, D., Singhal, M.: A low-overhead recovery technique using quasi-synchronous checkpointing. In: International Conference on Distributed Computing Systems, p. 100 (1996)
10. Vaidyanathan, K., Chai, L., Huang, W., Panda, D.K.: Efficient asynchronous memory copy operations on multi-core systems and i/oat. In: CLUSTER 2007: Proceedings of the 2007 IEEE International Conference on Cluster Computing, Washington, DC, USA, pp. 159–168. IEEE Computer Society Press, Los Alamitos (2007)
11. Goglin, B.: Improving message passing over ethernet with i/oat copy offload in open-mx. In: Proceedings of the 2008 IEEE International Conference on Cluster Computing, pp. 223–231. IEEE, Los Alamitos (2008)
12. Stricker, T., Gross, T.: Optimizing memory system performance for communication in parallel computers. In: ISCA 1995: Proceedings of the 22nd annual international symposium on Computer architecture, pp. 308–319. ACM, New York (1995)
13. Geoffray, P.: Opiom: Off-processor i/o with myrinet. Future Generation Comp. Syst. 18(4), 491–499 (2002)
14. Gabriel, E., Fagg, G.E., Bosilca, G., Angskun, T., Dongarra, J.J., Squyres, J.M., Sahay, V., Kambadur, P., Barrett, B., Lumsdaine, A., Castain, R.H., Daniel, D.J., Graham, R.L., Woodall, T.S.: Open MPI: Goals, concept, and design of a next generation MPI implementation. In: Proceedings of 11th uropean PVM/MPI Users' Group Meeting, Budapest, Hungary, pp. 97–104 (2004)
15. Bouteiller, A., Bosilca, G., Dongarra, J.: Retrospect: Deterministic replay of mpi applications for interactive distributed debugging. In: Proccedings of the 14th European PVM/MPI User's Group Meeting (EuroPVM/MPI), pp. 297–306 (2007)
16. Snell, Q.O., Mikler, A.R., Gustafson, J.L.: Netpipe: A network protocol independent performance evaluator. In: IASTED International Conference on Intelligent Information Management and Systems (1996)

Communication Target Selection for Replicated MPI Processes

Rakhi Anand, Edgar Gabriel, and Jaspal Subhlok

Department of Computer Science, University of Houston
{rakhi,gabriel,jaspal}@cs.uh.edu

Abstract. VolpexMPI is an MPI library designed for volunteer computing environments. In order to cope with the fundamental unreliability of these environments, VolpexMPI deploys two or more replicas of each MPI process. A receiver-driven communication scheme is employed to eliminate redundant message exchanges and sender based logging is employed to ensure seamless application progress with varying processor execution speeds and routine failures. In this model, to execute a receive operation, a decision has to be made as to which of the sending process replicas should be contacted first. Contacting the fastest replica appears to be the optimal local decision, but it can be globally non-optimal as it may slow-down the fastest replica. Further, identifying the fastest replica during execution is a challenge in itself. This paper evaluates various target selection algorithms to manage these trade-offs with the objective of minimizing the overall execution time. The algorithms are evaluated for the NAS Parallel Benchmarks utilizing heterogeneous network configurations, heterogeneous processor configurations and a combination of both.

1 Introduction

Idle desktops have been successfully used to run sequential and master-slave task parallel codes, most notably under Condor [1] and BOINC [2]. The distributed, heterogeneous and unreliable nature of these volunteer computing systems make the execution of parallel applications highly challenging. The nodes have varying compute, communication, and storage capacity and their availability can change frequently and without warning. Further, the nodes are connected with a shared network where available latency and available bandwidth can vary. Because of these properties, we refer to such nodes as *volatile* and parallel computing on volatile nodes is challenging.

The most popular approach for dealing with unreliable execution environments is to deploy a checkpoint-restart based mechanisms, as has been done (among others) by MPICH-V [3], OpenMPI [4] or RADIC-MPI [5]. A smaller number of projects are using replication based techniques, such as P2P-MPI [6] or rMPI [7]. We have recently introduced VolpexMPI [8], an MPI library which tackles the challenges mentioned above by deploying two or more copies of each MPI process, and utilizes a receiver based communication model with a sender side message logging. The design of VolpexMPI avoids an exponential increase

R. Keller et al. (Eds.): EuroMPI 2010, LNCS 6305, pp. 198–207, 2010.

in the number of messages with increasing degree of replication by ensuring that exactly one physical message is transmitted to each receiving process for each logical receive operation. On each process, a priority list of the available replicas corresponding to every other communicating process is maintained. Clearly, the ordering of processes in this list will have a fundamental impact on the overall performance of the application. We refer to this problem throughout this paper as the *target selection* problem.

This paper evaluates five different approaches for target selection: a process team based approach as described in [8], a network performance based approach, an algorithm based on the virtual time stamps of the messages, a timeout based algorithm, and a hybrid approach deploying both network parameters as well as virtual time stamps. The algorithms are evaluated for the NAS Parallel Benchmarks for heterogeneous network configurations, heterogeneous processor configurations and a combination of both.

The remainder of the paper is organized as follows: section 2 gives a brief overview of VolpexMPI. Section 3 describes the five target-selection algorithms in detail. The algorithms are than evaluated for various hardware configurations in section 4. Finally, section 5 summarizes the results and presents the ongoing work.

2   VolpexMPI

VolpexMPI is an MPI library designed to deal with the heterogeneity, unreliability and the distributed nature of volunteer computing environments. The key features of VolpexMPI are:

1. *Controlled redundancy:* A process can be initiated as two (or more) replicas. The fundamental goal for the execution model is to have the application progress at the speed of the fastest replica of each process.
2. *Receiver based direct communication:* The communication framework supports direct node to node communication with a *pull* model: the sending processes buffer data objects locally and receiving processes contact one of the replicas of the sending process to get the data object.
3. *Distributed sender based logging:* Messages sent are implicitly logged at the sender and are available for delivery to process instances that are lagging due to slow execution or recreation from a checkpoint.

The receiver based communication scheme along with the distributed sender based message logging allows a receiver process to contact any of the existing replicas of the sender MPI rank to request a message. Since different replicas can be in different execution states, a message matching scheme is employed to identify which message is being requested by a receiver. For example, it is not sufficient for process zero to request a message with a particular tag on communicator MPI_COMM_WORLD from process one, if the application would send, over its lifetime, multiple messages with this particular signature. To distinguish between the different incarnations of each message, VolpexMPI deploys a virtual timestamp to each message by counting the number of messages exchanged between pairs of processes. This virtual timestamp along with the tuple [communicator id, message tag, sender rank, receiver rank] uniquely identifies each message for the entire application execution. These timestamps are also used to monitor the progress of individual process replicas for resource management.

The overall sequence of operations for a point-to-point communication is as follows: Upon calling MPI_Send, the sending process only buffers the content of a message locally, along with the message envelope, which includes the virtual timestamp, in addition to the usual elements used for message matching. In case any of the replicas of the receiver process requests this particular message, the sender process will reply with the corresponding data.

The receiving process polls a potential sender and waits for the data item. Note, that there are two possibilities that have to be handled if a sender process does not currently have a matching message. First, the data might not be available yet, because the sender process lags the receiving process. This scenario is recognized by the sender by comparing the virtual time stamp of the request message with the most recent message having the same tuple [tag, communicator, sender rank, receiver rank]. In this case, the library will simply postpone the reply until the request message is available.

The second scenario is that the message is not available anymore, e.g., because the circular buffer used for sender side message logging has already overwritten the corresponding data item. This scenario only occurs, if the replica of the receiver process is lagging significantly behind the sender process. In this case the sender process will not be able to comply with the request. As of now we are not handling this situation. Thus, the lagging replica keeps on waiting for a particular message. However, there are various ways in which such a scenario can be handled, such as a time-out based mechanism, or an explicit reply indicating the inability to comply with the request. The long-term goal is to coordinate the size of the circular buffer with checkpoints of individual processes, which will allow guaranteed restarts with a bounded buffer size.

As different replicas of an MPI rank can be at significantly different stages of the execution, each process has to be able to prioritize the available replicas for each MPI rank. This is discussed in the next section.

3   Target Selection Algorithms

In order to meet the goal that the progress of an application correspond to the fastest replica for each process, the library has to provide an algorithm which allows a process to generate an order in which to contact the sender replicas. This is the main functionality provided by the target-selection module. The algorithm utilized by the target-selection module has to handle two seemingly contradicting goals: on one hand, it would be beneficial to contact the "fastest" replica from the performance perspective. On the other hand, the library does not want to slow-down the fastest replica by making it handle significantly larger number of messages, especially when a message is available from another replica. The specific goal, therefore, is to determine a replica which is "close" to the execution state of the receiver process.

The team based approach, the original algorithm implemented in VolpexMPI [8], divides the processes into teams at startup, with one replica of each process in every team. Processes communicate within their own teams and contact a process from another team only in case of failure. Since processes are communicating exclusively within a team, fast processes are bound to communicate with slow processes, causing the application to execute at the speed of slow processes. A slow process can be defined as a process running on a slow internet connection, having a slow processor speed, or busy in some other work. Thus, in order to advance application at the speed of fast processes, there should be a mechanism where each process can select their communicating partner. However, team based approach does not provide any such mechanism. In order to solve this problem, different algorithms based on network performance, timeout, virtual timeout, and hybrid approach were developed and implemented.

Network Performance Based Target Selection. Two key elements of network performance are latency and bandwidth. These parameters are used to establish an order among the replicas of an MPI rank on each process. For this, each process will use all of the known replicas of an MPI processes in a round-robin fashion for regularly occurring communication and time the operations. After receiving a fixed number of messages from each replica of the same rank, the receiver process calculates the latency and bandwidth corresponding to each sender process. Priority for the future is based on these parameters.

Note, that the algorithm has the ability to restart the evaluation process after a certain period of time, e.g. a certain number of messages to that process, or in case the estimated bandwidth value to the currently used replica changes significantly compared to the original evaluation. An important disadvantage of this approach is that, if one of the replicas used has an extremely slow network connectivity, the evaluation step will bring the fast running processes also to the speed of the slowest one for the duration of the evaluation.

Timeout Based Target Selection. In this approach, each process waits for the reply from a replica for a limited time. If the requested data is not available within the predefined time frame, the process switches to another replica and requests the same message. A technical challenge is how to deal with the data of the original abandoned replicas, since the user level message buffer should not be overwritten by that process anymore. Thus, the library has to effectively cancel the original request before moving to the next replica. If the data from the first (slow) replica comes in, the data will be placed into the unexpected message queue of the library instead of the user buffer, and can safely be purged from there. For this, VolpexMPI maintains a list of items that need to be removed from the unexpected message queue(s).

One drawback for this algorithm is that it is difficult to define a reasonable value for the timeout. If the timeout value is too small, processes change their target too frequently, whereas if the timeout value is large, processes will continue communication with slow processes for too long. Therefore, setting the correct threshold value plays a very important role. Another disadvantage is that all slow processes will try to contact fast processes making them handle more requests which may slow down the fast processes.

Virtual Time Stamp Based Target Selection. This algorithm employs the virtual timestamps of messages to compare the state of sender and receiver processes. As explained in section 2, a virtual timestamp is the message number for communication between a pair of processes. Each sender process attaches its most recent timestamp for the same message type, i.e. message with same tuple [communicator id, message tag, sender rank, receiver rank] when replying to a receiver request. The receiver compares the timestamps from different senders to determine the execution state of the processes.

The overall approach starts by using the teams as created by the team-based algorithm. If the difference between the timestamps of two processes exceeds a certain threshold value, a process can decide to switch to another replica for that particular rank. Ultimately, it will choose the replica closest to its own execution state. Similarly to the timeout based approach, the major difficulty in this algorithm is to decide on good values for the threshold, i.e. when to switch to next replica. This threshold value must not be too small to avoid frequent change in targets, nor too large to avoid long detection time of slow targets.

Hybrid Target Selection. This algorithm combines the network based algorithm and virtual timestamp based algorithm. Each process first sends a message to each replica and decides the preferred target based on the best network parameters. In order to determine the best target, pairwise communication is initiated during the initialization of the application. As a result each process is communicating with fastest communicating replica. This might make the fast running processes to slow down, since it has to serve potentially multiple instances of each MPI process. In a second step, the virtual timestamp based approach is used to separate slow replicas from faster ones, i.e., if a process is lagging far behind the sender process it changes its target to the slower one. The result is a reduced communication volume to faster processes. In the long term we envision the first step to be replaced by a more sophisticated process placement strategy of the `mpirun`, which can utilize proximity information obtained from the BOINC server and from internet metrics as presented in [9].

4   Performance Evaluation

This section describes the experiments with VolpexMPI library and the results obtained.The tests presented here have been executed on a regular dedicated cluster, in order to achieve reproducible results and understand characteristics of our algorithm. The cluster is composed of 29 compute nodes, 24 of them having a 2.2GHz dual-core AMD Opteron processors, and 5 nodes having two 2.2GHz quad-core AMD Opteron processors. Each node has 1GB main memory per core and network connected by 4xInfiniBand as well as a 48 port Linksys GE switch. For the subsequent analysis, only the Gigabit Ethernet interconnect has been utilized.

The NAS Parallel Benchmarks(NPBs) are executed for 8 process and the problem size B. For each target selection algorithm we record and compare the results obtained to the numbers obtained with the team based approach, the original target selection algorithm presented in [8]. Tests have been executed using two replicas per MPI process, henceforth denoted as double-redundancy, and three replicas per MPI process, also called triple redundancy runs. Note, that for triple redundancy runs only CG, EP and IS benchmarks have been executed. BT and SP would require 9 processes for that particular, which due to restrictions on the network configuration could not be executed. The triple redundancy tests of FT failed due to the memory requirement for each process.

4.1 Results for Heterogeneous Network Configurations

In this set of experiments we explore, the cluster switch has been configured such that the link bandwidth to eight nodes has been decreased from Gigabit Ethernet(1Gb/s) to Fast Ethernet (100Mb/s), creating a heterogeneous network configuration. For the double redundancy tests, processes have been distributed such that no two replicas of the same MPI rank are on same network, i.e. if process 0,A is on Fast Ethernet then process 0,B is on Gigabit Ethernet. Furthermore, all teams contain processes that use the Gigabit Ethernet and the Fast Ethernet network. For the triple redundancy tests, one team of processes is being executed on a single 8-core node, creating a third hierarchy level with respect to the quality of communication. Similarly to the double redundancy tests, teams have been initiated such that each process has one replica on each of the three hierarchy levels, and all teams contain some processes from all three hierarchy levels. The challenge for the algorithms are to modify the original teams provided at startup such that (ideally) one set of replicas only contain processes on fast nodes, and the other one only consists of slow nodes.

Fig. 1 shows the results obtained with all implemented algorithms for double(x2)(left) and triple(x3)(right) redundancy. The results indicate, that the execution time for the network based algorithm is almost equal to the execution time of the team based algorithm. The main reason for this somewhat surprising behavior is that in some instances the algorithm identifies the wrong replicas as

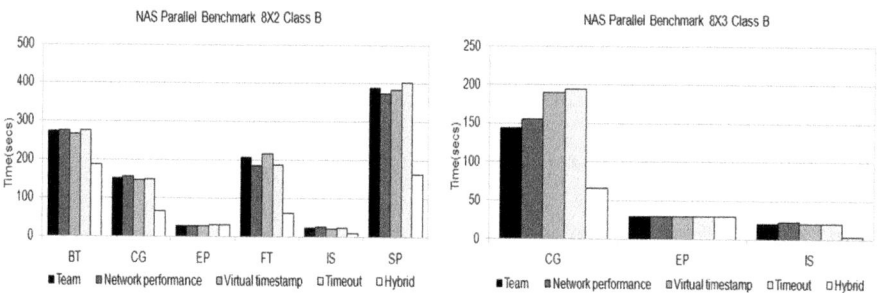

Fig. 1. Comparison of all target selection algorithms for heterogeneous networks

being the 'fastest' target. This happens if a fast process could not send response to the asking process because it is waiting itself for the result from a slow process, and which will falsify the measurements. Consider a simple example, where process 0,A and 1,B are running on fast nodes and processes 0,B and 1,A are running on slow nodes. In a situation when process 0,A sends a request for a message to 1,B, and 1,B is in turn waiting for a message from 0,B, its reply to process 1,A might be delayed. Note, that for artificial test case the algorithm worked as expected. However, for more realistic applications/benchmarks such as the ones used here, the MPI processes are too tightly coupled together in terms of send and receive operations, and the network performance based approach does not provide good performance results.

Next, we document the results for timeout based algorithm. The results for this algorithm are almost similar to the results obtained from network based algorithm for very similar reasons. As explained in section 3, the threshold value plays an important role in the overall performance of the algorithm. For the results presented here, we used a threshold value of 0.5 seconds for the double redundancy tests and 1.0 seconds for triple redundancy tests. We did perform experiments with various other threshold values, with lower values resulting in frequent switching of targets and higher threshold values preventing any process from switching to another replica.

Similarly to the other two algorithms, the virtual timestamp based algorithm does not show for the double and triple redundancy runs any difference to the team based approach, which is due to the fact that the synchronized communication patterns used in most of the NAS parallel benchmarks does not allow processes to 'drift apart'. Thus, the speed of fast processes is reduced due to the communication taking place with slow process and the application advances with the speed of slow processes. Note furthermore, that this algorithm in theory is designed to handle varying processor speeds and not necessarily varying network parameters for homogeneous processor configuration, and thus the result is not entirely unexpected.

Finally, the results obtained by using the hybrid approach in which processes are first grouped according to the network parameters and lagging processes are identified in a second step, is showing the best performance overall. The overall execution time is matching the performance that the same application would achieve when using Gigabit Ethernet network connections only. Thus, this algorithm is a significant improvement over the team-based approach utilized in [8]. Analyzing the results of this algorithm reveal, that the main difference comes from the fact, that the network parameters are not determined by timing the regularly occurring MPI messages of the application, but by introducing a pair-wise communication step that is executed in MPI_Init.

4.2   Results for Heterogeneous Processor Configurations

In order to analyze the behavior of the algorithms for systems comprised of heterogeneous processor, the frequency of 9 nodes has been reduced to 1.1 GHz, while all other nodes are running at full frequency, i.e 2.2 GHz. All nodes are

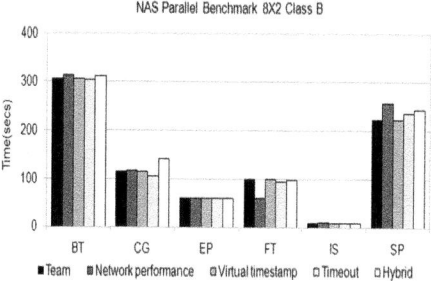

Fig. 2. Comparison of all algorithms for heterogeneous processor configurations

connected through Gigabit Ethernet, in order to eliminate network influences. Again we mixed the nodes running on slow frequency and nodes running on full frequency and compared the performances of different algorithms implemented.

Figure 2 shows the results obtained for this setting for all algorithms. The network based algorithm, timeout based algorithm and virtual timestamp based algorithm using double redundancy(left) runs are similar as discussed in the previous paragraph for similar reasons. In contrary to the previous section, the results obtained with the hybrid algorithm does not show any performance gain over other algorithms. This is due to the fact, that the network itself does not expose any hierarchies in this scenario, and therefore the pre-sorting of replicas and teams does not occur. In fact, nodes are grouped together without any proper order i.e each team consist of few processes running on slow nodes and other processes running on fast nodes.

4.3 Results for Combinations of Heterogeneous Network and Processor Configurations

For the last set of experiments, the network connection to 8 of the 29 nodes has been once again reduced to Fast Ethernet. Furthermore, the frequency of the same 8 nodes has been reduced to 1.1 GHz while all other nodes are running at full speed. Similarly to the previous tests, teams have been initiated such that each process has one replica on a slow and on a fast node, and all teams contain some processes from all three hierarchy levels. For the triple redundancy runs, a third configuration consisting of 8 processes running at full frequency, but located on a single 8-core processor are interleaved with the other two sets.

Figure 3 shows the results obtained for the double and triple redundancy runs. The results are similar to the results obtained in the previous sections, with the network based algorithm, timeout based algorithm and virtual timestamp based algorithm not being able to correctly identify the optimal configuration.

The hybrid approach however gives the performance numbers similar to the results as if all processes are running on fast nodes. Also, for triple redundancy runs where all processes from each initial team are mixed, the hybrid algorithm is clearly able to group processes as if all three teams are executing on separate

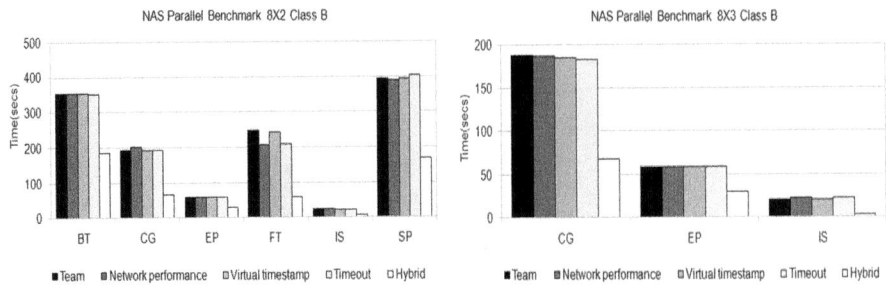

Fig. 3. Comparison of all algorithms for combination of heterogeneous network and processor configurations

Table 1. Performance Results for redundancy 3 runs on different networks

	Volpex Team A	Volpex Team B	Volpex Team C
CG	87.93	67.69	184.29
IS	3.56	8.15	23.82
EP	29.69	29.72	117.83

networks: Team A on shared memory, Team B on Gigabit Ethernet, and Team C on Fast Ethernet. This fact is highlighted by the results shown in table 1, which details the execution time observed by each individual team as identified by the hybrid target selection algorithm.

5 Summary

In this paper we presented and evaluated five different approaches for the target selection problem. The algorithms have been evaluated for the NAS Parallel Benchmarks for heterogeneous network configurations, heterogeneous processor configurations and a combination of both. The analysis reveals that the hybrid target selection algorithm shows a significant performance benefit over other algorithms for most (common) scenarios.

The ongoing work in this project includes a full evaluation of the new target selection method in a volunteer computing environment with a wider range of applications. On the algorithmic level we envision the initial placement to be driven by a more sophisticated process that can utilize proximity information obtained from the BOINC server and from internet metrics as presented in [9].

Acknowledgments. Partial support for this work was provided by the National Science Foundation's Computer Systems Research program under Award No. CNS-0834750. Any opinions, findings, and conclusions or recommendations expressed in this material are those of the authors and do not necessarily reflect the views of the National Science Foundation.

References

1. Thain, D., Tannenbaum, T., Livny, M.: Distributed computing in practice: the condor experience. Concurrency - Practice and Experience 17(2-4), 323–356 (2005)
2. Anderson, D.: BOINC: A system for public-resource computing and storage. In: Fifth IEEE/ACM International Workshop on Grid Computing (November 2004)
3. Bouteiller, A., Cappello, F., Herault, T., Krawezik, G., Lemarinier, P., Magniette, F.: MPICH-V2: a fault tolerant MPI for volatile nodes based on pessimistic sender based message logging. In: SC 2003: Proceedings of the 2003 ACM/IEEE conference on Supercomputing, Washington, DC, USA, p. 25. IEEE Computer Society, Los Alamitos (2003)
4. Hursey, J., Squyres, J.M., Mattox, T.I., Lumsdaine, A.: The design and implementation of checkpoint/restart process fault tolerance for Open MPI. In: Proceedings of the 21st IEEE International Parallel and Distributed Processing Symposium (IPDPS). IEEE Computer Society Press, Los Alamitos (March 2007)
5. Duarte, A., Rexachs, D., Luque, E.: An Intelligent Management of Fault Tolerance in Cluster Using RADICMPI. In: Mohr, B., Träff, J.L., Worringen, J., Dongarra, J. (eds.) PVM/MPI 2006. LNCS, vol. 4192, pp. 150–157. Springer, Heidelberg (2006)
6. Genaud, S., Rattanapoka, C.: Large-scale experiment of co-allocation strategies for peer-to-peer supercomputing in P2P-MPI. In: IEEE International Symposium on Parallel and Distributed Processing, IPDPS 2008, pp. 1–8 (2008)
7. Ferreira, K., Riesen, R., Oldfield, R., Stearly, J., Laros, J., Redretti, K., Kordenbrock, T., Brightwell, R.: Increasing fault resiliency in a message-passing environment. Technical report, Sandia National Laboratories (2009)
8. LeBlanc, T., Anand, R., Gabriel, E., Subhlok, J.: VolpexMPI: an MPI Library for Execution of Parallel Applications on Volatile Nodes. In: Ropo, M., Westerholm, J., Dongarra, J. (eds.) Recent Advances in Parallel Virtual Machine and Message Passing Interface. LNCS, vol. 5759, pp. 124–133. Springer, Heidelberg (2009)
9. Xu, Q., Subhlok, J.: Automatic clustering of grid nodes. In: Proceedings of the 6th IEEE/ACM Workshop on Grid Computing, Seattle, WA (November 2005)

Transparent Redundant Computing with MPI

Ron Brightwell, Kurt Ferreira, and Rolf Riesen

Sandia National Laboratories*
Albuquerque, NM USA
{rbbrigh,kbferre,rolf}@sandia.gov

Abstract. Extreme-scale parallel systems will require alternative methods for applications to maintain current levels of uninterrupted execution. Redundant computation is one approach to consider, if the benefits of increased resiliency outweigh the cost of consuming additional resources. We describe a transparent redundancy approach for MPI applications and detail two different implementations that provide the ability to tolerate a range of failure scenarios, including loss of application processes and connectivity. We compare these two approaches and show performance results from micro-benchmarks that bound worst-case message passing performance degradation. We propose several enhancements that could lower the overhead of providing resiliency through redundancy.

Keywords: Fault tolerance, Redundant computing, Profiling interface.

1 Introduction

It is widely accepted that future extreme-scale parallel computing systems will require alternative methods to enable applications to maintain current levels of uninterrupted execution. As the component count of future multi-petaflops systems continues to grow, the likelihood of a failure impacting an application grows as well. Current methods of providing resiliency for applications, such as checkpoint/restart, will become ineffective, largely due to the overhead required to checkpoint, restart, and recover lost work. As such, the research community is pursuing several alternative approaches aimed at providing the ability for an application to survive in the face of failures and to continue to make efficient computational progress.

One of the fundamental approaches for masking errors and providing fault tolerance is redundancy. Replicating state and repeating operations occurs in many parts of modern computing systems; e.g., RAID has become commonplace. In order for redundancy to be viable for parallel computing, the potential performance degradation has to be offset by significant benefits. An important consideration for existing parallel computing systems and applications is the

---

* Sandia is a multiprogram laboratory operated by Sandia Corporation, a Lockheed Martin Company, for the United States Department of Energy's National Nuclear Security Administration under contract DE-AC04-94AL85000.

R. Keller et al. (Eds.): EuroMPI 2010, LNCS 6305, pp. 208–218, 2010.

amount of invasiveness that will be required to provide fault tolerance. Incremental approaches that minimize modifications to applications, system software, and hardware are more likely to be adopted.

We are exploring approaches to providing redundancy for MPI applications. We are seeking to answer several important research questions: a) Can we employ redundant computing with MPI transparently? b) What missing functionality, if any, is needed? c) What is the worst-case overhead? d) Are there any possible software or hardware enhancements that could reduce this overhead?

In this paper, we present two approaches for providing transparent redundancy for MPI applications. Both of these approaches double the number of processes in the application but use different schemes for recognizing failed processes and lost messages.

2 Implementation

We implemented redundant computing as a library that resides between an application and an MPI implementation. The rMPI library is described in detail in [1]. Here we provide only a brief description and highlight the parts which are relevant for the remainder of this paper.

2.1 Design Choices

Future, large-scale machines where redundant computing may be of advantage [2] will have many nodes and will run MPI between these nodes to achieve the desired performance and scalability. Therefore, we implemented rMPI using the profiling interface of MPI. This provides us with portability across MPI implementations.

The second reason for implementing rMPI at the profiling layer is that we wanted to have a mechanism that is transparent to the application. Other than at job submission time when a user requests additional nodes for redundant computing, the application is not aware of the mechanism. It only sees, and interacts with, the active ranks and is unaware of the additional ranks and communication behind the scenes.

Using rMPI, we start an application on $n \ldots 2n$ nodes. During MPI_Init() we set up a new communicator for the first n active nodes and substitute that communicator whenever the application uses MPI_COMM_WORLD. Any nodes beyond n become redundant nodes for active nodes in a one-to-one fashion. Each redundant node performs the exact same computation as its active partner. The rMPI library ensures that it sees the same MPI behavior as the active node. That means if the active node is rank x, then the redundant node will also be rank x. The rMPI library performs the necessary translations to the actual ranks used by the underlying MPI library. Both nodes would send to rank y, even though there might be actually two nodes that have been assigned rank y.

Maintaining consistency for receives using MPI_ANY_SOURCE or MPI_ANY_TAG requires a consistency protocol between active and redundant nodes. For example, MPI guarantees message order between node pairs, but rMPI must make sure that the message order seen on an active node is the same on its redundant node. Otherwise, computation on these two nodes could diverge.

2.2   Mirror Protocol

We started implementing rMPI with a straightforward protocol called *mirror*. As the name implies, it duplicates every message an application sends by transmitting it first to the original destination and then one more time to the destination's redundant partner, if it exists.

Each receiver posts two receives into the same buffer for every application receive, if the sending node has a redundant partner that will also send. When nodes fail, rMPI is notified and stops sending to disabled nodes or posting receives for messages that will no longer be sent.

Figure 1 illustrates the process. As long as either the active node or its redundant partner are still alive, the application can continue. Only when both nodes in a bundle die, or a node without a redundant partner dies, will the application be interrupted and must restart.

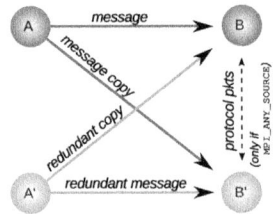

Mirroring message order on two independent nodes in case of a MPI_ANY_- SOURCE receive requires that only one node post the receive, and informs the second node of the actual receive order so it can post specific tag- and source-field receives to duplicate that order. If any MPI_ANY_- SOURCE receives are pending, the second node must queue all subsequent receives, without letting the MPI implementation see them, until all current MPI_ANY_SOURCE receives have been satisfied.

Fig. 1. In the mirror protocol each sender transmits the user messages twice and additional consistency protocol exchanges are needed in the case of MPI_ANY_SOURCE

2.3   Parallel Protocol

The mirror protocol consumes a lot of bandwidth when an application sends many large messages. To reduce this overhead we designed a second protocol named *parallel*. It is illustrated in Figure 2. Other than short protocol messages, rMPI only sends the original application messages between nodes. However, a larger number of protocol messages are now needed because the sender and its redundant partner must ensure that each of the destinations receive one copy of the message.

If one of the sender fails, the other must take over and send the additional copy. The parallel protocol somewhat resembles a transaction protocol where the two sending partners must ensure that both receivers get exactly one copy each of the application message. While this works well for large application messages where the overhead of a few additional short messages makes little difference, it is a problem for applications which send a lot of short messages. In that case the message rate that can be achieved by the application is reduced.

2.4   Issues

Initially we did not know whether transparent redundant computing could be
achieved at the MPI level. We have
shown [1] that it is possible, with relatively minor demands on the RAS system. Applications experience a performance impact that is in general less than
10%. Of course, micro-benchmarks clearly
show the overhead present in the two protocols we have described.

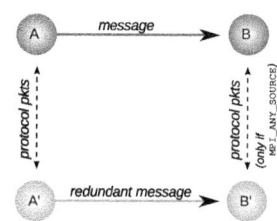

While moving lower than MPI in the
networking stack is not necessary, doing
so may have advantages. The purpose of
this paper is to further narrow the issues

Fig. 2. Message flow in the parallel protocol

with our current prototype of rMPI and propose new solutions. To start, we list
some issues that we have identified:
1. No redundant I/O. MPI-I/O and standard I/O are not currently handled by rMPI.
2. Missing integration with a RAS system. There is no standard way of doing this, but rMPI needs to know which nodes are alive, and the MPI implementation needs to survive the disappearance of individual nodes.
3. rMPI is an almost full re-implementation of the underlying MPI library.
4. Collective operations are reduced to point-to-point transmissions eliminating many of the optimization efforts performed by the underlying MPI library.
5. Mirroring message order on independent nodes for MPI_ANY_SOURCE receives causes consistency protocol overhead.
6. Delayed posting of MPI_ANY_SOURCE receives can increase the number of unexpected messages.
7. The mirror protocol consumes twice the bandwidth that the application needs and increases latency for small messages.
8. The parallel protocol is more frugal in its bandwidth consumption, but limits message rate.

In this paper we want to address items 3 through 8 and propose some ideas that could improve the performance of rMPI and limit some of its other shortcomings. In particular, we are interested in exploiting an intelligent network interface controller (NIC) and router designs to off-load some rMPI functionality. Furthermore, it would be interesting to design a solution that combines intra-node communication among the cores of a node with the necessary inter-node communication to reach redundant nodes which should be, for reliability purposes, physically as far away as possible inside the machine. The following sections describe our measurements and solutions.

3 Results

In this section, we present the performance impact of our two protocols using a latency, bandwidth, and a message rate microbenchmark. Latency and bandwidth tests are from the OSU MPI benchmark suite(OMB) [3] while the message rate test is from the Sandia MPI microbenchmark suite. Due to the protocols special handling of MPI_ANY_SOURCE, we created another microbenchmark similar to the OMB latency test which uses wild-card receives.

We conducted our tests on the Cray Red Storm system at Sandia National Laboratories. Each data points corresponds to the mean of five runs with error bars shown. In each of the following plots *native* refers to the performance of the benchmark without the *r*MPI library. *Base* for each of the protocols refers to performance with the *r*MPI library linked in but no redundant nodes used. The *parallel* and *mirror* lines are the performance of the application with a full set of replica nodes. Dashed lines show the overhead of keeping these replicas consistent.

Figures 3 and 4 illustrate the performance impact of the protocols on both bandwidth and message rate. Bandwidth in Figure 3 behaves as expected from the protocol descriptions in the previous section. The mirror protocol achieves about half of the observed bandwidth of native and the parallel protocol reaches nearly native bandwidth for large messages but for smaller messages the increased protocol message traffic hinders achievable bandwidth. Similarly, Figure 4 shows that for smaller messages mirror is able to achieve a higher message rate than parallel (with mirror's rate around half of that of native), but as message size increases parallel's rate approaches to within 10% of native.

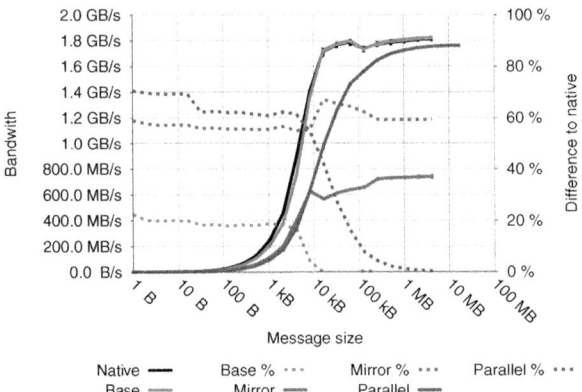

Fig. 3. Bandwidth measurements for the two protocols compared to native and baseline. Native is the benchmark without *r*MPI, base has *r*MPI linked in but does not use redundant nodes.

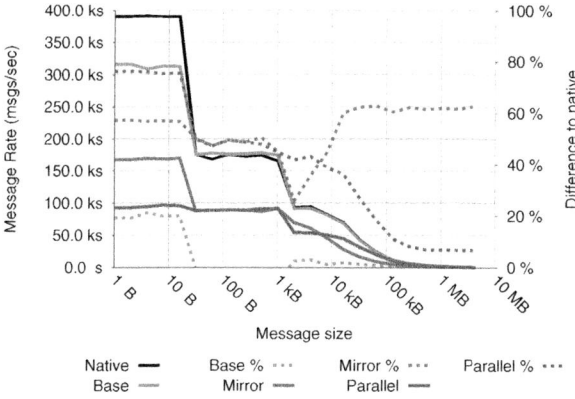

Fig. 4. Message rate measurements

We examine the protocols' impact on application latency next. Figure 3 shows the results of the latency microbenchmark without MPI_ANY_SOURCE receives, and Figure 3 repeats the experiment with MPI_ANY_SOURCE receives. Again, latency for parallel is significantly lower than mirror. This is due to the fact that mirror must either wait for both messages to arrive or receive one of the messages and cancel the other before proceeding. Each of these operations require much more time than the one receive that the parallel protocol must wait for. Note in Figure 3 that this difference in performance between the two protocols is smaller when an MPI_ANY_SOURCE receive is posted. This is because the overhead is dominated by the consistency protocol to enforce message order on the redundant nodes.

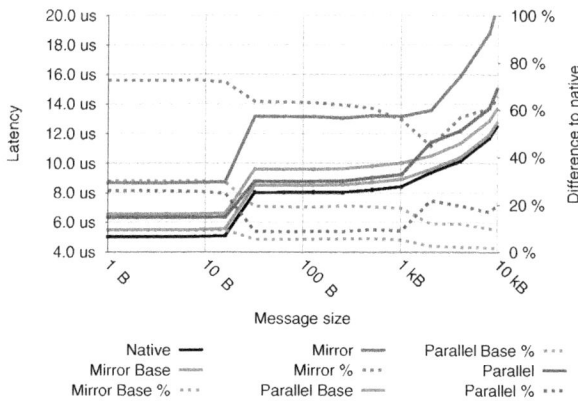

Fig. 5. Latency without MPI_ANY_SOURCE for the two protocols

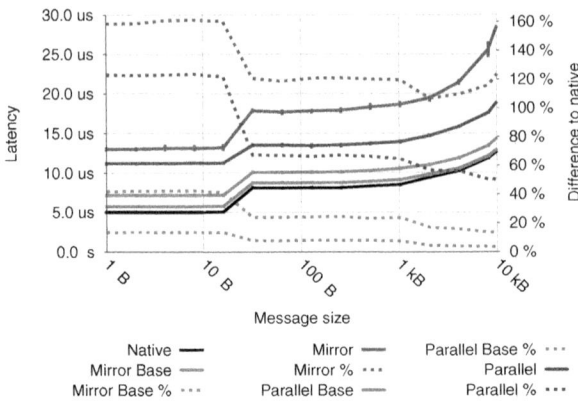

Fig. 6. Latency with `MPI_ANY_SOURCE` for the two protocols

These microbenchmark results show that parallel is better for bandwidth and latency sensitive applications, and mirror has the advantage of having a higher achievable message rate for smaller messages and places less demand on RAS system functionality [1]. In either case, it is clear that the performance of both protocols could be improved with some additional support from both the underlying hardware and the MPI implementation.

4   Accelerating Redundant Computing

In Section 2.4 we identified issues with the current implementation of rMPI and in Section 3 we measured some additional properties that are affected by the rMPI implementation and the protocols it uses. We are now ready to address issue items 3 through 8 from our list on page 211.

4.1   Integrating rMPI into an MPI Implementation

While implementing rMPI as a layer between the application and the MPI implementation allowed for quick prototyping, it has performance and code maintenance drawbacks that can be addressed by moving rMPI inside an existing MPI implementation. For example, this approach would allow for providing fully optimized collective operations.

4.2   Bandwidth and Latency Consumption

The mirror protocol sends each application message twice. Since the messages are identical, save for the different destinations, it would make sense to let the NIC duplicate the message and send two copies out, potentially reducing bandwidth consumption on the local NIC- to-memory connections. An even better approach would be to have the first router where the two message paths diverge, do this task. The message would have to be flagged as a redundant message and contain the address of, or routes to, both destinations. This mechanism would be an incremental increase in complexity inside a router or NIC. It would help the parallel protocol when it is operating in degraded mode.

Currently, *r*MPI operates below the collective operations and uses the MPI implementation underneath as a point-to-point transport layer. This leads to poor collective performance as can be seen in Figure 7. This figure shows the consistency protocol overhead for a barrier operation for both mirror and parallel. Similar slowdown can be seen for other collective operations [1]. Instead, *r*MPI should make use of the provided and optimized collective operations, and, for example, use one broadcast operation to deliver data to all nodes – active and redundant. This would help both protocols to take advantage of topology optimized MPI features.

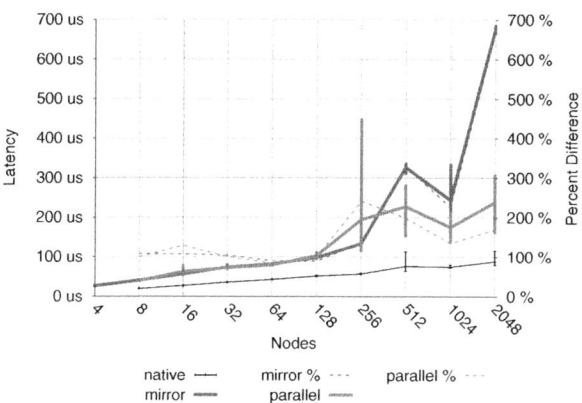

Fig. 7. MPI_Barrier() performance for the two protocols compared to native

4.3   Message Order Semantics in Case of MPI_ANY_SOURCE

Both protocols suffer when an application posts receives with the MPI_ANY_SOURCE or MPI_ANY_TAG marker. We cannot post the receive on the redundant node until we know in which order the message arrived at the active node. This causes overhead due to processing of unexpected messages. In addition, *r*MPI latency is degraded because of the protocol overhead to agree on a message order.

Implementation issues with MPI_ANY_SOURCE are well known and some researchers have advocated forbidding it, saying that well-written applications do not need it. *r*MPI can avoid most of the overhead when an application does not use MPI_ANY_SOURCE, since both nodes in a pair can detect the use of MPI_ANY_-SOURCE. However, the fundamental problem remains for a fully compliant MPI implementation. (*r*MPI currently handles MPI_ANY_SOURCE properly, but does not support message receives with both MPI_ANY_SOURCE and MPI_ANY_TAG specified.)

One option we have not evaluated yet, is a modification to the parallel protocol. In Figure 2, instead of node A' sending the redundant message, node B could send a copy to node B'. That way node B could control the message order B' sees. A redundant node would get all of its messages from its active partner. When sending, a protocol is needed to let the redundant node know that the message has been sent and that the redundant node can skip the transmission. If the active receiver (B in the example) fails, the active sender (A) would transmit to the redundant receiver (B'). If the active sender (A) fails, the redundant sender (A') would start sending to the active receiver (B).

We have not implemented this variation of the parallel protocol because we believe it would not significantly improve performance, and few high-performance applications use MPI_ANY_SOURCE in the critical path.

4.4 Use of One-Sided Operations

It might be possible to use one-sided operations to accelerate data delivery when MPI_ANY_SOURCE is used. This method would be useful for larger messages; temporary buffers and memory copies can be used for short messages. The active receiving node would inform the redundant receiver about the message order and let the redundant node get the data in the appropriate order using a pull protocol. This method would increase latency slightly, since the *get* request needs to be sent, and, depending on the architecture, a confirmation that the data has been picked up. This small overhead can be easily amortized for larger messages and the overhead to copy short messages into the correct user buffers is small.

5 Related Work

Although redundancy is one of the fundamental approaches to masking errors and providing resiliency, it has not been extensively explored or deployed in high-performance computing (HPC) environments. Since HPC applications can scale to consume any of the available resources in a system (e.g., compute power, memory), the cost of duplicating resources for resiliency has been perceived as being too high – especially given the reliability levels of current large-scale systems. However, there are several characteristics of future systems that are motivating the community to explore alternative approaches to resiliency. Recently, redundant computation has been suggested as a possible path [4,5] to resiliency, and the increasing probability of soft errors in future systems has also lead some to argue that higher levels of redundancy will also be needed [6].

There are several prior and ongoing research projects that are exploring resiliency and fault tolerance for MPI applications [7,8,9,10,11]. The MPI-3 Forum is also considering enhancements to the standard to enable fault-tolerant applications.

Most similar to rMPI is P2P-MPI [12,13] which provides fault tolerance for grid applications through replication. In contrast to rMPI, P2P-MPI does not ensure consistency when wild-cards are used. In addition, P2P-MPI is tied to

Java and requires a number of grid based services and protocols, which make this library inappropriate for an HPC environment. Furthermore, the failure analysis of P2P-MPI focuses on the probability of failures in the presence of replication, while in this work we focus on the impact of MTTI for the application which we believe is a more useful metric [2]. The approach we describe in this paper is very different from the other work on providing resiliency in the context of MPI applications, mostly due to the assumption that the cost of using resources for redundant computation will be acceptable for future large-scale systems.

There is some precedent for paying the resource cost of redundancy in high-performance computing. In [14] IBM describes a flow control protocol designed to efficiently manage limited buffer space for MPI unexpected messages on their BG/L system. Even though the network is reliable, an acknowledgment-based flow control protocol is used to ensure that unexpected messages do not overflow the limited amount of available memory on a node. This protocol can potentially slowdown all applications, but the authors argue that a factor of two increase in runtime is acceptable: *"Nevertheless, the main conclusion is that the overhead is never more than twice the execution time without memory problems, which is not a hight [sic] price to pay to make your application run without problems."*

References

1. Ferreira, K., Riesen, R., Oldfield, R., Stearley, J., Laros, J., Pedretti, K., Brightwell, R., Kordenbrock, T.: Increasing fault resiliency in a message-passing environment. Technical report SAND2009-6753, Sandia National Laboratories (2009)
2. Riesen, R., Ferreira, K., Stearley, J.: See applications run and throughput jump: The case for redundant computing in HPC. In: 1st International Workshop on Fault-Tolerance for HPC at Extreme Scale, FTXS 2010 (2010)
3. Network-Based Computing Laboratory, Ohio State University: OSU MPI benchmarks, OMB (2010), http://mvapich.cse.ohio-state.edu/benchmarks/
4. Schroeder, B., Gibson, G.A.: Understanding failures in petascale computers. Journal of Physics: Conference Series 78(1), 188–198 (2007)
5. Zheng, Z., Lan, Z.: Reliability-aware scalability models for high performance computing, In: Proceedings of the IEEE conference on Cluster Computing (2009)
6. He, X., Ou, L., Engelmann, C., Chen, X., Scott, S.L.: Symmetric active/active metadata service for high availability parallel file systems. Journal of Parallel and Distributed Computing (JPDC) 69(12), 961–973 (2009)
7. Fagg, G.E., Dongarra, J.: FT-MPI: Fault tolerant MPI, supporting dynamic applications in a dynamic world. In: Proceedings of the 7th European PVM/MPI Users' Group Meeting on Recent Advances in Parallel Virtual Machine and Message Passing Interface, pp. 346–353 (2000)
8. Gropp, W., Lusk, E.: Fault tolerance in message passing interface programs. International Journal of High Performance Computing Applications 18(3) (2004)
9. Bouteiller, A., Cappello, F., Herault, T., Krawezik, G., Lemarinier, P., Magniette, F.: MPICH-V2: a fault tolerant MPI for volatile nodes based on pessimistic sender based message logging. In: Proceedings of the ACM/IEEE International Conference on High Performance Computing and Networking (2003)
10. Hursey, J., Squyres, J., Mattox, T., Lumsdaine, A.: The design and implementation of checkpoint/restart process fault tolerance for Open MPI. In: Proceedings of the IEEE International Parallel and Distributed Processing Symposium (2007)

11. Santos, G., Duarte, A., Rexachs, D., Luque, E.: Providing non-stop service for message-passing based parallel applications with RADIC. In: Luque, E., Margalef, T., Benítez, D. (eds.) Euro-Par 2008. LNCS, vol. 5168, pp. 58–67. Springer, Heidelberg (2008)
12. Genaud, S., Rattanapoka, C.: P2P-MPI: A peer-to-peer framework for robust execution of message passing parallel programs on grids. J. Grid Comput. 5(1), 27–42 (2007)
13. Genaud, S., Jeannot, E., Rattanapoka, C.: Fault-management in P2P-MPI. Int. J. Parallel Program. 37(5), 433–461 (2009)
14. Farreras, M., Cortes, T., Labarta, J., Almasi, G.: Scaling MPI to short-memory MPPs such as BG/L. In: Proceeding of the International Conference on Supercomputing, pp. 209–218 (2006)

Checkpoint/Restart-Enabled Parallel Debugging

Joshua Hursey[1], Chris January[2], Mark O'Connor[2], Paul H. Hargrove[3],
David Lecomber[2], Jeffrey M. Squyres[4], and Andrew Lumsdaine[1]

[1] Open Systems Laboratory, Indiana University
{jjhursey,lums}@osl.iu.edu*
[2] Allinea Software Ltd.
{cjanuary,mark,david}@allinea.com
[3] Lawrence Berkeley National Laboratory
PHHargrove@lbl.gov**
[4] Cisco Systems, Inc.
jsquyres@cisco.com

Abstract. Debugging is often the most time consuming part of software development. HPC applications prolong the debugging process by adding more processes interacting in dynamic ways for longer periods of time. Checkpoint/restart-enabled parallel debugging returns the developer to an intermediate state closer to the bug. This focuses the debugging process, saving developers considerable amounts of time, but requires parallel debuggers cooperating with MPI implementations and checkpointers. This paper presents a design specification for such a cooperative relationship. Additionally, this paper discusses the application of this design to the GDB and DDT debuggers, Open MPI, and BLCR projects.

1 Introduction

The most time consuming component of the software development life-cycle is application debugging. Long running, large scale High Performance Computing (HPC) parallel applications compound the time complexity of the debugging process by adding more processes interacting in dynamic ways for longer periods of time. Cyclic or iterative debugging, a commonly used debugging technique, involves repeated program executions that assist the developer in gaining an understanding of the causes of the bug. Software developers can save hours or days of time spent debugging by checkpointing and restarting the parallel debugging session at intermediate points in the debugging cycle. For Message Passing Interface (MPI) applications, the parallel debugger must cooperate with the MPI implementation and Checkpoint/Restart Service (CRS) which account for the network state and process image. We present a design specification for this cooperative relationship to provide Checkpoint/Restart (C/R)-enabled parallel debugging.

* Supported by grants from the Lilly Endowment; National Science Foundation EIA-0202048; and U.S. Department of Energy DE-FC02-06ER25750^A003.
** Supported by the U.S. Department of Energy under Contract No. DE-AC02-05CH11231

R. Keller et al. (Eds.): EuroMPI 2010, LNCS 6305, pp. 219–228, 2010.

The C/R-enabled parallel debugging design supports multi-threaded MPI applications without requiring any application modifications. Additionally, all checkpoints, whether generated with or without a debugger attached, are usable within a debugging session or during normal execution. We highlight the *debugger detach* and *debugger reattach* problems that may lead to inconsistent views of the debugging session. This paper presents a solution to these problems that uses a thread suspension technique which provides the user with a consistent view of the debugging session across repeated checkpoint and restart operations of the parallel application.

2   Related Work

For HPC applications, the MPI [1] standard has become the *de facto* standard message passing programming interface. Even though some parallel debuggers support MPI applications, there is no official standard interface for the interaction between the parallel debugger and the MPI implementation. However, the MPI implementation community has informally adopted some consistent interfaces and behaviors for such interactions [2,3]. The MPI Forum is discussing including these interactions into a future MPI standard. This paper extends these debugging interactions to include support for C/R-enabled parallel debugging.

C/R rollback recovery techniques are well established in HPC [4]. The checkpoint, or snapshot, of the parallel application is defined as the state of the process and all connected communication channels [5]. The state of the communication channels is usually captured by a C/R-enabled MPI implementation, such as Open MPI [6]. Although C/R is not part of the MPI standard, it is often provided as a transparent service by MPI implementations [6,7,8,9]. The state of the process is captured by a Checkpoint/Restart Service (CRS), such as Berkeley Lab Checkpoint/Restart (BLCR) [10]. The combination of a C/R-enabled MPI and a CRS provide consistent global snapshots of the MPI application. Often global snapshots are used for fault recovery, but, as this paper demonstrates, can also be used to support reverse execution while debugging the MPI application. For an analysis of the performance implications of integrating C/R into an MPI implementation we refer the reader to previous literature on the subject [9,11].

Debugging has a long history in software engineering [12]. Reverse execution or back-stepping allows a debugger to either step backwards through the program execution to a previous state, or step forward to the next state. Reverse execution is commonly achieved though the use of checkpointing [13,14], event/message logging [15,16], or a combination of the two techniques [17,18]. When used in combination, the parallel debugger restarts the program from a checkpoint and replays the execution up to the breakpoint. A less common implementation technique is the actual execution of the program code in reverse without the use of checkpoints [19]. This technique is often challenged by complex logical program structures which can interfere with the end user behavior and applicability to certain programs.

Event logging is used to provide a deterministic re-execution of the program while debugging. Often this allows the debugger to reduce the number of processes involved in the debugging operation by simulating their presence through replaying events from the log. This is useful when debugging an application with a large number of processes. Event logging is also used to allow the user to view a historical trace of program execution without re-execution [20,21].

C/R is used to return the debugging session to an intermediary point in the program execution without replaying from the beginning of execution. For programs that run for a long period of time before exhibiting a bug, C/R can focus the debugging session on a smaller period of time closer to the bug. C/R is also useful for program validation and verification techniques that may run concurrently with the parallel program on smaller sections of the execution space [22].

3   Design

C/R-enabled parallel debugging of MPI applications requires the cooperation of the parallel debugger, the MPI implementation, and the CRS to provide consistently recoverable application states. The debugger provides the interface to the user and maintains state about the parallel debugging session (e.g., breakpoints, watchpoints).

The C/R-enabled MPI implementation marshals the network channels around C/R operations for the application. Though the network channels are often marshaled in a fully coordinated manner, this design does not require full coordination. Therefore the design is applicable to other checkpoint coordination protocol implementations (e.g., uncoordinated).

The CRS captures the state of a single process in the parallel application. This can be implemented at the user or system level. This paper requires an MPI application transparent CRS, which excludes application level CRSs. If the CRS is not transparent to the application, then taking the checkpoint would alter the state of the program being debugged, potentially confusing the user.

One goal of this design is to create *always usable checkpoints*. This means that regardless of whether the checkpoint was generated with the debugger attached or not, it must be able to be used on restart with or without the debugger. To provide the *always usable checkpoints* condition, the checkpoints generated by the CRS with the debugger attached must be able to exclude the debugger state. To achieve this, the debugger must detach from the process before a checkpoint and reattach, if desired, after the checkpoint has finished, similar to the technique used in [17]. Since we are separating the CRS from the debugger, we must consider the needs of both in our design.

In addition to the *always usable checkpoints* goal, this technique supports multi-threaded MPI applications without requiring any explicit modifications to the target application. Interfaces are prefixed with MPIR_ to fit the existing naming convention for debugging symbols in MPI implementations.

3.1   Preparing for a Checkpoint

The C/R-enabled MPI implementation may receive a checkpoint request internally or externally from the debugger, user, or system administrator. The MPI

```
volatile int MPIR_checkpoint_debug_gate = 0;
volatile int MPIR_debug_with_checkpoint = 0;
int MPIR_checkpoint_debugger_detach(void) { return 0; } // Detach Function
void MPIR_checkpoint_debugger_waitpoint(void) { // Thread Wait Function
    // MPI Designated Threads are released early,
    // All other threads enter the breakpoint below
    MPIR_checkpoint_debug_gate = 0;
    MPIR_checkpoint_debugger_breakpoint();
}
void MPIR_checkpoint_debugger_breakpoint(void) { // Debugger Breakpoint Func.
    while( MPIR_checkpoint_debug_gate == 0 ) { sleep(1); }
}
void MPIR_checkpoint_debugger_crs_hook(int state) { // CRS Hook Callback Func.
    if( MPIR_debug_with_checkpoint ) {
        MPIR_checkpoint_debug_gate = 0;
        MPIR_checkpoint_debugger_waitpoint();
    } else { MPIR_checkpoint_debug_gate = 1; }
}
```

Fig. 1. Debugger MPIR_ function pseudo code

implementation communicates the checkpoint request to the specified processes (usually all processes) in the MPI application. The MPI processes typically prepare for the checkpoint by marshaling the network state and flushing caches before requesting a checkpoint from the CRS.

If the MPI process is under debugger control at the time of the checkpoint, then the debugger must allow the MPI process to prepare for the checkpoint uninhibited by the debugger. If the debugger remains attached, it may interfere with the techniques used by the CRS to preserve the application state (e.g., by masking signals). Additionally, by detaching the debugger before the checkpoint, the implementation can provide the *always usable checkpoints* condition by ensuring that it does not inadvertently include debugger state in the CRS generated checkpoint.

The MPI process must inform the debugger of when to detach since the debugger is required to do so before a checkpoint is requested. The MPI process informs the debugger by calling the MPIR_checkpoint_debugger_detach() function when it requires the debugger to detach. This is an empty function that the debugger can reference in a breakpoint. It is left to the discretion of the MPI implementation when to call this function while preparing for the checkpoint, but it must be invoked before the checkpoint is requested from the CRS.

The period of time between when the debugger detaches from the MPI process and when the checkpoint is created by the CRS may allow the application to run uninhibited, we call this the *debugger detach problem*. To provide a seamless and consistent view to the user, the debugger must make a best effort attempt at preserving the exact position of the program counter(s) across a checkpoint operation. To address the *debugger detach problem*, the debugger forces all threads

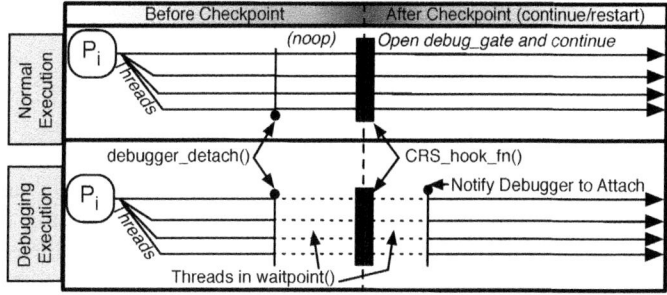

Fig. 2. Illustration of the design for each of the use case scenarios

into a waiting function (called `MPIR_checkpoint_debugger_waitpoint()`) at the current debugging position before detaching from the MPI process. By forcing all threads into a waiting function the debugger prevents the program from making any progress when it returns from the checkpoint operation.

The waiting function must allow certain designated threads to complete the checkpoint operation. In a single threaded application, this would be the main thread, but in a multi-threaded application this would be the thread(s) designated by the MPI implementation to prepare for and request the checkpoint from the CRS. The MPI implementation must provide an "early release" check for the designated thread(s) in the waiting function. All other threads directly enter the `MPIR_checkpoint_debugger_breakpoint()` function which waits in a loop for release by the debugger. Designated thread(s) are allowed to continue normal operation, but must enter the breakpoint function after the checkpoint has completed to provide a consistent state across all threads to the debugger, if it intends on reattaching. Figure 1 presents a pseudo code implementation of these functions.

The breakpoint function loop is controlled by the `MPIR_checkpoint_debug_gate` variable. When this variable is set to 0 the gate is *closed*, keeping threads waiting for the gate to be *opened* by the debugger. To *open* the gate, the debugger sets the variable to a non-zero value, and steps each thread out of the loop and the breakpoint function. Once all threads pass through the gate, the debugger then *closes* it once again by setting the variable back to 0.

3.2   Resuming after a Checkpoint

An MPI program either *continues* after a requested checkpoint in the same program, or is *restarted* from a previously established checkpoint saved on stable storage. In both cases the MPI designated thread(s) are responsible for recovering the internal MPI state including reconnecting processes in the network.

If the debugger intends to attach, the designated thread(s) must inform the debugger when it is safe to attach after restoring the MPI state of the process. If the debugger attaches too early, it may compromise the state of the checkpoint or the restoration of the MPI state. The designated thread(s) notify the debugger that it is safe to attach to the process by printing to stderr the hostname and PID of each recovered process. The message is prefixed by "MPIR_debug_info)" as to distinguish it from other output. Afterwards, the designated thread(s) enter the breakpoint function.

The period of time between when the MPI process is restarted by the CRS and when the debugger attaches may allow the application to run uninhibited, we call this the *debugger reattach problem*. If the MPI process was being debugged before the checkpoint was requested, then the threads are already being held in the breakpoint function, thus preventing them from running uninhibited. However, if the MPI process was not being debugged before the checkpoint then the user may experience inconsistent behavior due to the race to attach the debugger upon multiple restarts from the same checkpoint.

To address this problem, the CRS must provide a *hook* callback function that is pushed onto the stack of all threads before returning them to the running state. This technique preserves the individual thread's program counter position at the point of the checkpoint providing a best effort attempt at a consistent recovery position upon multiple restarts. The MPI implementation registers a hook callback function that will place all threads into the waiting function if the debugger intends to reattach. The intention of the debugger to reattach is indicated by the MPIR_debug_with_checkpoint variable. Since the hook function is the same function used when preparing for a checkpoint, the release of the threads from the waiting function is consistent from the perspective of the debugger.

If the debugger is not going to attach after the checkpoint or on restart, the hook callback does not need to enter the waiting function, again indicated by the MPIR_debug_with_checkpoint variable. Since the checkpoint could have been generated with a debugger previously attached, the hook function must release all threads from the breakpoint function by setting the MPIR_checkpoint_debug_gate variable to 1. The structure of the hook callback function allows for checkpoints generated while debugging to be used without debugging, and vice versa.

3.3   Additional MPIR_ Symbols

In addition to the detach, waiting, and breakpoint functions, this design defines a series of support variables to allow the debugger greater generality when interfacing with a C/R-enabled MPI implementation.

The MPIR_checkpointable variable indicates to the debugger that the MPI implementation is C/R-enabled and supports this design when set to 1. The MPIR_debug_with_checkpoint variable indicates to the MPI implementation if the debugger intends to attach. If the debugger wishes to detach from the program, it sets this value to 0 before detaching. This value is set to 1 when the debugger attaches to the job either while running or on restart.

The MPIR_checkpoint_command variable specifies the command to be used to initiate a checkpoint of the MPI process. The output of the checkpoint command must be formatted such that the debugger can use it directly as an argument to the restart command. The output on stderr is prefixed with "MPIR_checkpoint_handle)" as to distinguish it from other output.

The MPIR_restart_command variable specifies the restart command to prefix the output of the checkpoint command to restart an MPI application.
The MPIR_controller_hostname variable specifies the host on which to execute the MPIR_checkpoint_command and MPIR_restart_command commands.
The MPIR_checkpoint_ listing_command variable specifies the command that
lists the available checkpoints on the system.

4   Use Case Scenarios

To better illustrate how the various components cooperate to provide C/R-
enabled parallel debugging we present a set of use case scenarios. Figure 2
presents an illustration of the design for each scenario.

Scenario 1: No Debugger Involvement. This is the standard C/R scenario
in which the debugger is neither involved before a checkpoint nor afterwards. A
transition from the upper-left to upper-right quadrants in Figure 2. The MPI
processes involved in the checkpoint will prepare the internal MPI state and
request a checkpoint from the CRS then continue free execution afterwards.

Scenario 2: Debugger Attaches on Restart. In this scenario, the debugger
is attaching to a restarting MPI process from a checkpoint that was generated
without the debugger. A transition from the upper-left to lower-right quadrants
in Figure 2. This scenario is useful when repurposing checkpoints originally generated for fault tolerance purposes instead for debugging. The process of creating
the checkpoints is the same as in Scenario 1.

On restart, the hook callback function is called by the CRS in each thread
to preserve their program counter positions. Once the MPI designated thread(s)
have reconstructed the MPI state, the debugger is notified that it is safe to
attach. Once attached, the debugger walks all threads out of the breakpoint
function and resumes debugging operations.

Scenario 3: Debugger Attached While Checkpointing. In this scenario,
the debugger is attached when a checkpoint is requested of the MPI process. A
transition from the lower-left to lower-right quadrants in Figure 2. This scenario is
useful when creating checkpoints while debugging that can be returned to in later
iterations of debugging cycle or to provide backstepping functionality while debugging. The debugger will notice the call to the detach function and call the waiting
function in all threads in the MPI process before detaching. The MPI designated
checkpoint thread(s) are allowed to continue through this function in order to request the checkpoint from the CRS while all other threads wait there for later release. After the checkpoint or on restart the protocol proceeds as in Scenario 2.

Scenario 4: Debugger Detached on Restart. In this scenario, the debugger
is attached when the checkpoint is requested of the MPI process, but is not when
the MPI process is restarted from the checkpoint. A transition from the lower-left
to upper-right quadrants in Figure 2. This scenario is useful when analyzing the
uninhibited behavior of an application, periodically inspecting checkpoints for

validation purposes or, possibly, introducing tracing functionality to a running program. The process of creating the checkpoint is the same as in Scenario 3. By inspecting the MPIR_debug_with_checkpoint variable, the MPI processes know to let themselves out of the waiting function after the checkpoint and on restart.

5   Implementation

The design described in Section 3 was implemented using GNU's GDB debugger, Allinea's DDT Parallel Debugger, the Open MPI implementation of the MPI standard, and the BLCR CRS. Open MPI implements a fully coordinated C/R protocol [6] so when a checkpoint is requested of one process all processes in the MPI job are also checkpointed. We note again that full coordination is not required by the design, so other techniques can be used at the discretion of the MPI implementation.

5.1   Interlayer Notification Callback Functions

Open MPI uses the Interlayer Notification Callback (INC) functions to coordinate the internal state of the MPI implementation before and after checkpoint operations. After receiving notification of a checkpoint request, Open MPI calls the INC_checkpoint_prep() function. This function quiesces the network, and prepares various components for a checkpoint operation [11]. Once the INC is finished it designates a checkpoint thread, calls the debugger detach function, and then requests the checkpoint from the CRS, in this case BLCR.

After the checkpoint is created (called the *continue* state) or when the MPI process is restarted (called the *restart* state), BLCR calls the hook callback function in each thread (See Figure 1). The thread designated by Open MPI (in the INC_checkpoint_prep() function) is allowed to exit this function without waiting, while all other threads must wait if the debugger intends on attaching. The designated thread then calls the INC function for either the continue or restart phase depending on if the MPI process is continuing after a checkpoint or restarting from a checkpoint previously saved to stable storage.

If the debugger intends on attaching to the MPI process, then after reconstructing the MPI state, the designated thread notifies the debugger that it is safe to attach by printing the "MPIR_debug_info)" message to stderr as described in Section 3.2.

5.2   Stack Modification

In Section 3.1, the debugger was required to force all threads to call the waiting function before detaching before a checkpoint in order to preserve the program counter in all threads across a checkpoint operation. We explored two different ways to do this in the GDB debugger. The first required the debugger to force the function on the call stack of each thread. In GDB, we used the following command for each thread:

```
void MPIR_checkpoint_debugger_signal_handler(int num) {
    MPIR_checkpoint_debugger_waitpoint();
}
```

Fig. 3. Open MPI's SIGTSTP signal handler function to support stack modification

`call MPIR_checkpoint_debugger_waitpoint()`

Unfortunately this became brittle and corrupted the stack in GDB 6.8.

In response to this, we explored an alternative technique based on signals. Open MPI registered a signal callback function (See Figure 3) that calls the `MPIR_checkpoint_debugger_waitpoint()` function. The debugger can then send a designated signal (e.g., `SIGTSTP`) to each thread in the application, and the program will place itself in the waiting function.

Though the signal based technique worked best for GDB, other debuggers may have other techniques at their disposal to achieve this goal.

6   Conclusions

Debugging parallel applications is a time-consuming part of the software development life-cycle. C/R-enabled parallel debugging may be helpful in shortening the time required to debug long-running HPC parallel applications. This paper presented a design specification for the interaction between a parallel debugger, C/R-enabled MPI implementation, and CRS to achieve C/R-enabled parallel debugging for MPI applications. This design focuses on an abstract separation between the parallel debugger and the MPI and CRS implementations to allow for greater generality and flexibility in the design. The separation also enables the design to achieve the *always usable checkpoints* goal.

This design was implemented using GNU's GDB debugger, Allinea's DDT Parallel Debugger, Open MPI, and BLCR. An implementation of this design will be available in the Open MPI v1.5 release series. More information about this design can be found at the link below:

`http://osl.iu.edu/research/ft/crdebug/`

References

1. Message Passing Interface Forum: MPI: A Message Passing Interface. In: Proc. of Supercomputing 1993, pp. 878–883 (1993)
2. Cownie, J., Gropp, W.: A standard interface for debugger access to message queue information in MPI. In: Margalef, T., Dongarra, J., Luque, E. (eds.) PVM/MPI 1999. LNCS, vol. 1697, pp. 51–58. Springer, Heidelberg (1999)
3. Gottbrath, C.L., Barrett, B., Gropp, B., Lusk, E., Squyres, J.: An interface to support the identification of dynamic MPI 2 processes for scalable parallel debugging. In: Mohr, B., Träff, J.L., Worringen, J., Dongarra, J. (eds.) PVM/MPI 2006. LNCS, vol. 4192, pp. 115–122. Springer, Heidelberg (2006)

4. Elnozahy, E.N.M., Alvisi, L., Wang, Y.M., Johnson, D.B.: A survey of rollback-recovery protocols in message-passing systems. ACM Computing Surveys 34, 375–408 (2002)
5. Chandy, K.M., Lamport, L.: Distributed snapshots: determining global states of distributed systems. ACM Transactions on Computer Systems 3, 63–75 (1985)
6. Hursey, J., Squyres, J.M., Mattox, T.I., Lumsdaine, A.: The design and implementation of checkpoint/restart process fault tolerance for Open MPI. In: Proceedings of the IEEE International Parallel and Distributed Processing Symposium (2007)
7. Jung, H., Shin, D., Han, H., Kim, J.W., Yeom, H.Y., Lee, J.: Design and implementation of multiple fault-tolerant MPI over Myrinet (M^3). In: Proceedings of the ACM/IEEE Supercomputing Conference (2005)
8. Gao, Q., Yu, W., Huang, W., Panda, D.K.: Application-transparent checkpoint/restart for MPI programs over InfiniBand. In: International Conference on Parallel Processing, pp. 471–478 (2006)
9. Bouteiller, A., et al.: MPICH-V project: A multiprotocol automatic fault-tolerant MPI. International Journal of High Performance Computing Applications 20, 319–333 (2006)
10. Duell, J., Hargrove, P., Roman, E.: The design and implementation of Berkeley Lab's Linux Checkpoint/Restart. Technical Report LBNL-54941, Lawrence Berkeley National Laboratory (2002)
11. Hursey, J., Mattox, T.I., Lumsdaine, A.: Interconnect agnostic checkpoint/restart in Open MPI. In: Proceedings of the 18th ACM International Symposium on High Performance Distributed Computing, pp. 49–58 (2009)
12. Curtis, B.: Fifteen years of psychology in software engineering: Individual differences and cognitive science. In: Proceedings of the International Conference on Software Engineering, pp. 97–106 (1984)
13. Feldman, S.I., Brown, C.B.: IGOR: A system for program debugging via reversible execution. In: Proceedings of the ACM SIGPLAN/SIGOPS workshop on Parallel and Distributed Debugging, pp. 112–123 (1988)
14. Wittie, L.: The Bugnet distributed debugging system. In: Proceedings of the 2nd workshop on Making Distributed Systems Work, pp. 1–3 (1986)
15. Bouteiller, A., Bosilca, G., Dongarra, J.: Retrospect: Deterministic replay of MPI applications for interactive distributed debugging. In: Recent Advances in Parallel Virtual Machine and Message Passing Interface, pp. 297–306 (2007)
16. Ronsse, M., Bosschere, K.D., de Kergommeaux, J.C.: Execution replay and debugging. In: Proceedings of the Fourth International Workshop on Automated Debugging, Munich, Germany (2000)
17. King, S.T., Dunlap, G.W., Chen, P.M.: Debugging operating systems with time-traveling virtual machines. In: Proceedings of the USENIX Annual Technical Conference (2005)
18. Pan, D.Z., Linton, M.A.: Supporting reverse execution for parallel programs. In: Proceedings of the ACM SIGPLAN/SIGOPS workshop on Parallel and Distributed Debugging, pp. 124–129 (1988)
19. Agrawal, H., DeMillo, R.A., Spafford, E.H.: An execution-backtracking approach to debugging. IEEE Software 8(3), 21–26 (1991)
20. Undo Ltd.: UndoDB - Reversible debugging for Linux (2009)
21. TotalView Technologies: ReplayEngine (2009)
22. Sorin, D.J., Martin, M.M.K., Hill, M.D., Wood, D.A.: SafetyNet: Improving the availability of shared memory multiprocessors with global checkpoint/recovery. SIGARCH Computer Architecture News 30, 123–134 (2002)

Load Balancing for Regular Meshes on SMPs with MPI

Vivek Kale and William Gropp

University of Illinois at Urbana-Champaign, IL, USA
{vivek,wgropp}@illinois.edu

Abstract. Domain decomposition for regular meshes on parallel computers has traditionally been performed by attempting to exactly partition the work among the available processors (now cores). However, these strategies often do not consider the inherent system noise which can hinder MPI application scalability to emerging peta-scale machines with 10000+ nodes. In this work, we suggest a solution that uses a tunable hybrid static/dynamic scheduling strategy that can be incorporated into current MPI implementations of mesh codes. By applying this strategy to a 3D jacobi algorithm, we achieve performance gains of at least 16% for 64 SMP nodes.

1 Introduction

Much literature has emphasized effective decomposition strategies for good parallelism across nodes of a cluster. Recent work with hybrid programming models for clusters of SMPs has often focused on determining the best split of threads and processes, and the shape of the domains used by each thread [1,2]. In fact, these static decompositions are often auto-tuned for specific architectures to achieve reasonable performance gains. However, the fundamental problem is that this "static scheduling" assumes that the user's program has total access to all of the cores all of the time; these static decomposition strategies cannot be tuned easily to adapt in real-time to system noise (particularly due to OS jitter). The occasional use of the processor cores, by OS processes, runtime helper threads, or similar background processes, introduce noise that makes such static partitioning inefficient on a large number of nodes. For applications running on a single node, the general system noise is small though noticeable. Yet, for next-generation peta-scale machines, improving mesh computations to handle such system noise is a high priority. Current operating systems running on nodes of high-performance clusters of SMPs have been designed to minimally interfere with these computationally intensive applications running on SMP nodes [3], but the small performance variations due to system noise can still potentially impact scalability of an MPI application for a cluster on the order of 10,000 nodes. Indeed, to eliminate the effects of process migration, the use of approaches such as binding compute threads/processes to cores, just before running the application, is advocated [3]. However, this only provides a solution for migration and neglects overhead due to other types of system noise.

R. Keller et al. (Eds.): EuroMPI 2010, LNCS 6305, pp. 229–238, 2010.

In this work, we illuminate how the occasional use of the processor cores by OS processes, runtime helper threads, or similar background processes, introduce noise that makes such static schedules inefficient. In order to performance tune these codes with system noise in mind, we propose a solution which involves a partially dynamic scheduling strategy of work. Our solution uses ideas from task stealing and work queues to dynamically schedule tasklets. In this way, our MPI mesh codes work with the operating system running on an SMP node, rather than in isolation from it.

2   Problem Description

Our model problem is an exemplar of regular mesh code. For simplicity, we will call it a Jacobi algorithm, as the work that we perform in our model problem is the Jacobi relaxation iteration in solving a Poisson problem. However, the data and computational pattern are similar for both regular mesh codes (both implicit and explicit) and for algorithms that attempt to evenly divide work among processor cores (such as most sparse matrix-vector multiply implementations). Many MPI implmentations of regular mesh codes traditionally have a predefined domain decomposition, as can be seen in many libraries and microbenchmark suites [4]. This optimal decomposition is necessary to reduce communication overhead, minimize cache misses, and ensure data locality. In this work, we consider a slab decomposition of a 3-dimensional block implemented in MPI/pthreads hybrid model, an increasingly popular model for taking advantage of clusters of SMPs.

We use a problem size and dimension that can highlight many of the issues we see in real-world applications with mesh computations implemented in MPI: specifically, we use a 3D *block* with dimensions $64 \times 512 \times 64$ on each node for a fixed 1000 iterations. For our 7-point stencil computation, this generates a total of 1.6 GFLOPS per node With this problem size, we can ensure that computations are done out-of-cache so that it is just enough to excercise the full memory hierarchy. The block is partitioned into *vertical slabs* across processes along the X dimension. Each vertical slab is further partitioned into *horizontal slabs* across threads along the Y dimension. Each vertical slab contains a static section(top) and a dynamic section(bottom). We use this decomposition strategy because of its simplicity to implement and tune different parameters in our search space. A MPI *border exchange* communication occurs between left and right borders of blocks of each process across the YZ planes. The border exchange operation uses MPI_Isend and MPI_Irecv pair, along with an MPI_Waitall.We mitigate the issue of first-touch as noted in [5] by doing parallel memory allocation during the initialization of our mesh.

For such regular mesh computations, the communication between processes, even in an explicit mesh sweep, provides a synchronization between the processes. Any load imbalance between the processes can be amplified, even when using a good (but static) domain decomposition strategy. If even 1% of nodes are affected by system interference during one iteration of a computationally intensive MPI application on a cluster with 1000s of nodes, several nodes will be affected by noise during each iteration. Our solution to this problem is to use a partially dynamic scheduling strategy, and is presented in the section that follows.

3 Performance Tuning Experimentation

3.1 Performance Tuning Technique

The technique for supporting dynamic scheduling of computation was implemented with a queue that was shared among threads. Each element of the shared queue (we refer to it as a *tasklet*) contains the specification of what work the thread executing this tasklet is responsible for, and a flag indicating whether the tasklet has been completed by a thread. In order to preserve locality (so that in repeated computations the same threads can get the same work), we also maintain an additional tag specifying the last thread that ran this tasklet. In the execution of each iteration of the Jacobi algorithm, there are 3 distinct phases: MPI communication, statically scheduled computation, and dynamically scheduled computation. In phase 1, thread 0 does the MPI communication for border exchange. During this time, all other threads must wait at a thread barrier. In phase 2, a thread does all work that is statically allocated to it. Once a thread completes its statically allocated work it immediately moves to phase 3, where it starts pulling the next available tasklet from the queue shared among other threads, until the queue is empty. As in the completely static scheduled case, after threads have finished computation, they will need to wait at a barrier before continuing to the next iteration. The percentage of dynamic work, granularity/number of tasklets, and number of queues for a node, is specified as parameter. Through our experimental studies of tuning our dynamic scheduling strategy, we pose the following questions:

1. Does partially dynamic scheduling improve performance for mesh computations that have traditionally been completely statically scheduled?
2. What is the tasklet granularity we need to use for maintaining load balance of tasklets across threads?
3. In using such a technique, how can we decrease the overheads of synchronization of the work queues used for dynamic scheduling?
4. What is the impact of the technique for scaling to many nodes?

In the sections that follow, we first demonstrate the benefits of partial dynamic scheduling on one node in 3.2. Section 3.3 describes the effect of task granularity. Section 3.4 examines the impact on MPI runs with multiple nodes. Our experiments were conducted on a system with Power575 SMP nodes with 16 cores per node, and the operating system was IBM AIX. We assign a compute thread to each core, ensuring that the node is fully subscribed (ignoring the 2-way SMT available on these nodes as there are only 16 sets of functional units). If any OS or runtime threads need to run, they must take time away from one of our computational threads.

3.2   Reducing Impact of OS Jitter: Dynamic vs Static Scheduling

As mentioned above, threads first complete all static work assigned to it. Once a thread completes this stage, it moves to the dynamic phase, where it dequeues tasklets from the task queue. In the context of the stencil computation experimentation we do, each thread is assigned a horizontal slab from the static region at compile time. After a thread fully completes its statically allocated slab, it completes as many tasklets of the dynamic region as it can. The number of tasklets is a user-specified parameter. To explore the impact of using dynamic scheduling with locality preference, we enumerate 4 separate cases, based on the dynamic scheduling strategy.

1. 0% dynamic: Slabs are evenly partitioned, with each thread being assigned one slab. All slabs are assigned to threads at compile-time.
2. 100% dynamic + no locality: All slabs are dynamically assigned to threads via a queue.
3. 100% dynamic + locality: Same as 2, except that when a thread tries to dequeue a tasklet, it first searches for tasklets that it last executed in a previous jacobi iteration.
4. 50% static, 50% dynamic + locality: Each thread first does its static section, and then immediately starts pulling tasklets from the shared work queue. This approach is motivated by a desire to reduce overhead in managing the assignment of tasks to cores.

For the cases involving dynamic scheduling, we initially assume the number of tasklets to be 32, and that all threads within an MPI process share one work queue. We preset the number of iterations to be 1000 (rather than using convergence criteria) to allow us to more easily verify our results. In our experiments, we choose 1000 iterations as this adequately captures the periodicity of the jitter induced by the system services during a trial [3]. Figure 1 below shows the average performance we obtained over 40 trials for each of these cases. From the figure, we can see that the 50% dynamic scheduling gives significant performance benefits over the traditional static scheduling scheduling case. Using static scheduling, the average execution time we measure was about 7.00 seconds of wall-clock time. We make note that of the 40 trials we did, we obtained 6 *lucky* runs where the best performance we got was in the range 6.00 - 6.50 seconds. The remaining 34 runs were between 7.00 - 8.00 seconds. Using fully dynamic scheduling with no locality, performance was slightly worse than the statically scheduled case. For this case, there were some small performance variations (within 0.2 seconds) across the 40 trials; these were most probably due to the varying number of cache misses, in addition to system service interference. Using locality with fully dynamic scheduling, the performance variations over 40 trials here were even lower (within 0.1 seconds). Using the 50% dynamic scheduling strategy, the execution time was 6.53 seconds, giving us over 7% performance gain over our baseline static scheduling. Thus, we notice that just by using a reasonable partially dynamic scheduling strategy, performance variation can be reduced and overall performance can be improved. In all cases using

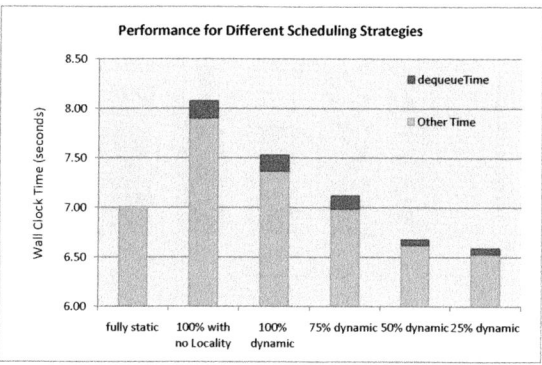

Fig. 1. The performance of different scheduling strategies used with the Jacobi Computation with 64 by 512 by 64 size block

dynamically scheduling, the thread idle times(not shown here) contribute to the largest percentage overhead. The high overhead in case 2 is likely attributed to the fact that threads suffer from doing non-local work. Because some threads suffer cache misses while others do not, the overall thread idle time (due to threads waiting at barriers) could be particularly high.

3.3   Tuning Tasklet Granularity for Reduced Thread Idle Time

As we noticed in the previous section, the idle times account for a large percentage of the performance. Total thread idle time (summed across threads) can be high because of load imbalance. Our setup above used 32 tasklets. However, the tasklets may have been too coarse grained (each tasklet has a 16-plane slab). With very coarse granularity, performance suffers because threads must wait (remain idle) at a barrier, until all threads have completed their dynamic phase. As a first strategy, we varied the number of tasklets, using 16, 32, 64, 96, and 128 tasklets as our test cases. The second strategy, called skewed workloads, addresses the tradeoff between fine-grain tasklets and coarse-grain tasklets. In this strategy, we use a work queue containing variable sized tasklets, with larger tasklets at the front of the queue and smaller tasklets towards the end. Skewed workloads reduce the contention overhead for dequeuing tasklets (seen when using fine-granularity tasklets) and also reduce the idle time of threads (seen when using coarse-grain tasklets). In figure 2, we notice that as we increase number of tasklets from 16 to 64 tasklets (decreasing tasklet size) we obtain significant performance gains, and the gains come primarily from the reduction in idle times. Overall, we notice that the performance increases rapidly in this region. As we increase from 64 to 128 tasklets, performance starts to decrease, primarily due to the contention for the tasklets and the increased dequeue overhead. We also see that performance of the skewed strategy (especially with 50% dynamic scheduling) is comparable to that of 64 tasklets, which has the best performance. In this way, a skewed strategy can yield competitive performance without needing

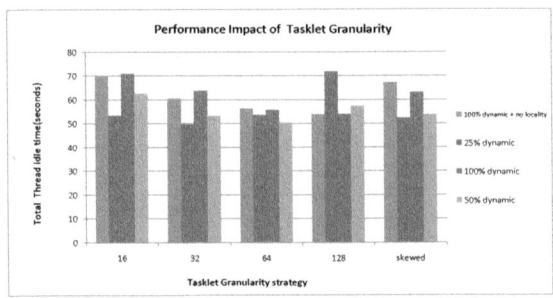

Fig. 2. Increasing task granularity to helps improve performance, particularly because of reduced thread idle times. However, at 128 tasklets the performance starts to degrade due to increasing contention for tasklets.

to predefine the tasklet granularity. To understand how tuning with a skewed workload benefits performance, figure 3 shows the distribution of timings for each of the 1000 iterations of the jacobi algorithm, comparing between static scheduling, 50% dynamic scheduling with fixed size tasklets, and 50% dynamic scheduling with skewed workloads. Using static scheduling, the maximum iteration time was 9.5 milliseconds(ms), about 40% larger than the average time of all iterations. Also, the timing distribution is bimodal, showing that half the iterations ran optimally as tuned to the architecture(running in about 6 ms), while the other half were slowed down by system noise(running in about 7.75 ms). Using 50% dynamic scheduling, the maximum iteration time is reduced to 8.25 ms, but it still suffers due to dequeue overheads, as can be seen by the mean of 7.25 ms. By using a skewed workload strategy, we see that the max is also 8.25 ms. However, the mean is lower (6.75 ms) than that seen when using fixed size tasklets, because of the lower dequeue overhead that this scheme provides. The skewed workloads provided 7% performance gains over the simple 50% dynamic scheduling strategy, which uses fixed-size coarse-grain tasklets of size 32. Furthermore, the reduced max time when using dynamic scheduling indicates that our dynamic scheduling strategy better withstands perturbations caused by system noise.

3.4   Using Our Technique to Improve Scalability

For many large MPI applications (especially with barriers) running on many nodes of a cluster, even a small system service interruption on a core of a node can accumulate to offset the entire computation, and degrade performance. In this way, the impact of a small load imbalance across cores is amplified for a large number of processes. This reduces the ability for application scalability, particularly for a cluster with a very large number of nodes (and there are many machines with more than 10000 nodes). To understand how our technique can be used to improve scalability, we tested our skewed workload with a 50% dynamic scheduling strategy on 1, 2, 4, 8, 16, 32, and 64 nodes of a cluster. One core of a

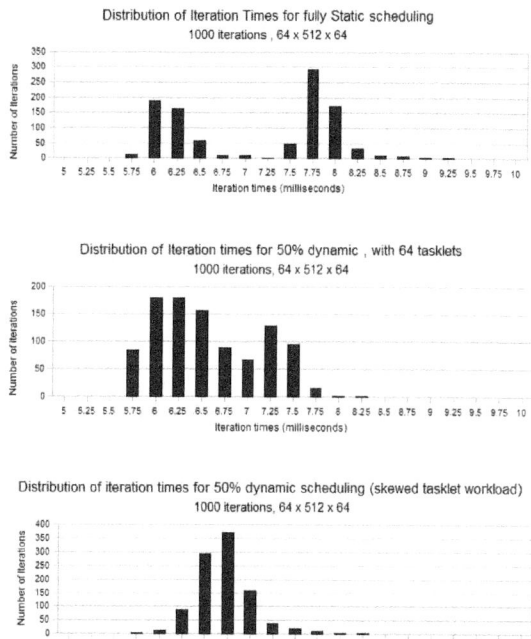

Fig. 3. Histogram view showing the performance variation of iterations for static scheduling, 50% dynamic scheduling with fixed-size tasklet granularity, and 50% dynamic scheduling with skewed workload strategy

node was assigned as a message thread to invoke MPI communication (for border exchanges) across nodes. We used the hybrid MPI/pthread programming model for implementation. Figure 4 shows how as we increase the number of nodes, using 50% dynamic scheduling always outperforms the other strategies and scales well. At 64 nodes, the 50% dynamic scheduling gives us on average a 30% performance improvement over the static scheduled case. As we can see for the case with static scheduling, a small overhead due to system services is amplified at 2 nodes and further degrades as we move up to 64 nodes. In contrast, for the 50% dynamic scheduling strategy using skewed workloads, the performance does not suffer as much when increasing the number of nodes, and our jitter mitigation techniques' benefits are visible at 64 nodes. To see the reasons for better scalability, we consider the iteration time distributions for 1 node in our 64 node runs, as shown in figure 5 (the distributions across all nodes were roughly the same). Compared to the top left histogram of figure 3, the histogram in figure 5 shows that the distribution has shifted significantly to the right for static scheduling. This makes sense since each node's jitter occurs at different times. The chain of dependencies through MPI messaging for border exchanges compounds the delay across nodes in consecutive iterations.With dynamic scheduling, the distribution has not shifted as much. For example, the mode(the tallest line) only shifted from 6.75 ms to 7.00 ms. This is because in each iteration, the node that

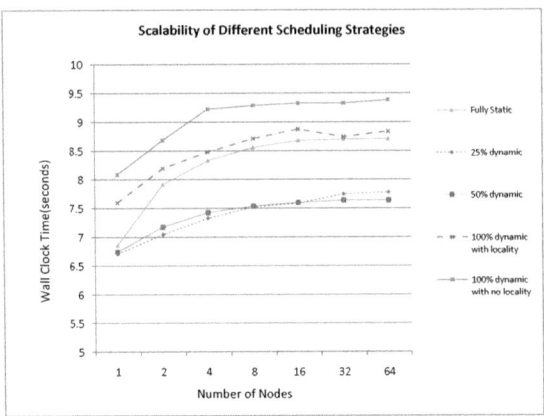

Fig. 4. Scalability results show that the 50% dynamic scheduling strategy performs better and also scales well compared to the traditional static scheduling approach

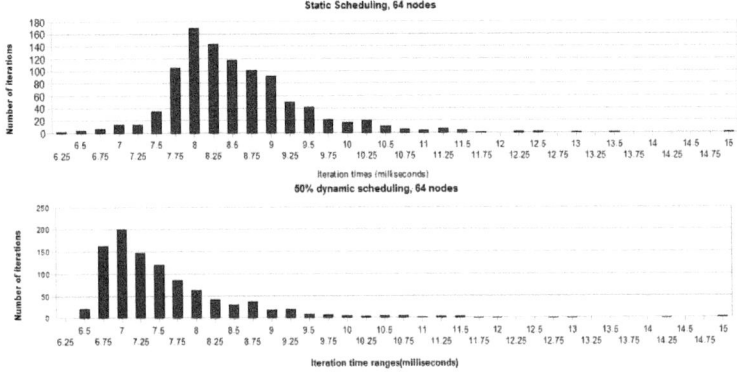

Fig. 5. The histograms (taken from node 0) in a 64 node run are shown. The left histogram corresponds to the static scheduling technique, while the right histogram corresponds to the 50% dynamic scheduling technique.

experiences noise mitigates its effect by scheduling delayed tasklets to its other threads.

4   Related Work

The work by [5, 1] shows how regular mesh (stencil) codes can be auto-tuned onto a multi-core architecture by enumerating different parameters and using sophisticated machine learning techniques to search for the best parameter configurations. In their work, the search space is based on the architectural parameters. In our work, we suggest another issue that one should be aware of for tuning codes: the random system noise incurred by OS-level events.

Cilk is a programming library [6] intended to enhance performance of many multi-core applications, and uses the ideas of shared queues and work stealing to dynamically schedule work. While the implementation of our dynamic strategy is similar to the Cilk dynamic scheduling strategy, we propose using a dynamic scheduling strategy for just the last fraction of the computation, rather than all of it. Furthermore, our method is locality-aware and allows one to tune this fraction of dynamic scheduling to the inherent system noise. We believe this can be particularly beneficial to scientific codes that are already optimally partitioned across nodes and tuned for the architecture. In [7] dynamic task scheduling with variable task sizes is used as a method for optimizing ScaLaPack libraries. Our work uses predefined, but tuned, task sizes that mitigate the system noise, without incurring dynamic scheduling overhead.

The work in [8] identifies, quantifies, and mitigates sources of OS jitter mitigation sources on large supercomputer. This work suggests different methodologies for handling each type of jitter source. This study suggests primarily modifying the operating system kernel to mitigate system noise. Specific methods for binding threads to cores [9] have been shown to have effect in reducing system interference (particularly process migration and its effects on cache misses) for high-performance scientific codes. However, these approaches cannot mitigate all system noise such as background processes or periodic OS timers. Our approach involves tuning an MPI application to any system noise, rather than modifying the operating system kernel to reduce its interference. In addition, the techniques we present can be used in conjunction with thread binding or other such techniques, rather than as an alternative.

5 Conclusions and Future Work

In this work, we introduced a dynamic scheduling strategy that can be used to improve scalability of MPI implementations of regular meshes. To do this, we started with a pthread mesh code that was tuned to the architecture of a 16-core SMP node. We then incorporated our partially dynamic scheduling strategy into the mesh code to handle inherent system noise. With this, we tuned our scheduling strategy further, particularly considering the grain size of the dynamic tasklets in our work queue. Finally, we added MPI for communication across nodes and demostrated the scalability of our approach. Through proper tuning, we showed that our methodology can provide good load balance and scale to a large number of nodes of our cluster of SMPs, even in the presence of system noise.

For future work, we plan to apply our technique to larger applications such as MILC [4]. We will also incorporate more tuning parameters (we are currently examining more sophisticated work-stealing techniques). In addition, we will tune our strategy so that it works alongside other architectural tuning strategies and other basic jitter mitigation techniques. We also plan to test on clusters with different system noise characteristics. With this, we hope to develop auto-tuning methods for such MPI/pthread code in the search space we have presented.

Acknowledgments. We thank Franck Cappello, as well as students at IN-RIA, for sharing their experience with OS jitter on IBM Power-series machines. Parts of this research are part of the Blue Waters sustained-petascale computing project, which is supported by the National Science Foundation (award number OCI 07-25070) and the state of Illinois.

References

1. Williams, S., Carter, J., Oliker, L., Shalf, J., Yelick, K.A.: Optimization of a lattice Boltzmann computation on state-of-the-art multicore platforms. Journal of Parallel and Distributed Computing (2009)
2. Cappello, F., Etiemble, D.: MPI versus MPI+OpenMP on IBM SP for the NAS benchmarks. In: Supercomputing 2000: Proceedings of the 2000 ACM/IEEE conference on Supercomputing (CDROM), Washington, DC, USA. IEEE Computer Society, Los Alamitos (2000)
3. Mann, P.D.V., Mittaly, U.: Handling OS jitter on multicore multithreaded systems. In: IPDPS 2009: Proceedings of the 2009 IEEE International Symposium on Parallel and Distributed Processing, Washington, DC, USA. IEEE Computer Society Press, Los Alamitos (2009)
4. Shi, G., Kindratenko, V., Gottlieb, S.: The bottom-up implementation of one MILC lattice QCD application on the Cell blade. International Journal of Parallel Programming 37 (2009)
5. Kamil, S., Chan, C., Williams, S., Oliker, L., Shalf, J., Howison, M., Bethel, E.W.: A generalized framework for auto-tuning stencil computations. In: Proceedings of the Cray User Group Conference (2009)
6. Blumofe, R.D., Joerg, C.F., Kuszmaul, B.C., Leiserson, C.E., Randall, K.H., Zhou, Y.: Cilk: An efficient multithreaded runtime system. Journal of Parallel and Distributed Computing (1995)
7. Song, F., YarKhan, A., Dongarra, J.: Dynamic task scheduling for linear algebra algorithms on distributed-memory multicore systems. In: SC 2009: Proceedings of the Conference on High Performance Computing Networking, Storage and Analysis. ACM, New York (2009)
8. Petrini, F., Kerbyson, D.J., Pakin, S.: The case of the missing supercomputer performance: Achieving optimal performance on the 8,192 processors of ASCI Q. In: SC 2003: Proceedings of the 2003 ACM/IEEE conference on Supercomputing, Washington, DC, USA, IEEE Computer Society Press, Los Alamitos (2003)
9. Klug, T., Ott, M., Weidendorfer, J., Trinitis, C., Müchen, T.U.: Autopin, automated optimization of thread-to-core pinning on multicore systems (2008)

Adaptive MPI Multirail Tuning for Non-uniform Input/Output Access

Stéphanie Moreaud, Brice Goglin, and Raymond Namyst

Université de Bordeaux, INRIA – LaBRI
351, cours de la Libération F-33405 Talence cedex, France
{smoreaud,goglin,namyst}@labri.fr

Abstract. Multicore processors have not only reintroduced Non-Uniform Memory Access (NUMA) architectures in nowadays parallel computers, but they are also responsible for non-uniform access times with respect to Input/Output devices (NUIOA). In clusters of multicore machines equipped with several network interfaces, performance of communication between processes thus depends on which cores these processes are scheduled on, and on their distance to the Network Interface Cards involved. We propose a technique allowing multirail communication between processes to carefully distribute data among the network interfaces so as to counterbalance NUIOA effects. We demonstrate the relevance of our approach by evaluating its implementation within OPEN MPI on a MYRI-10G + INFINIBAND cluster.

Keywords: Multirail, Non-Uniform I/O Access, Hardware Locality, Adaptive Tuning, OPEN MPI.

1 Introduction

Multicore processors are widely used in high-performance computing. This architecture trend is increasing the complexity of the compute nodes while introducing non-uniform memory topologies. A careful combined placement of tasks and data depending on their affinities is now required so as to exploit the quintessence of modern machines. Furthermore, access to the networking hardware also becomes non-uniform since interface cards may be closer to some processors. This feature has been known to impact networking performance for a long time [1] but current MPI implementations do not take it into account in their communication strategies.

One way to take affinities between processes and communication into account is to modify the process placement strategy so as to offer a privileged networking access to communication-intensive processes. In this article, we look at an orthogonal idea: optimizing communication with a predefined process placement. We study multirail configurations (scattering messages across multiple network interfaces) and show that MPI implementations should not blindly split messages in halves when sending over two rails. *Non-Uniform I/O Access* should be involved in this splitting strategy so as to improve the overall performance.

R. Keller et al. (Eds.): EuroMPI 2010, LNCS 6305, pp. 239–248, 2010.

The remaining of the paper is organized as follows. Section 2 presents modern architectures and the affinities of Network Interface Cards (*NICs*). Our proposal and implementation of a NUIOA-aware multirail MPI is then described in Section 3 while its performance is presented in Section 4.

2   Background and Motivation

In this section, we describe the architecture of modern cluster nodes and introduce *Non-Uniform Input/Output Access* as a consequence that must definitely be taken into account in the design of communication algorithms and strategies.

2.1   Multicore and NUMA Architectures

Multicore processors have represented over 90 % of the 500 most powerful computing systems in the world[1] for the last five years. This hardware trend is currently leading to an increasing share of NUMA architectures (*Non-Uniform Memory Access*), the vast majority of recent cluster installations relying on INTEL NEHALEM or AMD OPTERON processors that have introduced scalable but non-uniform memory interconnects. AMD HYPERTRANSPORT and, more recently, INTEL QPI were designed towards this goal by attaching a memory node to each processor socket, as depicted on Figure 1. These increased complexity and hierarchical aspects, from multiple hardware threads, cores, shared caches to distributed memory banks, raise the need for carefully placing tasks and data according to their affinities. Once tasks are distributed among all the cores, an additional step is to optimize communication and synchronization between tasks depending on their topological distance within the machine [2].

In addition, the increasing number of cores in machines causes network interfaces and I/O buses to become potential bottlenecks. Indeed, concurrent requests from all cores may lead to contention and may thus reduce the overall application performance significantly [3]. Regarding this problem, multirail machines are now commonly considered as a workaround since their multiple NICs scale better with the number of cores. However, such complex architectures, interconnecting numerous hardware components, also raise the need to take affinities and locality into account when scheduling network processing [4].

2.2   Non-Uniform Input/Output Access

NUMA architectures have been the target of numerous research projects in the context of high-performance computing, from affinity-based OPENMP thread scheduling to MPI process placement [5,6]. The impact of process placement in NUMA machines on high-speed networking has been known for several years already. As shown on Figure 1, network interfaces may be closer to some NUMA nodes and processors than to the others, causing their data transfer performance

---

[1]   Top500, http://www.top500.org.

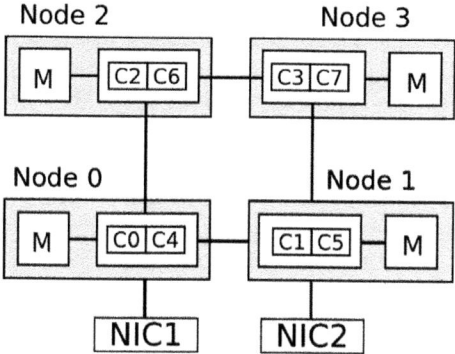

Fig. 1. Quad-socket dual-core host with two I/O chipsets connected to NUMA nodes #0 and #1

to vary. However, this property is almost only taken into account for microbenchmarks by binding processes as close as possible to the network interface.

As depicted by Figure 2, we demonstrated with previous OPTERON architectures and several network technologies that the actual throughput is dramatically related to process placement with regards to network interfaces [1]. While latency depends only slightly on process placement (usually less than 100 nanoseconds), we observed up to 40% throughput degradation when using multiple high-performance interconnects. This behavior, called *Non-Uniform Input/Output Access* (NUIOA), induces the need for a careful placement of tasks depending on their communication intensiveness.

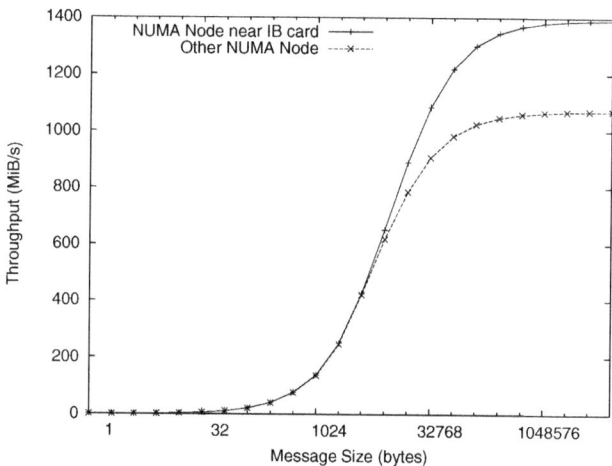

Fig. 2. Influence of the locality of processes and network cards on the RDMA Write throughput between 2 dual-OPTERON 265 machines with INFINIBAND DDR cards

This characteristic has been observed for various memory interconnects. It still appears with latest INTEL NEHALEM processors and QPI architectures. It is sometimes also referred to as *Non-Uniform Network Access* but it is actually not specific to network devices. Indeed, we observed DMA throughput degradation by up to 42% when accessing a NVIDIA GPU from the distant NUMA node of a dual-XEON 5550 machine. Moreover, in the presence of multiple devices, it becomes important to carefully distribute the workload among devices. Since these NUMA architectures are now spreading into high-performance computing, we intend to look at adapting the MPI implementation to these new constraints.

3   NUIOA-Aware Multirail

We introduce in this section our proposal towards a MPI implementation that adapts multirail communication to Non-Uniform I/O Access. We then describe how we implemented it in OPEN MPI.

3.1   Proposal

Dealing with affinities inside NUMA machines usually requires to place tasks with intensive inter-communication or synchronization inside the same NUMA node or shared cache. Meanwhile, distributing the workload across the whole machine increases the available processing power and memory bandwidth. Finding a tradeoff between these goals is difficult and depends on the application requirements. Adding the locality of network interfaces to the problem brings new constraints since some cores may have no I/O devices near them. It leads to the idea of keeping these cores for tasks that are not communication-intensive. Other processes may be given a privileged access to all or only some of the interfaces (they may be close to different cores). Moreover, detecting which tasks are communication-intensive may be difficult. And numerous MPI applications have uniform communication patterns since most developers try to avoid irregular parallelism so as to exploit all the processing cores.

While binding communication-intensive tasks near the network interfaces is not easy, we look at an orthogonal problem: to optimize the implementation of communication within a predefined process placement. This placement may have been chosen by the MPI process launcher depending on other requirements such as affinities between tasks [6]. Given a fixed distribution of processes on a NUIOA architecture, we propose to adapt the implementation of MPI communication primitives to better exploit multiple network interfaces.

3.2   Distributing Message Chunks According to NICs Localities

Several MPI implementations may use multiple network interfaces at the same time. For throughput reasons, large messages are usually split across all available *rails* and reassembled on the receiver side. OPEN MPI [7] and MPICH2/NEWMAD [8] may even use different models of interfaces and wires and dynamically adapt their utilization depending on their relative performance. For instance, a 3 MiB message would be sent as a 1 MiB chunk on a DDR IN-FINIBAND link and another 2 MiB chunk on a QDR link. As explained above, the actual throughput of these network rails depends on the process location. We propose to adapt the size of the chunks to the distance of each process from network interfaces.

We implemented this idea in OPEN MPI 1.4.1. Each network interface is managed by a BTL component (*Byte Transfer Layer*) that gathers its expected bandwidth by looking at its model and current configuration. By default, these bandwidths are accumulated in the BML (*BTL Multiplexing Layer*) so as to compute a *weight* for each BTL. Sending a large message then results in one chunk per BTL that connects the processes, and a *splitting ratio* is determined so that each chunk size is proportional to the BTL weight, as depicted on Figure 3.

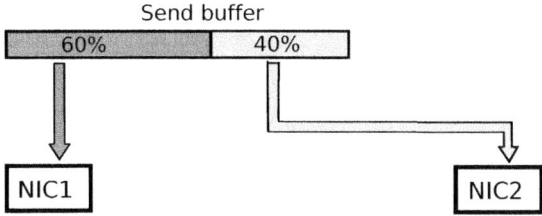

Fig. 3. Multirail using a 60% splitting ratio. 100% means that the whole message is sent through the NIC1.

We modified the R2 BML component so as to take hardware locality into account when computing these weights. The expected bandwidth of each BTL is adjusted by looking at the current process and BTL physical device locations. This change is specific to each process since it depends on its actual binding. It must thus be done late in the initialization phase, usually after the `paffinity` component (in charge of processor affinity) has done the actual binding of processes.

3.3 Gathering NIC and Process Locality Information

Our NUIOA-aware tuning of BTL weights relies on the knowledge of the location of each networking device in the underlying hardware topology. Most BIOSes tell the operating system which NUMA node is close to each I/O bus. It is thus easy to determine the affinity of each PCI device. However, user-level processes do not manipulate PCI devices, they only know about software handles such as INFINIBAND devices in the VERBS interface. Some low-level drivers offer a way to derive these software handles into hardware devices thanks to `sysfs` special files under LINUX. For other drivers such as MYRICOM MX, a dedicated command had to be added to retrieve the locality of a given MX endpoint.

We implemented in the HWLOC library (*Hardware Locality*[2]) the ability to directly return the set of cores near a given INFINIBAND software device.

---

[2] http://www.open-mpi.org/projects/hwloc/

OPEN MPI will switch to using HWLOC for process binding in the near future since it offers an extensive set of features for high-performance computing [9]. The BML will thus easily know where OPEN MPI binds each process. Until this is implemented, we let the `paffinity` component bind processes and later have the BML retrieve both process and network device locations through HWLOC. Once this information has been gathered, the BML adjusts the splitting ratio according to the actual performance of each BTL in this NUIOA placement.

4   Performance

We present in this section the performance evaluation of NUIOA effects, from single rail ping-pong to multirail MPI collective operations.

4.1   Experimentation Platform

The experimentation platform is composed of several quad-socket hosts with dual-core OPTERON 8218 processors (2.6 GHz). As depicted by Figure 1, this NUMA architecture contains four NUMA nodes, two of them being also connected to their own I/O bus. Each bus has a PCIe 8x slot where we plug either a MYRICOM MYRI-10G NIC or a MELLANOX MT25418 CONNECT-X DDR IN-FINIBAND card. These hosts run the *Intel MPI Benchmarks* (IMB) on top of our modified OPEN MPI 1.4.1 implementation, using the MX or OPENIB BTLs.

4.2   Single-Rail Micro-Benchmark

Figure 4 summarizes NUIOA effects on our experimentation platform by presenting the MPI throughput of a ping-pong between two hosts depending on the process binding. It confirms that NUIOA effects are indeed significant on our platform, whenever messages contain dozens of kilobytes.

However the actual impact depends a lot on the underlying networking hardware. Indeed, the throughput over MX with MYRI-10G cards varies only very slightly while the INFINIBAND throughput decreases by 23% when the process is not bound near the network interface. The raw INFINIBAND throughput being larger, it may be more subject to contention, but we do not feel that this fact would induce such a difference. Instead we think that these NICs may be using different DMA strategies to transfer data inside the host, causing the congestion to differ. Indeed, if INFINIBAND uses smaller DMA packets, more packets are in-flight at the same time on the HYPERTRANSPORT bus, causing more saturation of the HYPERTRANSPORT request and response buffers.

Moreover, we also observed that increasing the NUMA distance further does not further decrease the INFINIBAND throughput: once the process is not near the card, the throughput does not vary anymore when binding it farther away.

4.3   Point-to-Point Multirail

Figure 5 now presents the multirail throughput for large messages depending on process placement when an INFINIBAND NIC is connected to each I/O bus.

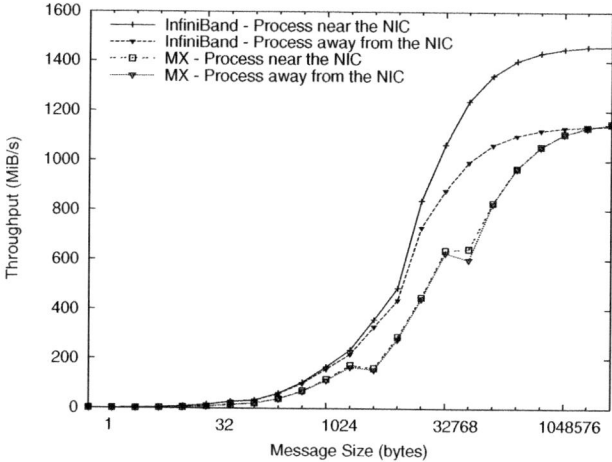

Fig. 4. Influence of the locality of processes and network cards on a single-rail IMB ping-pong with OPEN MPI

It clearly shows that when the process is bound near one of the cards (NUMA node #0 or #1), the MPI implementation should privilege this card by assigning about 58% of the message size to it. This ratio is actually very close to the ratio between monorail throughputs that we may derive from Figure 4 (56.6%). Such a tuning offers 15% better throughput than the usual half-size splitting. Moreover, when the process is not close on any NIC (NUMA node #2 or #3), the messages should be split in half-size chunks as usual. Again, this could have been derived from Figure 4 since monorail throughput does not vary from with the binding when not close to the NIC.

When MYRI-10G NICs are used, the closest NIC should only be very slightly privileged (51%). It is also expected since MYRI-10G performance varies only slightly with process placement. Finally, combining one MYRI-10G and one IN-FINIBAND NIC also matches our expectations: when a process is close to the INFINIBAND NIC, it should privilege it significantly (57%), while a process near the MYRI-10G NIC should only privilege it slightly (51%).

These results confirm that combining monorail throughputs from the given process bindings is an interesting way to approximate the optimal multirail ratio for point-to-point operations, as suggested in earlier works [10].

4.4   Contention

We now look at the impact of contention on the memory bus on our NUIOA-aware multirail. Indeed, previous results used idle hosts where data transfers between NUMA nodes and the NICs were optimal. In the case of real applications, some processors may access each others' memory, causing contention on the HYPERTRANSPORT links.

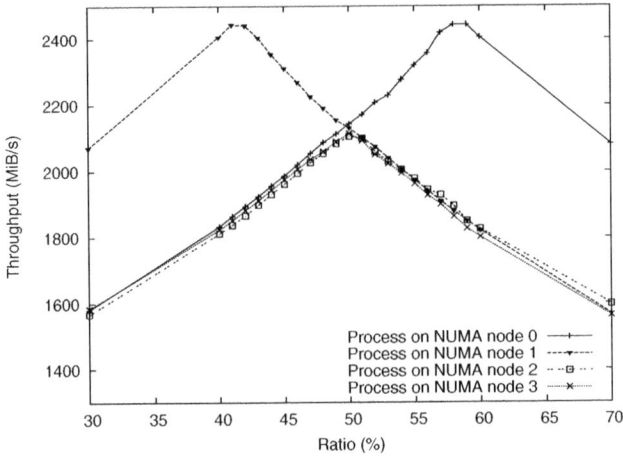

Fig. 5. Multirail IMB ping-pong throughput for 1 MiB messages, depending on the ratio between two INFINIBAND cards and on the process placement

We added such contention on some HYPERTRANSPORT links during our afore-mentioned multirail IMB ping-pong. This reduces the network throughput but does not modify the splitting ratio. This is a surprising result since we had carefully chosen which HYPERTRANSPORT link should get contention so as to disturb the data path towards a single NIC and not the other (thanks to the HYPERTRANSPORT routing table). If the ratio does not change, then it means that our contention on the single link reduces the overall memory bandwidth instead of only the bandwidth on this link.

4.5 Collective Operations

Previous sections showed that an interesting splitting of large point-to-point mes-sages across multiple rails may be derived from each rail NUIOA throughput. We now look at collective MPI operations. Since processes are now communicating from all NUMA nodes at the same time, we should now find the splitting ratio of each running process simultaneously. Figure 6 presents the performance of the all-to-all operation depending on the splitting ratios. We distinguish processes depending on whether they are close to one NIC or not.

The optimal tuning that we obtained first reveals that processes that are not close to any NIC should send one half of each message on each NIC. This result matches our earlier point-to-point observations since NUIOA effects do not matter once processes are far from NICs, but it seems less significant here. Then, the interesting result is that processes close to one NIC should only use this NIC. This result contradicts the previous section since contention now appears as critical for performance. We think that contention in this all-to-all benchmark were more intensive than in the previous section, causing the ratio to vary. In the end, this all-to-all tuning outperforms the default splitting strategy by 5%.

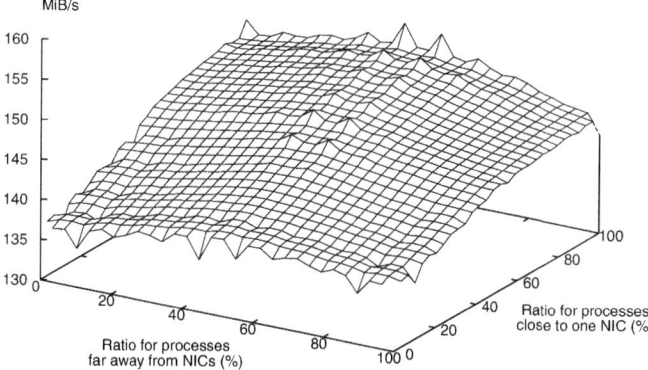

Fig. 6. IMB all-to-all, throughput per process between 16 processes on 2 hosts with 2 INFINIBAND NICs, depending on multirail splitting ratios

When using one INFINIBAND and one MYRI-10G NIC, we obtained similar results. However, when using two MYRI-10G NICs we again observed less NUIOA effects since the ratio almost does not matter (the variation of results among multiple runs is larger than the variation due to splitting ratios). Looking at other collectives, we observed that communication intensive operations, such as allgather, tend towards all-to-all ratios while other operations show barely noticeable variations with changes of splitting ratios.

5   Conclusion and Future Works

The increasing number of cores and the widespread use of NUMA architectures in cluster computing nodes leads to the multiplication of connected hardware components. It raises the need to take affinities and localities into account in the design of communication strategies. Indeed, the performance of communication over high-speed networks is directly related to the relative location of processes and network interfaces.

In this paper, we propose to optimize the implementation of MPI primitives by adapting the use of multiple network interfaces to their locations with regards to processes. Thanks to the knowledge of process/NIC affinities in HWLOC, we determine a splitting ratio that increases the throughput by up to 15% over the standard multirail strategy. Communication-intensive patterns such as all-to-all even show that processes that are close to one NIC should not use other NIC so as to avoid contention on the memory bus.

Combined with per-core sampling of the interfaces, or even auto-tuning [11], this approach lets us envision a far better utilization of multiple rails. We plan to integrate our work in the mainline OPEN MPI implementation and further experiment with it on real applications. We also intend to integrate this knowledge about NIC affinities in collective algorithms, where the role of each process may differ (e.g. local root of a reduction) and thus where each process would certainly use different thresholds.

References

1. Moreaud, S., Goglin, B.: Impact of NUMA Effects on High-Speed Networking with Multi-Opteron Machines. In: The 19th IASTED International Conference on Parallel and Distributed Computing and Systems (PDCS 2007), Cambridge, Massachussetts (2007)
2. Buntinas, D., Goglin, B., Goodell, D., Mercier, G., Moreaud, S.: Cache-Efficient, Intranode Large-Message MPI Communication with MPICH2-Nemesis. In: Proceedings of the 38th International Conference on Parallel Processing (ICPP-2009), Vienna, Austria, pp. 462–469. IEEE Computer Society Press, Los Alamitos (2009)
3. Narayanaswamy, G., Balaji, P., Feng, W.: Impact of Network Sharing in Multi-core Architectures. In: Proceedings of the IEEE International Conference on Computer Communication and Networks (ICCCN), St. Thomas, U.S. Virgin Islands (2008)
4. Jang, H.C., Jin, H.W.: MiAMI: Multi-core Aware Processor Affinity for TCP/IP over Multiple Network Interfaces. In: Proceedings of the 17th Annual Symposium on High-Performance Interconnects (HotI 2009), New York, NJ, pp. 73–82 (2009)
5. Rabenseifner, R., Hager, G., Jost, G.: Hybrid MPI/OpenMP Parallel Programming on Clusters of Multi-Core SMP Nodes. In: Proceedings of the 17th Euromicro International Conference on Parallel, Distributed, and Network-Based Processing (PDP 2009), Weimar, Germany, pp. 427–436 (2009)
6. Mercier, G., Clet-Ortega, J.: Towards an Efficient Process Placement Policy for MPI Applications in Multicore Environments. In: Ropo, M., Westerholm, J., Dongarra, J. (eds.) Recent Advances in Parallel Virtual Machine and Message Passing Interface. LNCS, vol. 5759, pp. 104–115. Springer, Heidelberg (2009)
7. Gabriel, E., Fagg, G.E., Bosilca, G., Angskun, T., Dongarra, J.J., Squyres, J.M., Sahay, V., Kambadur, P., Barrett, B., Lumsdaine, A., Castain, R.H., Daniel, D.J., Graham, R.L., Woodall, T.S.: Open MPI: Goals, concept, and design of a next generation MPI implementation. In: Proceedings of 11th European PVM/MPI Users' Group Meeting, Budapest, Hungary, pp. 97–104 (2004)
8. Mercier, G., Trahay, F., Buntinas, D., Brunet, É.: NewMadeleine: An Efficient Support for High-Performance Networks in MPICH2. In: Proceedings of 23rd IEEE International Parallel and Distributed Processing Symposium (IPDPS 2009), Rome, Italy. IEEE Computer Society Press, Los Alamitos (2009)
9. Broquedis, F., Clet-Ortega, J., Moreaud, S., Furmento, N., Goglin, B., Mercier, G., Thibault, S., Namyst, R.: hwloc: a Generic Framework for Managing Hardware Affinities in HPC Applications. In: Proceedings of the 18th Euromicro International Conference on Parallel, Distributed and Network-Based Processing (PDP 2010), Pisa, Italia. IEEE Computer Society Press, Los Alamitos (2010)
10. Aumage, O., Brunet, E., Mercier, G., Namyst, R.: High-Performance Multi-Rail Support with the NewMadeleine Communication Library. In: Proceedings of the Sixteenth International Heterogeneity in Computing Workshop (HCW 2007), held in conjunction with IPDPS 2007, Long Beach, CA (2007)
11. Pellegrini, S., Wang, J., Fahringer, T., Moritsch, H.: Optimizing MPI Runtime Parameter Settings by Using Machine Learning. In: Ropo, M., Westerholm, J., Dongarra, J. (eds.) Recent Advances in Parallel Virtual Machine and Message Passing Interface. LNCS, vol. 5759, pp. 196–206. Springer, Heidelberg (2009)

Using Triggered Operations to Offload Collective Communication Operations

K. Scott Hemmert[1], Brian Barrett[1], and Keith D. Underwood[2]

[1] Sandia National Laboratries*
P.O. Box 5800, MS-1110
Albuquerque, NM, 87185-1110
kshemme@sandia.gov, bwbarre@sandia.gov
[2] Intel Corporation
Hillsboro, OR, USA
keith.d.underwood@intel.com

Abstract. Efficient collective operations are a major component of application scalability. Offload of collective operations onto the network interface reduces many of the latencies that are inherent in network communications and, consequently, reduces the time to perform the collective operation. To support offload, it is desirable to expose semantic building blocks that are simple to offload and yet powerful enough to implement a variety of collective algorithms. This paper presents the implementation of barrier and broadcast leveraging triggered operations — a semantic building block for collective offload. Triggered operations are shown to be both semantically powerful and capable of improving performance.

1   Introduction

Although the vast majority of data volume that is transferred within science and engineering applications is in relatively localized, point to point communications, these applications also include some number of global communications known as collectives. Many collective communications are inherently less scalable, as they involve communications all the way across the machine and contributions from every node. As system sizes increase, it becomes increasingly difficult to implement fast collectives across the entire system. One approach to improving collective performance is to offload collective operations to the network.

Many prior approaches to offloading collective operations have offloaded the entire collective operation, including the communication setup and computation [1]. While this eliminates the host overhead, it creates a more complicated offload function that is harder to adapt over time. As an alternative, a similar level of offload can be achieved with more elementary building blocks that are both easier to implement in hardware and less subject to change. When building

---

* Sandia is a multiprogram laboratory operated by Sandia Corporation, a Lockheed Martin Company, for the United States Department of Energy's National Nuclear Security Administration under contract DE-AC04-94AL85000.

R. Keller et al. (Eds.): EuroMPI 2010, LNCS 6305, pp. 249–256, 2010.

blocks are provided, the host library (e.g., MPI) is able to more readily adopt new collective algorithms as they are being developed. Alternatively, the host can more easily tune the algorithm based on the size of the system, the layout of the job on the system, and the size of the collective.

Portals 4 [2] introduced a set of semantic building blocks that included triggered operations and counting events that were explored for MPI_ALLREDUCE in [3]. Triggered operations allow an application to schedule a new network operation to occur in the future when a counting event reaches a specified threshold. This paper illustrates the breadth of triggered operations for implementing collective algorithms ranging from tree based barriers to dissemination barriers to bulk data broadcasts. Simulation results show the performance improvements that can be achieved using offload through triggered operations.

2   Related Work

Offload of collective operations has been an active area of research for many years. Custom engineered systems like the Cray T3D provided hardware barrier synchronization [4] and IBM's BG/L provided a dedicated collective network. Similarly, research into hardware support for collective operations on commodity hardware began in the mid-1990's [5]. This work became more prevalent with the arrival of programmable network interfaces like Myrinet and Quadrics. Barrier [1,6] and broadcast [7] are particularly popular targets.

Offloading collective operations onto a Myrinet NIC requires significant enhancements to the control program running on the NIC processor. Because this requires significant effort for each collective operation offloaded, mechanisms to provide more dynamic offloaded capability were proposed [8]. Unlike Myrinet, the Quadrics Elan network supported a user-level thread on the NIC. Because this user-level thread has direct access to the address space of the process that created it, it is easier to create extended functionality to offload collectives. The programming environment for the Elan adapters provides some key functionality. For example, Elan event functions can increment a counter by a user specified amount when an operation, such as a DMA transfer, completes. Events could be chained to allow the triggering of one event to trigger others. Elan events are very similar to the counting events that were added to the Portals [2] API.

3   Triggered Operations in Portals 4

Triggered operations and counting events were introduced into Portals 4 [2] as semantic building blocks for collective communication offload. Triggered operations provide a mechanism through which an application can schedule message operations that initiate when a condition is met. Triggered versions of each of the Portals data movement operations were added (e.g., PtlTriggeredPut(), PtlTriggeredGet(), and PtlTriggeredAtomic()) by extending the argument list to include a counting event on which the operation will trigger and a threshold at which it triggers. In turn, counting events are the lightweight semantic provided to track the completion of network operations. Counting events are opaque objects containing an integer that can be allocated, set to a value, or incremented by a value through the Portals API. In addition, they can be attached to various Portals structures and configured to count a variety of network operations, such as the local or remote completion of a message as well as the completion of incoming operations on a buffer (e.g., the completion of a PtlPut() or PtlAtomic() to a local buffer).

A triggered operation is issued by the application and then initiated by the network layer when a counting event reaches a threshold. Through careful use of counting events and triggered operations, an almost arbitrary sequence of network operations can be setup by the application and then allowed to progress asynchronously. A discussion of how reduction operations can be implemented using triggered operations is presented in [3].

4   Evaluation Methodology

The Structural Simulation Toolkit (SST) v2.0 [9] was used to simulate both host-based and offloaded versions of several collective algorithms. SST provides a component-based simulation environment, designed for simulating large-scale HPC environments. It simultaneously provides both cycle-accurate and event-based simulation capabilities. Here, we present both the algorithms simulated and a description of the parameters used for simulation.

4.1   Collective Algorithms

Three barrier algorithms were simulated for both a host based and a triggered operation based implementation. The first algorithm was a binomial tree (not shown) with the experimentally determined optimal radix chosen for both the host and the triggered cases. The tree algorithm is similar to what was explored for Allreduce in prior work [3]. The second algorithm used was the recursive doubling algorithm (also not shown), which is a simplified variant of the Allreduce in [3], since no data movement is required.

The final algorithm explored is the dissemination barrier [10]. In the radix-2 version, the dissemination barrier has a series of rounds, R, where each node, N, sends a message to node $(N + 2^R)$ mod P. A message in a given round can only be sent after messages for all prior rounds have been received. Because some nodes can proceed through the rounds faster than others, a node must receive a specific set of messages before proceeding. This is synonymous with receiving the message for this round and having completed the previous round, which is how the algorithm in Figure 1 is structured. Figure 1 is also extended to show a higher radix dissemination barrier algorithm.

A binomial tree algorithm is used for broadcast. Figure 2 shows how triggered operations can be leveraged for a rendezvous style protocol implementing a tree. At communicator creation, each node creates a descriptor to receive messages from their "parent" in the tree. When the collective is initiated, children determine who their parent will be based on the root and issue a triggered get.

```
//Round 0 message from self when we enter
for (j = 1; j < radix; j++) PtlPut(user_md_h, (id+j) % num_nodes, 0);
//Signal round 1. Only receive radix−1 messages and not signal from previous round
PtlTriggeredCTInc(level_ct_hs[1], 1, level_ct_hs[0], radix−1);
PtlTriggeredCTInc(level_ct_hs[0], −(radix−1), level_ct_hs[0], radix−1);
for (i = 1, level = 0x2 ; level < num_nodes ; level <<= log2(radix), ++i) {
    for (j = 0; j < (radix−1); ++j) {
        remote = (id + level + i) % num_nodes;
        // Start round i when input from round i − 1 peer arrives and
        // communication to round i − 1 completes
        PtlTriggeredPut(md_h, remote, i, level_ct_hs[i], radix);
    }
    //Signal round i+1 that round i (and all previous rounds) is done
    PtlTriggeredCTInc(level_ct_hs[i+1], 1, level_ct_hs[i], radix);
    //Clean−up this iteration
    PtlTriggeredCTInc(level_ct_hs[i], −radix, level_ct_hs[i], radix);
}
// wait for completion and clean up last level
PtlCTWait(level_ct_h[levels], 1);
PtlTriggeredCTInc(level_ct_hs[levels], −1, level_ct_hs[levels], 1);
```

Fig. 1. Pseudo-code for the triggered dissemination barrier algorithm

When the data is available in the local buffer, the parent notifies the child, which increments the counting event that releases the triggered get. The algorithm is pipelined by issuing multiple triggered gets with offsets that trigger at different count thresholds. Short messages are sent using a puts into the bounce buffer, with a user-level copy on completion.

4.2   Simulation Model

The collective operation simulations utilize a cycle-based router and network model combined with an event driven model of the network interface and the host. A torus network of up to 32K nodes ($32 \times 32 \times 32$) was simulated. Simulations were run with and without simulated OS interference to determine the success of offloaded implementations in eliminating noise. The router simulation matched those used in earlier simulations [9]. In contrast, the node was modeled as a simple state machine. Message insertion rate, delays for copying data to the NIC, and delays associated with memory copies were all modeled as interrelated occupancies in a queuing model. The NIC used a similar set of occupancies to model the NIC level operations and fed data (packets) to the router model.

Key parameters were modeled for both the NIC and host processing times: bus delays, delays through the NIC, occupancy in the receive processing logic, and memory latencies. Parameters that were used corresponded to network latencies of 1 μs and 1.5 μs and are shown in Table 1. The one way message rates are limited by the highest latency, unpipelined processing stage. Since the hardware

```
if (my_root == my_id) {
    /* Notify children that all chunks are ready */
    for (j = 0 ; j < msg_size ; j += chunk_size)
        for (i = 0 ; i < num_children ; ++i) PtlPut(bounce_md, 0, 0, 0, child[i], 0);
} else {
    /* iterate over chunks */
    for (offset = 0 ; offset < msg_size ; offset += chunk_size) {
        /* when a chunk is ready, issue get. Local and remote offset are the same */
        PtlTriggeredGet(out_md, offset, chunk_size, my_root, offset,
                bounce_ct, j / chunk_size);
        /* then when the get is completed, send ready notice to children */
        for (i = 0 ; i < num_children ; ++i)
            PtlTriggeredPut(bounce_md, 0, 0, 0, child[i], 0, out_md_ct, offset / chunk_size);
    }
    /* reset 0−byte put received counter */
    PtlTriggeredCTInc(bounce_ct, −count, bounce_ct, count);
}
/* wait for children gets */
if (num_children > 0) PtlCTWait(out_me_ct, count);
/* wait for local gets to complete */
else PtlCTWait(out_md_ct, count);
```

Fig. 2. Pseudo-code for the long message triggered binomial tree broadcast

stages are pipelined, the message rate limiter is the software, which yields 5.7 million messages per second (Mmsgs/s) and 3.3 Mmsgs/s for 1 μs and 1.5 μs latency, respectively. The rate at which the NIC can issue triggered operations is limited by the 8 flit header and the 500 MHz clock to yield 62.5 Mmsgs/s (once the operations are queued on the NIC). To enqueue triggered operations, the software faces the same limitations as message transmits; therefore, triggered operations can be enqueued at 10 Mmsgs/s and 5 Mmsgs/s, based on the 100 ns and 200 ns TX software delays, respectively. As a final parameter, setup time for the collective operation — time needed by MPI before communication starts to setup the algorithm — is set to 200 ns. In addition to the baseline simulations, we run simulations representing the impact of OS noise. One "long noise at infrequent interval" signature (25 μs at 1 KHz) from a previous study [11] is used to represent OS noise.

Table 1. Summary of simulation parameters

Msg Latency	1000 ns	1500 ns	Msg Latency	1000 ns	1500 ns
TX Software Delay	100 ns	200 ns	RX Software Delay	175 ns	300 ns
TX Bus Delay	200 ns	300 ns	RX Bus Delay	200 ns	300 ns
TX NIC Delay	75 ns	100 ns	RX NIC Delay	150 ns	200 ns
Memory Latency	100 ns	100 ns	Read over Bus	400 ns	500 ns

5   Results

Figure 3 compares the performance of the three barrier algorithms with 1000 ns and 1500 ns latency for both host and triggered implementations. The triggered implementation has over a 2× advantage that is larger at higher latency, since the triggered operations experience lower *effective* latency and higher *effective* message rate when they actually issue from the NIC. Much of the real latency and message posting overheads are overlapped with other communications for triggered operations (i.e., they happen before the host implementation is free to initiate the messages). Note that we used a Radix-8 implementation for the triggered dissemination barrier, which gave it a significant advantage over the very similar recursive doubling algorithm using Radix-2. Radix-8 results for the host variant (not shown) were far worse than Radix-2.

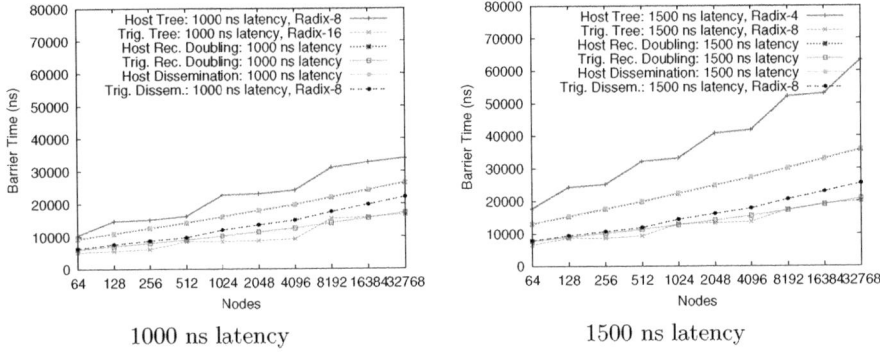

1000 ns latency   1500 ns latency

Fig. 3. Simulated barrier time

Figure 4 presents results of simulations with noise. While the barrier time increases substantially for host based implementations and shows growing impact as the number of nodes increases, the barrier time for implementations using triggered operations shows more modest impact from noise and the noise impact levels off at large node counts. In addition, note that the introduction of noise changes the "right" algorithm to use. Both dissemination and recursive doubling algorithms require processing by every node at every stage, but a tree based algorithm uses logarithmically fewer nodes at each stage. This carries over to triggered operations, since all nodes spend a significant amount of time injecting messages in the dissemination and recursive doubling barrier algorithms, but most nodes inject few messages in the binomial tree.

Broadcast performance is presented in Figure 4 for small messages (8 bytes) and in Figure 5 for a sweep over larger messages at 4096 nodes. At small messages, the triggered operations provide a 15–20% performance improvement over host based algorithms, in addition to substantially less noise sensitivity. At larger message sizes, however, broadcast using triggered operations has a smaller performance and noise sensitivity advantage. Serialization delay to transfer data

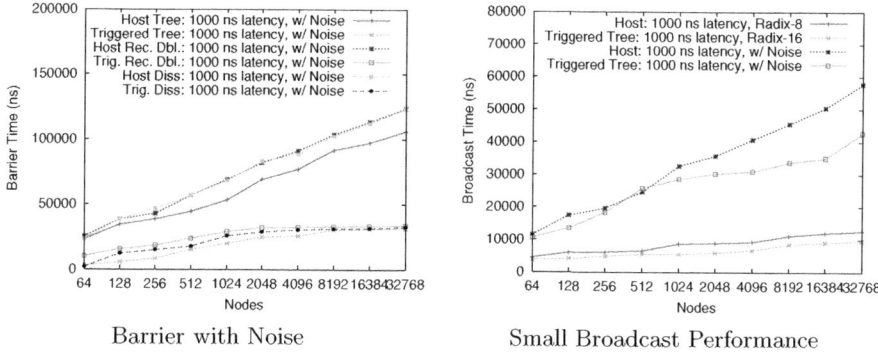

Barrier with Noise   Small Broadcast Performance

Fig. 4. Simulated time with Noise

dominates both noise and processor overheads, which can be seen by the convergence of the host and triggered results in Figure 5. The triggered technique shows promise for non-blocking collectives, however, as it offers similar performance to host-based collectives with minimal processor overhead after an initial setup period. The use of a multi-post triggered interface to save round-trip communication with the NIC when setting up the messaging pipeline would further reduce the (small) processor overhead experienced for triggered bcasts of large messages and will be examined in future work.

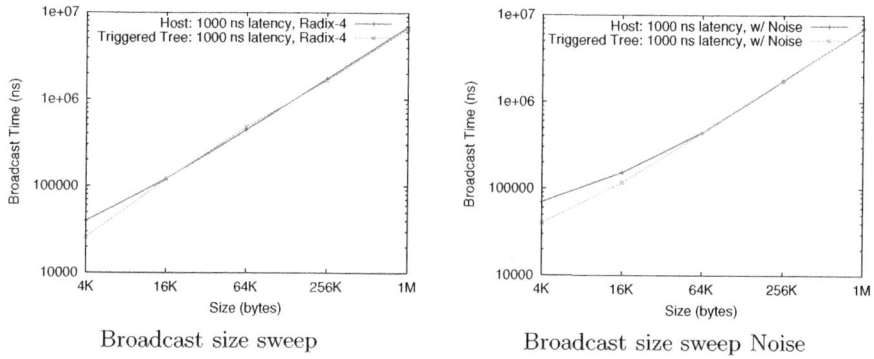

Broadcast size sweep   Broadcast size sweep Noise

Fig. 5. Sweep of broadcast size with and without noise

6   Conclusions

This paper has illustrated that triggered operations leveraging counting events are semantically sufficient to implement a variety of collective algorithms. Pseudo-code was shown for a rendezvous-like functionality for long broadcasts and pseudo-code was shown for higher radix dissemination barriers. Collectives based on triggered operations are shown to be both higher performing (by over 2×) and more resistant to interference from system noise.

256   K.S. Hemmert, B. Barrett, and K.D. Underwood

References

1. Buntinas, D., Panda, D.K., Sadayappan, P.: Fast NIC-based barrier over Myrinet/GM. In: Proceedings of the International Parallel and Distributed Processing Symposium (April 2001)
2. Riesen, R.E., Pedretti, K.T., Brightwell, R., Barrett, B.W., Underwood, K.D., Hudson, T.B., Maccabe, A.B.: The Portals 4.0 message passing interface. Technical Report SAND2008-2639, Sandia National Laboratories (April 2008)
3. Underwood, K.D., Coffman, J., Larsen, R., Hemmert, K.S., Barrett, B.W., Brightwell, R., Levenhagen, M.: Enabling flexible collective communication offload with triggered operations. Submitted to Proceedings of the 2010 IEEE International Conference on Cluster Computing (September 2010)
4. Scott, S.L., Thorson, G.: Optimized routing in the Cray T3D. In: Bolding, K., Snyder, L. (eds.) PCRCW 1994. LNCS, vol. 853, pp. 281–294. Springer, Heidelberg (1994)
5. Yih Huang, P.K.M.: Efficient collective operations with ATM network interface support. In: Proceedings of the International Conference on Parallel Processing, August 1996, pp. 34–43 (1996)
6. Yu, W., Buntinas, D., Graham, R.L., Panda, D.K.: Efficient and scalable barrier over Quadrics and Myrinet with a new NIC-based collective message passing protocol. In: Proceedings of the Workshop on Communication Architecture for Clusters (April 2004)
7. Buntinas, D., Panda, D.K., Duato, J., Sadayappan, P.: Broadcast/multicast over Myrinet using NIC-assisted multidestination messages. In: Proceedings of the Fourth International Workshop on Communication, Architecture, and Applications for Network-Based Parallel Computing (January 2000)
8. Wagner, A., Jin, H.-W., Panda, D.K., Riesen, R.: NIC-based offload of dynamic user-defined modules for Myrinet clusters. In: Proceedings of the 2004 IEEE International Conference on Cluster Computing, September 2004, pp. 205–214 (2004)
9. Underwood, K.D., Levenhagen, M., Rodrigues, A.: Simulating Red Storm: Challenges and successes in building a system simulation. In: 21st International Parallel and Distributed Processing Symposium (IPDPS 2007) (March 2007)
10. Hoefler, T., Mehlan, T., Mietke, F., Rehm, W.: Fast barrier synchronization for InfiniBand. In: 20th International Parallel and Distributed Processing Symposium, IPDPS 2006. (April 2006)
11. Ferreira, K.B., Bridges, P., Brightwell, R.: Characterizing application sensitivity to OS interference using kernel-level noise injection. In: SC 2008: Proceedings of the, ACM/IEEE conference on Supercomputing, pp. 1–12. IEEE Press, Piscataway (2008)

Second-Order Algorithmic Differentiation by Source Transformation of MPI Code

Michel Schanen[*], Michael Förster, and Uwe Naumann

LuFG Informatik 12: Software and Tools for Computational Engineering
RWTH Aachen University, Germany
{schanen,foerster,naumann}@stce.rwth-aachen.de
http://www.stce.rwth-aachen.de

Abstract. A source transformation tool for algorithmic differentiation is introduced, capable of transforming MPI-enabled code into second-order adjoint code. Our derivative code compiler (dcc) is used for the source transformation while a runtime library handles the adjoining of the MPI routines. This paper describes in detail the link between these two components in order to compute second derivatives. This process is illustrated by a simplified parallel implementation of Burgers' equation in a second-order optimization setting, for example, Newton's method.

1 Introduction

Burgers' equation $\frac{\partial u}{\partial t} + u\frac{\partial u}{\partial x} = \nu\frac{\partial^2 u}{\partial x^2}$ [1] is used in fluid dynamics to describe shock waves moving through gases. u is the velocity field of the fluid with viscosity ν. Without going into the physical details, such types of equations represent the basis of many numerical simulations. Suppose we have a one-dimensional discrete problem with n_x finite difference grid points. We simulate a physical process by integrating over n_t time steps for given initial conditions $u_i, 0 \leq i < n_x$. Since the initial conditions u_i often cannot be measured, they are replaced by guessed values. To improve their accuracy additional observed values u_{ob} are taken into account. The cost function

$$cost = \sum_{i=0}^{n_x} \sum_{j=0}^{n_t} \frac{(u[i][j] - u_{ob}[i][j])^2}{2} \tag{1}$$

then compares the observed values u_{ob} with the computed values u. This allows us to optimize the initial conditions by applying, for example, Newton's method [2] to perform a parameter estimation with Burgers' equation as constraints. The Newton step is repeated until the cost undercuts a certain threshold. Accurate second derivatives are highly desirable.

Modern large-scale simulations exploit parallelism. We assume that the message passing library MPI [3] is used in a Burgers simulation to implement a

---

[*] This work was supported by the Fond National de la Recherche of Luxembourg under grant PHD-09-145.

R. Keller et al. (Eds.): EuroMPI 2010, LNCS 6305, pp. 257–264, 2010.

standard reduction over the cost function. This paper introduces an extension to the derivative code compiler dcc [4] enabling the differentiation of MPI code based on a subset of C by accessing an external library for adjoining MPI calls [5].

In Sect. 2 we introduce algorithmic differentiation as implemented by dcc followed by a brief overview of the adjoint MPI library. The link between these two components is the wrapper introduced in Sect. 3. Section 4 discusses issues related to non-blocking communication. Finally, the Burgers test case is considered in Sect. 5.

2 Derivative Code Compiler

Suppose we have an implementation of a multivariate vector function $y = F(x)$: $\mathbb{R}^n \rightarrow \mathbb{R}^m$. There exist two distinct derivative models, differing in the order of application of the chain rule. Let ∇F be the Jacobian of F. The *tangent-linear* model of F computes the directional derivative $\dot{y} = \nabla F(x) \cdot \dot{x}$ of the outputs y with respect to the inputs x for a given direction $\dot{x} \in \mathbb{R}^n$. The runtime complexity for accumulating the whole Jacobian is $O(n) \cdot Cost(F)$, where $Cost(F)$ denotes the computational cost of a single function evaluation. By exploiting the associativity of the chain rule, the *adjoint model* of F computes *adjoints* $\bar{x}$

$$\bar{x} = (\nabla F(x))^T \cdot \bar{y} \tag{2}$$

of the outputs y with respect to the inputs x for given adjoints $\bar{y} \in \mathbb{R}^m$ leading to a runtime complexity of $O(m) \cdot Cost(F)$ for the accumulation of the entire Jacobian.

The generalized model for arbitrary orders of algorithmic differentiation (AD)[1] is presented in [6]. dcc semi-automatically generates such derivative models for computer programs written in a well-defined subset of C++. The jth-order derivative code is generated by reapplication of dcc to the $(j-1)$th-order derivative code. Moreover, dcc implements both the tangent-linear and adjoint modes of AD while preserving the reapplication feature. In this paper we use dcc to generate second-order adjoint code by reapplying it in tangent-linear mode to the adjoint model of the cost function (1). This process is illustrated in Fig. 1.

For second derivatives every variable $y, \bar{y}, x$, and $\bar{x}$ of the adjoint code is augmented with an additional directional derivative component $\dot{y}, \dot{\bar{y}}, \dot{x}$, and $\dot{\bar{x}}$, respectively. It follows that each incrementation of the differentiation order doubles the number of arguments.

When computing the cost of our parameter estimation problem by MPI's reduction operation we have to deal with a code that is similar to the code fragment of the function f in Listing (1). Without loss of generality we omit all other MPI_Reduce arguments besides the master process id root, the send buffer c, and the receive buffer cost. Additionally, we inserted a multiplication statement to illustrate the differentiation model.

---

[1] Refer to the AD community's web portal http://www.autodiff.org/ for information on research groups, AD tools, and an extensive bibliography.

$$F\left(\overset{\downarrow}{\underset{\downarrow}{x}}, y\right) \xrightarrow{dcc} \bar{F}\left(\overset{\downarrow}{x}, \underset{\downarrow}{\bar{x}}, y, \underset{\downarrow}{\bar{y}}\right) \xrightarrow{dcc} \dot{\bar{F}}\left(\overset{\downarrow}{x}, \overset{\downarrow}{\dot{x}}, \underset{\downarrow}{\bar{x}}, \underset{\downarrow}{\dot{\bar{x}}}, y, \dot{y}, \underset{\downarrow}{\bar{y}}, \underset{\downarrow}{\dot{\bar{y}}}\right) , \text{ where}$$

$$\dot{\bar{x}} = \dot{x}^T \cdot \nabla^2 F\left(x\right) \cdot \bar{y} + \nabla F\left(x\right)^T \cdot \dot{\bar{y}} \qquad\qquad \dot{y} = \nabla F\left(x\right) \cdot \dot{x}$$
$$\bar{x} = \nabla F\left(x\right)^T \cdot \bar{y} \qquad\qquad\qquad\qquad y = F\left(x\right)$$

Fig. 1. Second-order adjoint derivative model generated by dcc. Upper arrows mark the inputs of a program whereas base arrows mark the outputs.

```
void f(double *x, double *cost) {
    ...
    c=c*x[myid];
    MPI_Reduce(cost,c,root);
}
```

Listing 1. dcc input: c is a double; myid is the process id.

Application of dcc in adjoint mode (2) yields the function b1_f shown in Listing (2). All added code segments are highlighted. First the name of the function is prefixed with b1_ (b for bar used to denote adjoints and 1 for first derivatives). Each of the arguments (x,cost) is augmented with its respective adjoint (b1_x, b1_cost). The adjoint code is split into two separate runs indicated by the parameter m.

```
1    void b1_f(int m, double* x, double* b1_x, double* cost, double* b1_cost
          ) {
2        m=1; // forward run
3        ...
4        push(c); c=c*x[myid];
5        b1_MPI_Reduce(m, cost, b1_cost, c, b1_c,root);
6
7        m=2; // reverse run
8        b1_MPI_Reduce(m, cost, b1_cost, c, b1_c,root);
9        pop(c);
10       b1_x[myid]+=c*b1_c; b1_c+=x[myid]*b1_c;
11       ...
12   }
```

Listing 2. Adjoint dcc output

The *forward run* (m=1) computes cost as a function of x. When the value of the variable c is overwritten, it is pushed onto a stack for later use in the computation of the adjoints. Note that the current value of c enters the computation of the adjoint b1_x[myid] as the local partial derivative in line 10. Overwriting c in line 4 without saving the current value and restoring it in line 9 would result in an incorrect b1_x[myid].

After the forward run the *reverse run* is started computing the adjoints b1_x as a function of b1_cost. It is executed in reverse order of the original computation. Hence we first execute b1_MPI_Reduce. Here, b1_cost is actually broadcast to all processes where it is saved in b1_c.

Each MPI routine is either called in forward (m=1) or reverse (m=2) mode. Therefore b1_MPI_Reduce is called twice, once to reduce c to cost and a second time to broadcast the adjoint b1_cost to the adjoints b1_c of every process amounting to the adjoint reduction operation.

Since computing the adjoints involves the reversal of the whole program, this also implies the reversal of all communication inside MPI. A Send has to become a Receive, a Receive becomes a Send and so on. In our example, MPI_Reduce has to become a MPI_Broadcast in the reverse run (Fig. 2). A general approach is needed to transform each MPI routine into its adjoint counterpart. This logic is provided by a basic library written in C as presented in [5]. Each MPI routine is associated with a forward routine, essentially the same as the original MPI call and a backward routine, propagating the adjoints. We have implemented an adjoint MPI library (AMPI) which provides the most commonly used MPI routines in their forward and backward variants, including non-blocking communication.

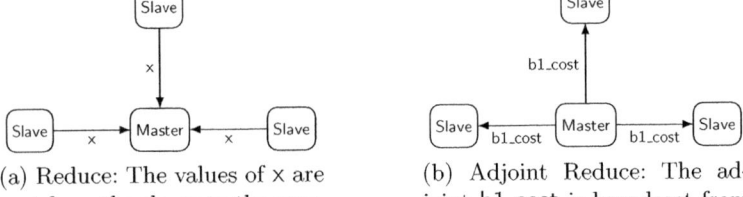

(a) Reduce: The values of x are sent from the slaves to the master.

(b) Adjoint Reduce: The adjoint b1_cost is broadcast from the master to the slaves.

Fig. 2. Adjoint MPI

The essential part of our work, is to link the generated b1_MPI calls with the underlying AMPI routines. The corresponding wrapper is the main novel contribution of this paper.

3   AMPI_dcc Wrapper

The AMPI_dcc wrapper library is the interface between AMPI and the adjoint code generated by dcc. Its sole responsibility is to ensure that values and adjoints are mapped properly from and to the AMPI routine calls. Hence, the relevant part of the AMPI calling signatures is the handling of values and adjoints. The *forward routines* need to be called with the original values as arguments in the forward run, whereas the *backward routines* need to be called with the adjoints as arguments in the reverse run. The logic of transforming the MPI communications is entirely handled by the AMPI library. Therefore AMPI_Reduce_b is actually a Broadcast as illustrated in Fig. 2.

Furthermore, we want to compute second derivatives by reapplying dcc in tangent-linear mode to the adjoint code. As explained in Fig. 1, the Reduce will be differentiated again yielding

d2_b1_MPI_Reduce(m,cost,**d2_cost**,b1_cost,**d2_b1_cost**,x,**d2_x**,b1_x,**d2_b1_x**).

The function prefix d2_b1 stands for the second-order adjoint code obtained by tangent-linear (d2_) over adjoint (b1_) mode. As in Fig. 1, every value and every adjoint get an additional tangent-linear component (highlighted) prefixed with d2_. Hence there are twice as many variables in the signature of d2_b1_MPI_Reduce than in the first-order adjoint routine b1_MPI_Reduce. Since the underlying AMPI library does not have specific second-order routines, we need to split the communication into two AMPI calls as shown in Fig. 3. This amounts to a doubling of the MPI communication. As in first-order adjoint mode we call d2_b1_MPI_Reduce in the forward run and in the backward run. In the forward run (m=1) d2_b1_MPI_Reduce calls the AMPI library twice: Once to reduce the value of x to cost and once more to reduce the value of the tangent-linear component of d2_x to d2_cost. The resulting MPI calls are two Allreduce calls. As in first-order adjoint mode, the Allreduces are necessary, since each process need the values of cost and d2_cost to compute the adjoints in the reverse run. In the reverse run (m=2), we propagate the adjoints in reverse order of the original code. Therefore the reverse AMPI Reduce routine AMPI_Reduce_b is called. Again this is done twice since the tangent-linear components (d2_b1_cost, d2_b1_x) of both adjoints and the adjoints (b1_cost, b1_x) themselves need to be set. The actual resulting MPI communications are the two Broadcasts of b1_cost and d2_b1_cost explained in Sect. 2. b1_x and d2_b1_x are computed based on the adjoint of the reduction operation.

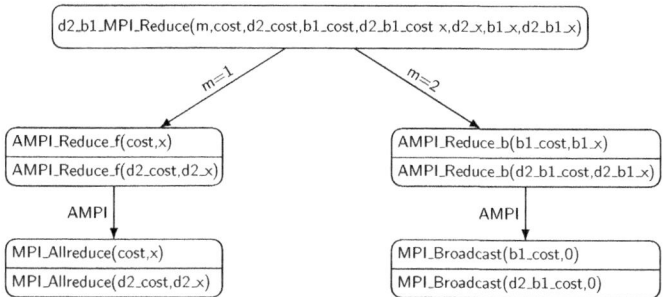

Fig. 3. d2_d1_MPI_Reduce

The procedure of linking each second-order adjoint MPI routine with the corresponding AMPI library routines has been implemented and tested for blocking Send/Receive, non-blocking Send/Receive, Reduce, Broadcast and Waitall. Issues arising from non-blocking communication patterns are explained in the following section.

4 Non-blocking Communication

Reversal of non-blocking communication (MPI_Wait, MPI_Isend, MPI_Irecv) is challenging, due to the undetermined state of the buffer variables between a

Wait and a Irecv or Isend. A solution to the request handling and non-blocking communication patterns has been presented in [7]. Additionally, our solution imposes certain restrictions on the memory management of the buffer.

The wait routine MPI_Wait is exceptional in the way that it does not have any variable as an argument. Let us assume that we have a function placing an Isend followed some time later by a Wait. When computing the adjoints in the reverse run, the adjoint Wait will be called before the adjoint Isend. The AMPI library treats this situation by providing a memory manager, handling all the adjoint buffers. It requires that the memory allocation of the adjoint variables b1_v is completed during the forward run. Unfortunately, this requirement contradicts our current reversal model.

As seen in Fig. 4(a), the forward run m=1 (double right arrow) and the reverse run m=2 (double left arrow) of a function b1_bar are currently separated into two subroutine calls except for the root routine foo. Therefore we call this reversal mode *split reversal* [6]. It implies that the adjoint buffers of b1_MPI are in different memory locations in the forward and reverse runs. As this contradicts the premiss of AMPI, it is impossible for b1_bar to call an AMPI_dcc routine. We could avoid this difficulty, if we merged the forward and reverse run into one function call.

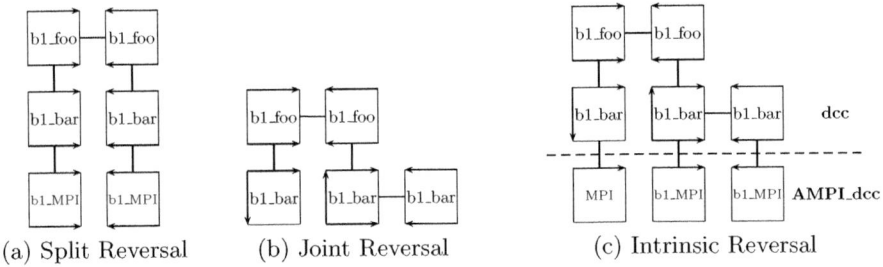

(a) Split Reversal   (b) Joint Reversal   (c) Intrinsic Reversal

Fig. 4. Reversal models

dcc uses *joint reversal* [8] shown in Fig. 4(b) as its data-flow reversal model by introducing three modes m=1,2,3. It is a trade-off between computational and memory complexity of the generated adjoint codes. Let foo be the root routine. The three modes are defined as follows:

- m=1: forward and reverse run
- m=2: store arguments + original routine call
- m=3: restore arguments + forward and reverse run

Only the root routine b1_foo is called with m=1. It starts its forward run (double right arrow) and calls b1_bar with m=2. b1_bar then stores its arguments (down arrow) and runs the undifferentiated original routine bar (single right arrow). b1_foo now starts its reverse run (double left arrow). To do this, it needs the adjoints of the reversed routine b1_bar. This is done by calling b1_bar with m=3.

b1_bar restores its arguments (up arrow) and performs its forward run (double right arrow) followed by the reverse run (double left arrow). We end up with all the adjoints when b1_foo finishes its reverse run.

Lets assume the routine bar has an MPI call in joint reversal. We then use *intrinsic reversal* for the called AMPI_dcc. Just like any intrinsic function the AMPI_dcc routines are differentiated through provision of their differentiated counterpart by the AMPI_dcc library and not by dcc. Thus we can separate the forward and the reverse run, since we have internal structures that link these two calls if necessary. In Fig. 4(c) we see that with joint reversal, the adjoint buffers are the same for both b1_MPI calls since they are called by the same instance of the routine b1_bar as opposed to split reversal in Fig. 4(a). Obviously this comes at a communication cost, since the original MPI routine has to be called during the undifferentiated call (m=2) of b1_bar.

The prime benefit of our approach to the reversal of non-blocking MPI communication is the avoidance of modification of dcc's source transformation algorithms. dcc handles all MPI calls similar to any arbitrary user-written subroutine by calling adjoint versions in modes m=1,2,3. Dealing with memory management issues is deferred entirely to the wrapper.

5   Test Case

Due to space restrictions we only present the code fragment covering the MPI reduction of cost over cost_mpi (Listing (3)) of the Burgers simulation introduced in Sect. 1. We initialize cost_mpi of each process to 0. Then we calculate the left (mpi_x) and right (mpi_nx−1) boundaries of the partitioned one-dimensional domain based on each process' id (myid). Each process then computes its local contribution to the cost cost_mpi by comparing at each time step j the velocity u with the observed velocity values uob. Last but not least, we reduce the values cost_mpi to cost with root being our master process.

```
cost_mpi=0;
mpi_i = (myid) * (nx/numprocs); mpi_nx = (myid+1) * (nx/numprocs);
while (mpi_i < mpi_nx) {
    j=0;
    while (j<n) {
        cost_mpi=cost_mpi+0.5*(u[mpi_i][j]−uob[mpi_i][j])*(u[mpi_i][j]−uob[
            mpi_i][j]);
        j=j+1;
    }
    mpi_i=mpi_i+1;
}
MPI_Reduce(cost, cost_mpi, root);
```

Listing 3. Reduction of cost in Burgers simulation

This code is entirely accepted by dcc and allows us to generate second-order adjoint code. We then compile the resulting code with gcc, while linking the MPI calls to our AMPI_dcc wrapper. The complete code is available on request. As anticipated, the results of the parallel computation exactly match those of the serial run without the MPI enhancements. While the correctness of our approach is proven, no conclusion about speed may be drawn yet. Ongoing work focusses on the application of the proposed method to practically relevant problems including simulations in oceanography and waterways engineering [9].

6   Summary and Conclusion

Our goal was to compute second derivatives of numerical simulations implemented as MPI-augmented code. For this purpose we wrote a wrapper AMPI_dcc capable of linking second-order adjoint calls generated by the derivative code compiler dcc [4] with the AMPI library [5]. The wrapper covers the most commonly used MPI calls. Nonetheless there are certain limits. As the reversal of a program is closely interwoven with memory management, there are certain caveats that need to be considered when dealing with non-blocking communication. The proposed approach is extensible to any order of differentiation. The corresponding wrappers could be implemented by hand. Alternatively, one could envision a model, where even the wrapper is generated for an arbitrary order of differentiation.

References

1. Zwillinger, D.: Handbook of Differential Equations, 3rd edn. Academic Press, Boston (1997)
2. Kelley, T.: Solving nonlinear equations with Newton's method. Fundamentals of Algorithms. SIAM, Philadelphia (2003)
3. Gropp, W., Lusk, E., Skjellum, A.: Using MPI: Portable Parallel Programming with the Message Passing Interface. MIT Press, Cambridge (1994)
4. Hannemann, R., Marquardt, W., Gendler, B., Naumann, U.: Discrete first- and second-order adjoints and automatic differentiation for the sensitivity analysis of dynamic models. In: Procedia Computer Science. Elsevier, Amsterdam (to appear, 2010)
5. Utke, J., Hascoët, L., Heimbach, P., Hill, C., Hovland, P., Naumann, U.: Toward Adjoinable MPI. In: Proceedings of the 23rd IEEE International Parallel & Distributed Processing Symposium, Washington, DC, USA. IEEE Computer Society Press, Los Alamitos (2009)
6. Griewank, A., Walter, A.: Evaluating Derivatives. Principles and Techniques of Algorithmic Differentiation, 2nd edn. SIAM, Philadelphia (2008)
7. Schanen, M., Naumann, U., Hascoët, L., Utke, J.: Interpretative adjoints for numerical simulation codes using mpi. Procedia Computer Science 1, 1819–1827 (2010); ICCS 2010
8. Utke, J., Naumann, U., Fagan, M., Tallent, N., Strout, M., Heimbach, P., Hill, C., Wunsch, C.: OpenAD/F: A modular, open-source tool for automatic differentiation of Fortran codes. ACM Transactions on Mathematical Software 34, 1–18 (2008)
9. Riehme, J., Kopmann, R., Naumann, U.: Uncertainty quantification based on forward sensitivity analysis in sisyphe. In: Proceedings of ECCOMAS-CFD 2010 (to appear, 2010)

Locality and Topology Aware Intra-node Communication among Multicore CPUs

Teng Ma, George Bosilca, Aurelien Bouteiller, and Jack J. Dongarra

Innovative Computing Laboratory,
University of Tennessee Computer Science Department
1122 Volunteer Blvd., Knoxville, TN 37996-3450, USA
{tma,bosilca,bouteill,dongarra}@eecs.utk.edu

Abstract. A major trend in HPC is the escalation toward *manycore*, where systems are composed of shared memory nodes featuring numerous processing units. Unfortunately, with scale comes complexity, here in the form of non-uniform memory accesses and cache hierarchies. For most HPC applications, harnessing the power of multicores is hindered by the topology oblivious tuning of the MPI library. In this paper, we propose a framework to tune every type of shared memory communications according to locality and topology. An implementation inside Open MPI is evaluated experimentally and demonstrates significant speedups compared to vanilla Open MPI and MPICH2.

1 Introduction

Because the emergence of thermic and power issues have prevented further performance improvements through the usual frequency scaling, CPU vendors have resorted to multicore architectures to deliver the expected level of performance progression. Unfortunately, incorporating more processing units does not give an instant and automatic speed boost to applications; programmers have to take into account numerous issues posed by the intrinsic parallel and heterogeneous nature of multicore chips. Although HPC developers have become proficient at harnessing the power of parallel systems, through the use of various programming models such as Single Process Multiple Data (SPMD) and tools like MPI or OpenMP, straight out applications of those paradigm on cluster of multicores types of architecture have exhibited disappointing performance [1]. To enable the integration of more cores inside a computing node, vendors are forced to expose more complex architectures, exhibiting Non Uniform Memory Accesses (NUMA) and several levels of partitioned cache hierarchies. Furthermore, design and implementation of multi-core architecture present a large diversity among vendors. As an example, Intel's Tigerton CPUs feature a SMP architecture, while AMD's Istanbul and Intel's Nehalem exhibit NUMA characteristics. In a node, some cores reside on different sockets, interconnected by network-style fast connections such as Intel's QuickPath and AMD's HyperTransport. Even inside a single die, one can encounter different L2 caches, shared between exclusive

R. Keller et al. (Eds.): EuroMPI 2010, LNCS 6305, pp. 265–274, 2010.

groups of cores, so that two cores of the same processor might or might not share the same level of cache, depending on their respective position on the die.

While hybrid approaches and novel programming models are being investigated, the large majority of applications available in the HPC ecosystem today are based on the message passing paradigm (using the MPI standard). Converting every and each of those applications to take into account the fine subtleties of the various and changing vendor implementations of multicore systems would impose a significant and lasting burden to the community. Among the issues preventing message passing from delivering performance on cluster of multicore systems is the use of a flat set of tuning parameters for all shared memory communications, regardless of the underlying hardware architecture, more precisely the distance to different levels of cache and memory and the physical topology imposed by the chips. In this paper, we propose to alleviate this issue by providing a topology aware framework inside the message passing middleware, in order to unleash legacy application performance on the most recent architectures without shifting the programming model. The prominent feature of this framework is to optimize intra-node communication by selecting the optimal tuning parameter set at runtime. Multiple communication parameter sets are provided and can be selected, according to the run-time placement of the MPI processes and considering the topology of the underlying hardware.

The rest of this paper is organized as follows: Section 2 introduces the related work on multi-core intra-node communication. Section 3 formulates and outlines the extent of the problem when considering modern multicore processors. Then Section 4 describes our framework designed to combine locality and topology information with intra-node communication, and its implementation in a leading MPI implementation. A performance study is presented in the Section 5, substantiating the benefits of this approach when compared to the Open MPI and MPICH2 implementations. Finally, Section 6 concludes the paper with a discussion of the results and future directions.

2   Related Work

MPICH2 [2] and OpenMPI [3,4] are the two major implementations of the MPI standard. Both feature an optimized device to handle shared memory communications: Nemesis [5] for MPICH2 and the SM BTL for Open MPI. In both MPI implementations, large messages are divided into fragments to establish a pipeline. The smallest message to use the pipeline protocol as well as the fragment size are examples of crucial parameters to reach maximum bandwidth without sacrificing latency. The OPTO tool [6] has been proposed to optimize the run-time parameters of the Open MPI environment. It uses a brute-force searching of the parameter space by evaluating benchmarks such as NetPipe, for point-to-point communication, and SkaMPI, for collective communication. This set of *tuned* parameters is then used for every communication of any application.

In regular MPI shared memory implementations, any transfer actually involves two memory copies: one from the user buffer to the shared memory buffer

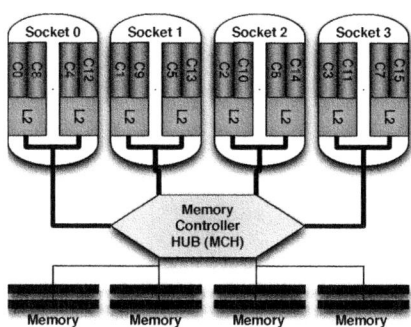
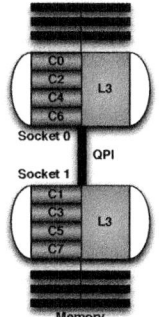

(a) Four sockets Intel Tigerton node   (b) Two sockets Intel Nehalem node

Fig. 1. Architecture comparison between 2 generations of Intel multicore CPUs

and another to the destination user buffer. The LiMIC [7] kernel module can decrease the number of necessary memory copies to one by doing the memory movement with kernel access rights. KNEM [8] is a similar kernel module that also features DMA (Direct Memory Access) copy by using Intel I/O acceleration technique (I/OAT). DMA copy can decrease cache pollution and CPU noise from communication. However, DMA performance suffers when multiple communications overload a single DMA device, which is very likely with the current trend to increase the number of cores. In that context, rather than easing the tuning process, using kernel-based DMA methods is another parameter that changes according to the communication workload.

Several efforts have proposed to embrace the hierarchical nature of Grid systems network [9,10,11]. These papers propose different approaches to map the collective communication topology to the actual network topology, an idea that applies as well to multicore processors. Yet, our work focuses on the optimization of point-to-point message, and its indirect improvement on collective communication performance. The optimization and tuning of the collective algorithm itself, according to the hardware topology, is left for future works.

While shared memory communication tuning has been an active research area, using the underlying hardware topology to define different tuning parameter sets, as we propose in our framework, has never been attempted.

3   Multicore and Multifarious Hierarchies

Modern CPUs exhibit several levels of cache, with non uniform memory accesses. The communication distance between two cores of a single node varies depending on those hierarchies. Furthermore, each vendor exhibits different characteristics that tends to radically change between successive CPU generations. Figure 1(a) and Figure 1(b) illustrate such differences by describing two typical cluster nodes featuring different generations of Intel multicore CPUs. Figure 1(a) shows the architecture of a node with four Intel's Tigerton CPUs (16 cores). In this machine, all sockets are interconnected by one memory controller. Thus, the apparent distance to the memory is the same for every core. However, three different communication path exist with distinct costs. The first one is between core 0 and core 8 which are on the same die and share the L2 cache. While core 0 and core 4 do not share L2 cache, communication inside the same socket are significantly faster than resorting to the FSB (front side bus). Between core 0 and core 2, hosted in different sockets, the FSB is the only option. Figure 1(b) describes the architecture of a node with two Intel Nehalem CPUs (8 cores). Each processor has an independent memory controller, which makes it a NUMA architecture. While all cores of a socket share a common L3 cache, the Nehalem architecture also exhibits different communication performance whether the cores are on the same socket or not.

SPMD had been a very successful programming model for single core architectures. Consequently, numerous applications and libraries have been developed following this approach, and have benefited from its easy portability across different vendors, and excellent level of performance. However, in the context of multicore CPUs, not considering the locality and topology of the cores distribution inside the CPUs dreadfully affects the overall application experienced communication performance.

Nowadays MPI implementations provide a specific optimized device to handle shared memory communications. This device is usually applied directly to core-to-core communications with a single set of tuning parameter oblivious of the topology between the sender and receiver cores. As an example, in the Nemesis device of MPICH2, when the message size is smaller than PIPELINE_THRESHOLD (128KBytes), the copy_limit, defining the pipeline size, is set to 16KBytes. For larger messages, this parameter is changed to MPID_MEM_COPY_BUF_LEN which is often 32KBytes. Open MPI also has a set of similar communication parameters (btl_eager_limit and btl_max_send_size) which are used for protocol switch point and pipeline size respectively. Unlike MPICH2, in Open MPI users can tune those parameters without recompiling the MPI library. Despite this added flexibility, users rarely have the expertise and time to properly tune parameters for the communication pattern of their application, leading most runs to use the default parameters. Furthermore, for any run of an application, a single set of tuning parameters can be used. Therefore, it is impossible to apply a different set of tuning parameters to communications to account for different characteristics of the links between cores. The experimental section of this paper (5) provides an evaluation of the extent of the performance issue induced.

4   Multi-tuning Framework

Because Open MPI is based on a modular and component model while at the same time retaining outstanding performance, it is a very convenient vessel to investigate new features. Thus, while the principles presented are generic, our topology aware multicore communication framework is implemented into Open MPI. The framework is composed of three main components: the rule discovery module, the machine topology discovery module and the runtime communication tuning module.

Table 1. An example of rule discovery table

CPU	locality	btl_eager_limit	pipe_size	use_knem	DMA_min
Tigerton	no shared L2 cache	2096	0.5 * L1 cache	1	2196608
Nehalem EP	no shared L2 cache	4192	0.5 * L1 cache	0	null
Tigerton	shared L2 cache	2096	L1 cache	1	4804864

The rule discovery module is used offline to construct a table storing knowledge about best tuning parameters for a particular architecture. In this table, we store common knowledge about the relationship between CPU architecture, locality, topology and tuning parameters. Table 1 is an example of a generated rule table, where the tuning of various pipeline length and the use of a DMA engine is governed by the sharing of an L2 cache. Rules are inferred from a mathematical model taking into account the size of the L1 data cache, L2 cache and the sharing of cache hierarchies between cores. As an example, if two cores share the L2 cache, the heuristic is to use half of L1 data cache size as the pipeline size. Most rules depends on the cache reuse policy and the snoopy cache protocol. Different sets of rules have been defined for different families of processors. While more experimental evaluations are needed to assert the soundness of the models proposed, the results presented in the experimental section of this paper are encouraging. Should some architecture be difficult to describe with a mathematical model, the previously discussed parameter exhaustive OPTO tool could also be used to build the rule table.

The machine topology discovery module discover, once for every run, all information about the cache hierarchies, core mapping and proximity between each pair of cores. This module is based on the Portable Hardware Locality (hwloc) [12] project. It provides a portable abstraction (OS, versions, architectures, etc.) of the hierarchical topology of modern CPUs, including NUMA memory nodes, sockets, shared caches, cores and hyper-threading. It also gathers various system attributes such as cache and memory information.

Based on the tables generated by the two discovery modules, the runtime communication tuning modules instantiates several distinct SM BTL with adequate tuned parameters for each type of communications. Then, among the available instantiation of the SM BTL, for every message, the best one [13] is selected to actually transfer the data, based on the rules applied to this type and size of message and the source to destination core distance.

5   Experimental Evaluation

Experimental Conditions. Our experimental setup includes two different Intel based machines. The first one is based on four Intel Xeon E7340 at 2.4GHz (Tigerton), as described by Figure 1(a). Its L1 data cache is 32KB and L2 data cache is 4MB. The four sockets are interconnected each other by the front side bus. The second system is based on two Intel Xeon E5520 at 2.27GHz (Nehalem), as described by Figure 1(b). Its L1 data cache is 32KB, and each core has an

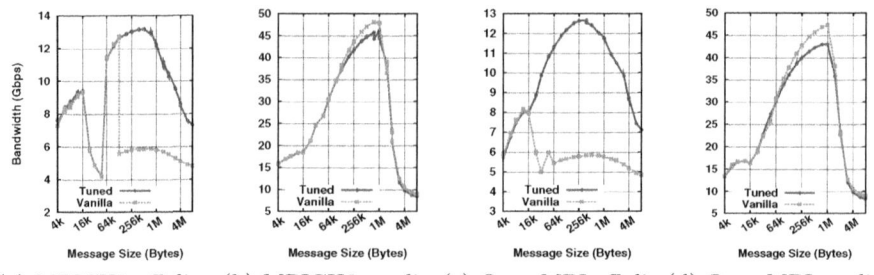

(a) MPICH2 off-die   (b) MPICH2 on-die   (c) Open MPI off-die (d) Open MPI on-die

Fig. 2. Impact on bandwidth of pipeline fragment size tuning according to core distance on the Tigerton machine for MPICH2 and Open MPI

independent L2 data cache whose size is 256KB. CPUs are interconnected by Intel QuickPath. The same operating system (Linux 2.6.30) is deployed on both machines. MPICH2-1.2.1 and Open MPI trunk (r22930) are used. We used Net-PIPE [14], Intel MPI benchmarks [15] and the NAS parallel benchmarks [16] to evaluate the performance of our tuning framework. All benchmarks are compiled with gcc 4.1.2, with the -O3 flag.

Assessment of the Severity of the Performance Issues. The first set of experiments evaluates the performance loss incurred by using a single set of tuning parameters, regardless of core locality. Figure 2 presents the performance comparison between a vanilla version of MPICH-2 and Open MPI with a similar version where pipeline size has been hand-tuned for maximum inter-socket bandwidth. In vanilla MPICH-2, for messages larger than 128KB, the pipeline size switches from 16KB to 32KB. As the steep bandwidth drop illustrates in Figure 2(a), this is a very inappropriate tuning for communications between cores located in different sockets. The hand tuned version, that retains the original 16KB pipeline for larger messages, is capable of sustaining a higher bandwidth, up to a very significant 2.5 times improvement. Open MPI exhibit the same behavior (Figure 2(c)), illustrating that the issue is not implementation specific. However, as illustrated by the Figures 2(b) and 2(d), when communicating inside the same die, the default parameters are perfectly tuned and perform slightly better. While the hand tuned parameters yield significant benefits in certain cases (inter-socket communications), using them in certain cases decrease performance, illustrating the need for using *simultaneously* different sets of tuning parameters for different types of communications.

Effectiveness of the Multi-tuning Framework. The four Figures 3 presents the comparison between vanilla MPICH2, vanilla Open MPI and multi-tuned Open MPI in the NetPIPE ping-pong benchmark for a variety of machines and core distributions. The bandwidth values for message smaller than 4KB have been removed for clarity, as the performance of the three versions were similar. On the Tigerton machine, the multi-tuned version outperforms both vanilla Open MPI and MPICH2 for inter-socket communications (Figure 3(a)), thanks to using

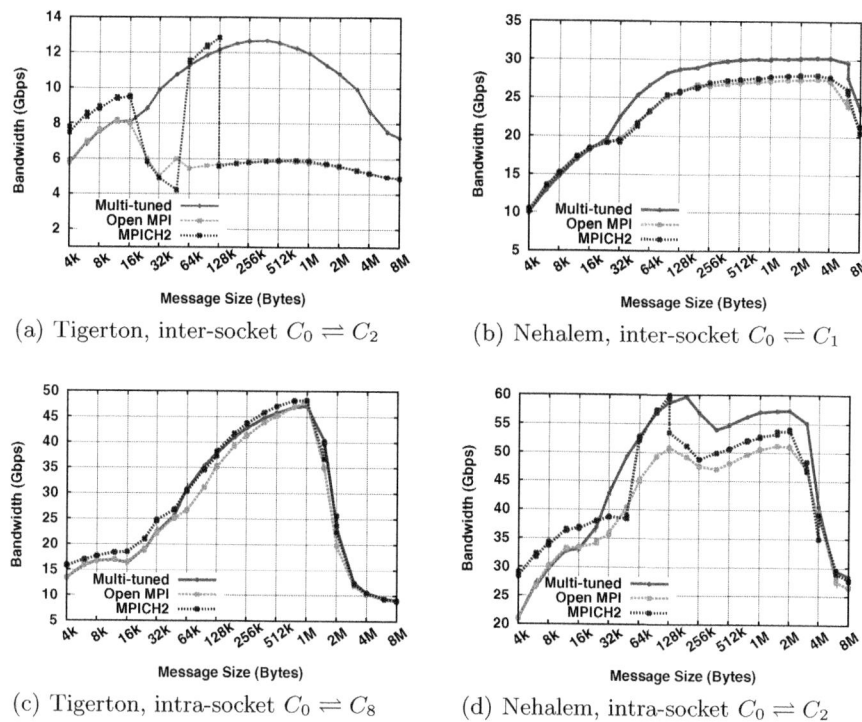

(a) Tigerton, inter-socket $C_0 \rightleftharpoons C_2$

(b) Nehalem, inter-socket $C_0 \rightleftharpoons C_1$

(c) Tigerton, intra-socket $C_0 \rightleftharpoons C_8$

(d) Nehalem, intra-socket $C_0 \rightleftharpoons C_2$

Fig. 3. Bandwidth of the ping-pong test for vanilla MPICH2, vanilla OpenMPI and multi-tuned Open MPI

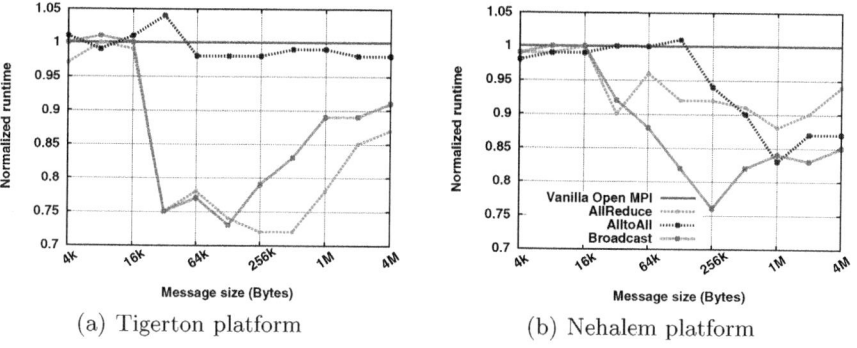

(a) Tigerton platform

(b) Nehalem platform

Fig. 4. Run time of the IMB collective tests of the multi-tuned Open MPI (normalized to the vanilla Open MPI performance, lower is better)

better tuning. Contrarily to the naive hand-tuned version presented in the previous experiment (Figure 2(d)), it does not suffer from any performance degradation for intra-socket communication (Figure 3(c)). The same holds with the Nehalem processor; though initially MPICH2 performs better than Open MPI, the multi-tuned Open MPI version outperforms both vanilla MPI for intra-socket

Table 2. Run time of the NAS benchmarks

name	Nehalem (8 cores)			name	Tigerton (16 cores)		
	Open MPI	Tuned	Speedup		Open MPI	Tuned	Speedup
IS.C	3.72s	3.48s	6.9%	IS.C	6.27s	6.22s	0.8%
FT.B	15.81s	15.31s	3.3%	FT.B	24.53s	24.45s	0.32%
LU.C	278.34s	276.81s	0.55%	LU.C	209.07s	204.64s	2.16%
MG.B	2.60s	2.57s	1.2%	MG.B	4.94s	4.91s	0.61%
CG.C	46.42s	46.34s	0.17%	CG.C	97.31s	96.21s	0.93%

communications. Though all cores are on the same die, they don't share L2 cache, which is the prominent performance affecting factor.

Collective Communications. Figure 4(a) and Figure 4(b) present the run time of multi-tuned OpenMPI for Broadcast, AlltoAll, and AllReduce collective communication, on the Tigerton and Nehalem machines, normalized to the run time of the similar algorithm in vanilla Open MPI. In this experiment, the multi-tuning only takes place at the the point-to-point communication level underlying the collective algorithm; the collective algorithm itself is left unchanged. For messages smaller than 4KB, the physical page size, multi-tuned and vanilla Open MPI always use the same eager protocol, exhibiting equal performance (thus not presented on the graph); yet significant performance improvement are visible for larger message sizes.

In the AlltoAll test, while multi-tuned Open MPI reduces the execution time by only 2% on the Tigerton platform, it yields up to 17% improvement on the Nehalem platform. This result can be explained by the communication pattern of the shared memory AlltoAll collective operation: it does not use a tree topology. On the Tigerton platform, the cross bandwidth of the FSB is consequently easily saturated by this naive algorithm, negating the inter-socket bandwidth benefit achieved on the simple point-to-point benchmark. As multi-tuning does not yield much gains for the intra-socket communication on the Tigerton architecture, the overall benefit is small. On the contrary, the Nehalem platform features only two sockets connected through the much faster QPI interface and adapts better to cross-traffic. Moreover, intra-socket point-to-point bandwidth is also improved on this system, which transfers as well to the collective performance.

In the AllReduce test, multi-tuned OpenMPI benefits from a 12% run time reduction on Nehalem and up to 23% for the very common 256KB message size on Tigerton. Both platform exhibit a close to 25% performance improvement for some message sizes on the Broadcast collective operation. In these two collectives, the shared memory collective algorithm uses a tree topology which does not saturate the inter-socket link; therefore, benefits of multi-tuning on point-to-point performance are reflected in the collective performance. For the entire range of message sizes, multi-tuned collective operations compare favorably to vanilla Open MPI, except for 32KB on the Tigerton platform.

Applications. Additionally, we used application benchmarks to evaluate the performance of our framework. We used IS, FT, LU, MG and CG from the NAS benchmarks. Table 2 shows the comparison between the runtime of the multi-tuned version of Open MPI and vanilla Open MPI for these benchmarks. Compared with regular MPI, the multi-tuned approach always decreases the overall application runtime. As communication are using the extremely fast shared memory device, the communication to computation ratio is balancing toward computation bound performance. As a consequence, the overall impact of communication performance on the application runtime is small, an effect more pronounced on the benchmark achieving good communication overlap by computations such as MG and CG. Though the CG benchmark is communication intensive, only its latency bound communications are difficult to overlap, an area where tuning is already adequate by default. The FT benchmark, which uses an all-to-all collective communication, exhibits a similar performance profile as the AlltoAll test, with almost no gain on the Tigerton platform but some improvement on the Nehalem. The maximum performance improvement of multi-tuning is achieved in the communication intensive IS benchmark on the Nehalem platform, with close to 7% application run time improvement.

6 Conclusion and Future Work

In this paper, we studied the problem of intra-node communication inside multicore CPUs. Our experiments show that ignoring the locality and topology information in the MPI software stack is an obstacle to harness the optimal communication performance on multicore systems. We then introduced a framework to 1) build tuning rules for different models of CPUs, 2) discover the run-time information such as CPU type, cache size, locality and etc. and 3) take advantage of this knowledge to finely tune the internals of the MPI library for different kind of communications. Our experiments show that the multi-tuned Open MPI version based on this framework, always exhibits better application performance, thanks to improved communication speed for both point-to-point and collective communication patterns. With the future increase in the number of cores per node this benefit is expected to be magnified.

Future Works. While fine tuning of point-to-point communication to adapt to the multicore topology has proven beneficial indirectly to the collective communications, tuning the collective itself according to the same information has the potential to further increase performance. We expect to witness the same hardware locality dependent tuning for the selection and parametrization of the collective algorithm itself, an area where we believe our multi-tuning approach will be valuable.

References

1. Rabenseifner, R., Hager, G., Jost, G.: Hybrid MPI/OpenMP parallel programming on clusters of multi-core SMP nodes. In: Parallel, Distributed and Network-based Processing, pp. 427–436 (2009)
2. Gropp, W., Lusk, E., Doss, N., Skjellum, A.: A high-performance, portable implementation of the MPI message passing interface standard. Parallel Computing 22, 789–828 (1996)

3. Gabriel, E., Fagg, G.E., Bosilca, G., Angskun, T., Dongarra, J.J., Squyres, J.M., Sahay, V., Kambadur, P., Barrett, B., Lumsdaine, A., Castain, R.H., Daniel, D.J., Graham, R.L., Woodall, T.S.: Open MPI: Goals, concept, and design of a next generation MPI implementation. In: Proceedings, 11th European PVM/MPI Users' Group Meeting, Budapest, Hungary, pp. 97–104 (2004)
4. Graham, R.L., Woodall, T.S., Squyres, J.M.: Open MPI: A flexible high performance MPI. In: Proceedings of 6th Annual International Conference on Parallel Processing and Applied Mathematics, Poznan, Poland (2005)
5. Buntinas, D., Mercier, G., Gropp, W.: Design and evaluation of Nemesis, a scalable, low-latency, message-passing communication subsystem. In: Sixth IEEE International Symposium on Cluster Computing and the Grid, vol. 1, pp. 10–20 (2006)
6. Chaarawi, M., Squyres, M., Gabriel, J., Feki, E.,, S.: A tool for optimizing runtime parameters of Open MPI. In: Lastovetsky, A., Kechadi, T., Dongarra, J. (eds.) EuroPVM/MPI 2008. LNCS, vol. 5205, pp. 210–217. Springer, Heidelberg (2008)
7. Jin, H.W., Sur, S., Chai, L., Panda, D.: LiMIC: support for high-performance MPI intra-node communication on linux cluster. In: International Conference on Parallel Processing, ICPP 2005, pp. 184–191 (2005)
8. Buntinas, D., Goglin, B., Goodell, D., Mercier, G., Moreaud, S.: Cache-Efficient, Intranode Large-Message MPI Communication with MPICH2-Nemesis. In: Proceedings of the 38th International Conference on Parallel Processing (ICPP 2009), pp. 462–469. IEEE Computer Society Press, Vienna (2009)
9. Kielmann, T., Hofman, R.F.H., Bal, H.E., Plaat, A., Bhoedjang, R.A.F.: Magpie: Mpi's collective communication operations for clustered wide area systems. In: Proceedings of the 1999 ACM SIGPLAN Symposium on Principles and Practice of Parallel Programming (PPOPP 1999), pp. 131–140 (1999)
10. Karonis, N.T., de Supinski, B.R., Foster, I., Gropp, W., Lusk, E., Bresnahan, J.: Exploiting hierarchy in parallel computer networks to optimize collective operation performance. In: The 14th International Parallel and Distributed Processing Symposium, p. 377 (2000)
11. Filgueira, R., Singh, D.E., Pichel, J.C., Isaila, F., Carretero, J.: Data Locality Aware Strategy for two-phase Collective I/O. In: Palma, J.M.L.M., Amestoy, P.R., Daydé, M., Mattoso, M., Lopes, J.C. (eds.) VECPAR 2008. LNCS, vol. 5336, pp. 137–149. Springer, Heidelberg (2008)
12. Broquedis, F., Clet Ortega, J., Moreaud, S., Furmento, N., Goglin, B., Mercier, G., Thibault, S., Namyst, R.: hwloc: a Generic Framework for Managing Hardware Affinities in HPC Applications. In: The 18th Euromicro International Conference on Parallel, Distributed and Network-Based Computing (2010)
13. Shipman, G.M., Woodall, T.S., Bosilca, G., Graham, R.L., Maccabe, A.B.: High performance RDMA protocols in HPC. In: Mohr, B., Träff, J.L., Worringen, J., Dongarra, J. (eds.) PVM/MPI 2006. LNCS, vol. 4192, pp. 76–85. Springer, Heidelberg (2006)
14. Snell, Q.O., Mikler, A.R., Gustafson, J.L.: NetPIPE: A network protocol independent performance evaluator. In: IASTED International Conference on Intelligent Information Management and Systems (1996)
15. Intel: Intel MPI benchmarks 3.2 (2010), http://software.intel.com/en-us/articles/intel-mpi-benchmarks/
16. Bailey, D.H., Barszcz, E., Barton, J.T., Browning, D.S., Carter, R.L., Fatoohi, R.A., Frederickson, P.O., Lasinski, T.A., Simon, H.D., Venkatakrishnan, V., Weeratunga, S.K.: The NAS parallel benchmarks. Technical report. The International Journal of Supercomputer Applications (1991)

Transparent Neutral Element Elimination in MPI Reduction Operations

Jesper Larsson Träff

Department of Scientific Computing, University of Vienna
Nordbergstrasse 15C, A-1090 Vienna, Austria
traff@par.univie.ac.at

Abstract. We describe simple and easy to implement MPI library internal functionality that enables MPI *reduction operations* to be performed more efficiently with increasing sparsity (fraction of neutral elements for the given operator) of the input (and intermediate result) vectors. Using this functionality we give an implementation of the MPI_Reduce collective operation that completely transparently to the application programmer exploits sparsity of both input and intermediate result vectors. Experiments carried out on a 64-core Intel Nehalem multi-core cluster with InfiniBand interconnect show considerable and worthwhile improvements as the sparsity of the input grows, about a factor of three with 1% non-zero elements which is close to best possible for the approach. The overhead incurred for dense vectors is negligible when compared to the same implementation not exploiting sparsity of input and intermediate results. The implemented SPS_Reduce function is for both very small and large vectors faster than the native MPI_Reduce of the used MPI library, indicating that the improvements reported are not artifacts of suboptimal reduction algorithms.

1 Introduction

Many of the MPI built-in, binary reduction operators (like e.g. MPI_SUM [8, Section 5.9.1]) have a *neutral (zero) element* for many of the basic MPI datatypes. Thus, there is an obvious, input dependent potential for improving implementations of the MPI reduction collectives [8, Section 5.9] by *elimination* of the neutral elements in partial result vectors, and by performing intermediate applications of the binary reduction operator directly with such compressed vectors. We present elimination schemes and simple, library internal functionality that makes it straightforward to incorporate this improvement in (*any*) point-to-point communication based MPI library implementation of the reduction collectives. By the design we take care that the elimination can *never* (bar small overhead for scanning parts of input and intermediate result vectors) be worse than implementations using standard functionality for sending, receiving and local operator application. We back this claim up by giving text-book implementations of the MPI_Reduce collective operation by means of binomial trees and linear pipelines, and perform experiments that show that significant performance improvements

R. Keller et al. (Eds.): EuroMPI 2010, LNCS 6305, pp. 275–284, 2010.

can be achieved as sparsity (fraction of neutral elements) increases, with very little additional overhead for dense input vectors. We suggest that the optimization of elimination of neutral elements comes with so little cost as to be useful as per default MPI library implementations of the reduction collectives.

Message compression to improve communication performance is nothing new for MPI libraries. A number of papers have explored the use of general purpose message compression in MPI, see e.g. [1,6,7] and [2] and further references there. Such general schemes cannot exploit sparsity for operations with additional semantics, like the collective operations and in particular the MPI reduction operations. Elimination of neutral elements for improvements in both message complexity and computational load was recently somewhat explored in [5] that use run-length encodings of floating-point numbers for zero-element elimination and a simple, linear pipeline reduction implementation. We use a different neutral-element elimination scheme that can cover all basic MPI datatypes (and in principle be extended also to user-defined types and reduction operators), give the accompanying, general reduction operator application functionality, and thus demonstrate how neutral-element elimination can easily be incorporated in any MPI library implementation of the collective reduction operations. We believe that actual applications do exhibit cases of MPI reductions performed on large vectors with a substantial fraction of zeroes, but this claim needs to be backed up with concrete examples. Another source of inspiration is the sparse exchange problem investigated in [4]. One of the proposed algorithms entails a reduce-scatter computation of p-element vectors that will typically have a very large number of zeroes. For this protocol the optimization presented here will be immediately applicable.

2   Neutral Element Elimination

The idea of the neutral element elimination optimization is to represent the neutral elements of input (and intermediate) vectors for the MPI reduction functions only implicitly. MPI offers a number (12) of built-in binary reduction operators that work on the MPI built-in datatypes. The neutral element, if defined, is different for different combinations of operator and datatype, thus a function SPS_Op_get_neutral(op,type,zero) that returns either a (pointer to) the neutral (zero) element of the operator op for datatype type, or NULL is needed, and can easily be written. This is sufficient for most of the MPI operators, but not for the logical operation MPI_LAND where the neutral element is anything that is not (integer) 0. To cater for such cases, variants of the discussed functionality are needed, but it is obvious that these cases can be handled, and we will not describe such any further.

The aim is to provide functionality that can replace the send, receive and local reduction functions normally used in point-to-point based implementations of the reduction collectives. To this end send and receive functions that take an extra parameter specifying the neutral element are introduced. A (non)blocking SPS_Send_elim(sendbuf,...,dest,tag,zero,comm) function with an extra void pointer to the zero element of the datatype is introduced, that (as described in the next section) only sends the non-zero elements of sendbuf.

Conversely, SPS_Recv_fill(recvbuf,...,source,tag,comm,request) also with a void pointer to the zero element of the datatype, expands the received result (that must have been sent with SPS_Send_elim) by filling in again the eliminated zero elements. We will actually *not* use this function, but try to maintain vectors in the compressed format as long as possible.

Ideally, both functions should be implemented such that no extra space than allocated for the send and receive buffers is used internally. At least the count and datatype arguments should always give an upper bound on the amount of extra space that need to be allocated. In order to guarantee that the new reduction collective implementations are never worse than their native counterparts it is important than no more data be ever sent than would have been done with a standard MPI_Send call, that is as specified by the count and datatype arguments.

3   Implicit Representation of Neutral Elements

A simple representation of vectors with many neutral elements is as a list of indices of non-neutral elements paired with a list of the corresponding, actual values. This simple format will be used as a starting point. Indices and values are kept separate to minimize effects of bad word alignments (e.g. indices being integers and values bytes).

With this representation the heart of SPS_Send_elim would look as follows, assuming here that auxiliary space has been allocated for the index and value arrays. This version does not do neutral element elimination in-place, but has crucial the property that no more data are sent than an MPI_Send operation would have done.

```
MPI_Type_size(type,&typesize); space = 0; j = 0;
for (i=0; i<count; i++) {
  if (sendbuf[i]!=*zero) {
    space += sizeof(int)+typesize; index[j++] = i;
  }
  if (space>=cutoff) break;
}
if (space>=cutoff) {
  // no compression possible, send normal
  MPI_Send(sendbuf,total,MPI_BYTE,dest,DENSETAG,comm);
} else {
  value = (type*)(index+j); // values after indices
  for (i=0; i<j; i++) value[i] = sendbuf[index[i]];
  MPI_Send(index,space,MPI_BYTE,dest,SPARSETAG,comm);
}
```

The overhead for this format is at most one scan of the input, with a break as soon as the size of the compacted data reaches a certain cutoff size, which could for instance be a fixed fraction of the size of the send buffer (in the experiments cutoff is set to the size of the send buffer). Both the loops for setting the index vector and the copy of the non-zero values into the value array are regular and well suited for vectorization. Tags DENSETAG and SPARSETAG are used to distinguish between implicitly represented, compressed vectors and normal, uncompressed buffers. Data are sent as uninterpreted MPI_BYTEs; an actual library implementation of the functionality would be more careful and send non-compressed data with their actual MPI datatype. For this simple format, the tag suffices to distinguish between compressed and uncompressed formats; for the compressed format the start of the list of values can be computed from the amount of data received, as shown in the next section.

For inputs where the non-zero elements come in consecutive blocks (banded vectors), the index-value format is obviously wasteful, and a representation based for instance on runs would be more space-efficient. It is easy to switch format as soon as a run of more than two successive indices is seen. Each such run would be represented by an index, a run-length, and the block of values. Again, the indices and run-lengths should be kept separate from the values, and the two formats could be mixed as follows. A positive index indicates a block of size 1, the corresponding value is the next value in the array of values. A negative index indicates a larger block, and is followed (in the index array) by the length of the run. The values for the block are found consecutively in the values array. One word is needed for the size of the index array, since this now cannot be computed from the number of bytes in index and value array. This format would be typed by a third special tag, RUNTAG. The overhead in preparing index array, and in copying the values is higher for this format, and might overall the change feasibility of using neutral element elimination. The format has not (yet) been implemented.

4   Implementations of Collective Reduction Algorithms

With the two functions for sending with elimination of neutral element and receiving with expansion, any (point-to-point) based implementation of any of the MPI reduction functions can be modified to employ neutral element elimination. This is the benefit gained from (possibly more powerful) general purpose message compression. However, expansion is wasteful and should be avoided. By introducing an extra utility function for doing local reductions on the index-value arrays this can easily be accomplished.

In addition to the MPI 2.2 functionality MPI_Reduce_local we therefore introduce a function SPS_Reduce_local_index(index,value,inout,count,type,op), taking an index-value pair as input argument; the count argument in this case is the number of such pairs. The implementation for MPI_SUM would look as follows.

```
for (i=0; i<count; i++) inout[index[i]] = in[i]+inout[index[i]];
```

If commutativity shall not be exploited, more local reduction functions for left/right reduction are convenient to minimize the amount of copying between intermediate buffers. For user-defined operators we would have to call MPI_-Reduce_local in each iteration.

Now, the reception and local reduction part of a typical (pipelined) implementation of, say, MPI_Reduce would look as follows.

```
MPI_Recv(partbuf,blocksize,MPI_BYTE,child,MPI_ANY_TAG,
          comm,&status);
if (status.MPI_TAG==DENSETAG) {
  MPI_Reduce_local(partbuf,recvbuf,count,type,op);
} else if (status.MPI_TAG==SPARSETAG) {
  MPI_Get_count(&status,MPI_BYTE,&sparsecount);
  sparsecount = sparsecount/(sizeof(int)+typesize);
  index = (int*)partbuf; value = (void*)(index+sparsecount);
  SPS_Reduce_local_index(index,value,recvbuf,sparsecount,type,op);
} else {
  // impossible for data sent with SPS_Send_elim
}
```

where child is some process to receive from. The right local reduction operator application function is selected based on the tag of the message (DENSETAG or SPARSETAG).

Using this scheme we have here implemented a binomial tree for short vectors, and a linear pipeline for long vectors. There is no claim that these are best possible algorithms, and any other reduction algorithm could have been chosen, e.g. the butterfly based algorithms described in [9].

4.1 Algorithm Performance

The functionality proposed so far would simply replace the send, receive and local reduction operations as outlined. Since no more data are sent than in the original algorithm, performance should, apart from the small overhead in scanning through the vectors to be sent (until the size cutoff is reached) never be worse; for pipelined algorithms the *same* number of blocks is being transmitted. This nice property is also the limiting factor in the performance improvement that can be achieved, with the best possible performance being bounded by the cost of sending only empty blocks (corresponding to vectors of zeroes). Alternative functionality would make it possible to send/receive variable parts of the input (intermediate) vectors, compressing as many non-zeroes as possible into the pipeline buffer. This alternative would thus pipeline blocks of the same size as in the original algorithm, but possibly fewer blocks. There would be some slight implementation difficulties by children in tree based algorithms going out of sync, but these are not major obstacles. Nevertheless, incorporating this variant into existing MPI libraries would definitely require more reimplementation work. We note that for both variants, estimating performance and in particular finding best possible pipeline block sizes in an on-line computation is not easily possible, since the actual amount of work in sending data and doing local reductions is input and intermediate result dependent (non-oblivious).

5   Experimental Setup and Evaluation

Experiments with the SPS_Reduce function were conducted at the University of Vienna on a newly installed 64-core Intel Nehalem based cluster, consisting of 8 compute nodes each with two Xeon quad-core X5550 2.66GHz CPUs, interconnected by a QDR InfiniBand. The system runs the mvapich2-1.2 MPI implementation. Performance of SPS_Reduce was measured with a benchmark written according to the guidelines in [3]. We report the effective reduction bandwidth (number of reduced MBytes per second) calculated from minimum completion time of the slowest MPI process to finish over a small number of iterations (25). Some care is taken that reduction buffers are not in cache when SPS_Reduce is called. The communicator is MPI_COMM_WORLD in which the MPI processes are distributed consecutively over the cores.

Although critical, it has not been attempted to highly tune the choice of pipeline blocksize for the linear pipeline algorithm. A fixed, (surprisingly) small pipeline blocksize of somewhat less than 10,000 Bytes seemed to give good results. The switch point from binomial tree to linear pipeline is at 100,000 Bytes, independently of the number of MPI processes. These choices were made for the sake of the experiments, and are certainly not overall best for production quality MPI_Reduce implementations.

With the experiments we aim to illustrate the possible improvements through neutral element elimination, estimate the overhead compared to algorithms not employing the optimization, and show differences for different datatypes. All experiments reported here have been performed with all 64 cores of the cluster.

Figure 1 shows the effect of varying the density (fraction of non-neutral elements) of the input vectors. All input vectors have the same structure and consist of blocks of one non-neutral element followed by a stride of neutral elements. Thus, intermediate results computed by the reduction algorithms have the same structure and density. The operator is MPI_SUM and the datatype MPI_INT. As vector size increase, the improvement increases, becoming up to a factor three

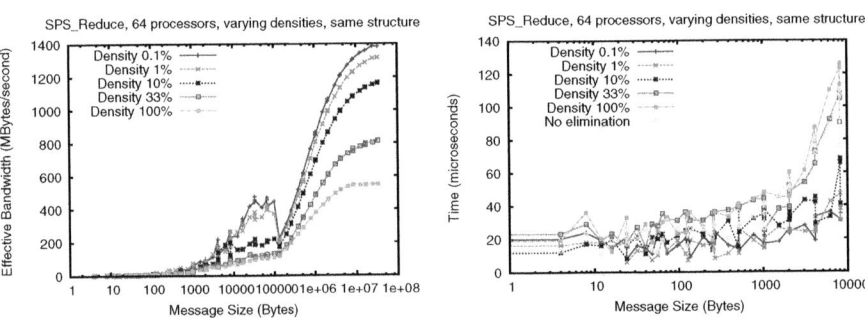

Fig. 1. Varying input density, same block structure of input and intermediate vectors, 0.1%, 1%, 10%, 33%, 100% non-neutral elements. Effective reduction bandwidth (left), latency (time) for input vectors up to 10,000Bytes (right).

for density 0.1% compared to fully dense vectors, but not much better than the improvement obtained for 1% density. For a density of 33%, corresponding to a reduction in message size to 2/3 of the input vectors, the improvement is proportional to the reduction in size, but improvements much larger than a factor three is not be possible with the machinery of this paper: the difference between sending a single integer and sending a full pipeline buffer (slightly less than 10KBytes) over the InfiniBand network is about a factor of three (2.9μ seconds vs. 9μ seconds). For small vectors up to a few thousand elements, the scan overhead for dense input does make a difference, up to 50%, but this quickly trails off; in a production setting the cutoff should probably be set such that neutral element elimination is only performed for vectors larger than a few hundred (or thousand) elements. For the vectors with low density some interesting effects are seen in the medium vector size regime. For 10,000 to 100,000 Bytes the effective message can become small enough that a different point-to-point message protocol is used, leading to a disproportional increase in effective reduction bandwidth. This is seen for densities of 1% and 0.1%, again with little difference between the two.

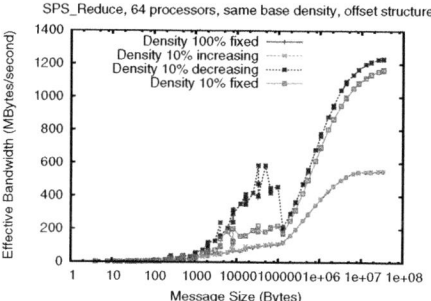

Fig. 2. Increasing and decreasing densities of intermediate results, compared to base density of 10% and 100%

The structure of the input vectors used in the experiments of Figure 1 was such that the intermediate results (of reducing any two vectors) again have the same structure: no neutral new elements appear or disappear. The effect of gradually *increasing density* as further non-zero elements are generated by intermediate reductions; and of *decreasing density* as neutral elements are created by intermediate reductions is shown in Figure 2. A density of 10% is taken as base, and vectors with the block structure described above but offset by the rank of the calling process are used as input. Addition of these input vectors will gradually lead to fully dense vectors (depending on the order in which partial results are combined). With input elements chosen as -1 for even ranks and 1 for odd ranks, neutral elements are sometimes generated (again, depending on reduction order) leading to less dense intermediate results.

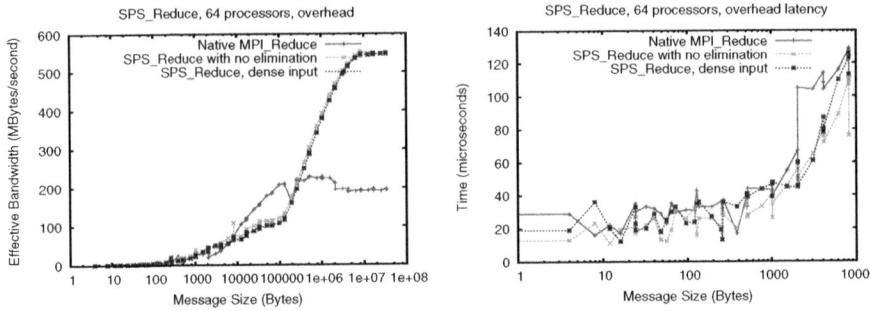

Fig. 3. Native MPI_Reduce compared to SPS_Reduce with no elimination compared to SPS_Reduce with dense input, bandwidth (left) and latency (time) for input vectors up to 10,000Bytes (right)

A crucial factor for the usability of the optimization is the overhead in performing elimination compared to SPS_Reduce implementations that do not attempt to do this (including MPI_Reduce), and thus do not have to scan the input. This is measured with the experiment documented in Figure 3 which compares the same basic algorithms with and without neutral element elimination. For large vectors there is virtually no overhead incurred in having to scan for zeroes; interestingly about a factor three improvement over mvapich's MPI_Reduce implementation is observed with the simple linear pipeline. For very small vectors up to a few hundred bytes the overhead can sometimes be significant (50%). It therefore may make sense to use the optimization only beyond some small threshold of some hundred bytes up to a few thousand bytes.

Finally, to illustrate that different MPI datatypes will have different compression and thus different improvements, results of running SPS_Reduce on other datatypes than MPI_INT are shown in Figure 4. The benchmark was run with density 1% and density 100% for MPI_SIGNED_CHAR and MPI_DOUBLE. For

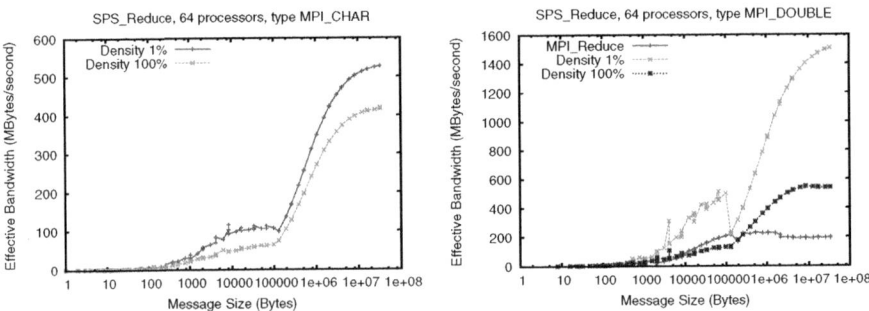

Fig. 4. Effects of neutral element elimination for MPI_SIGNED_CHAR (left) and MPI_DOUBLE, density of 1% compared to density 100%

`MPI_SIGNED_CHAR` the improvement for the sparse input is only about 20%, while for for `MPI_DOUBLE` it is again well over a factor of three. For this type it was possible to compare against the native MPI_Reduce of `mvapich`, which again was almost a factor three slower than the linear pipeline implemented here.

6   MPI Standardization

As described in this paper, elimination of neutral elements applies to the (commutative) built-in operators of MPI. The machinery could be used for user-defined operators as well, but MPI lacks functionality for users to declare neutral elements for user-defined reduction operators. There is therefore no portable way for an MPI library to determine whether a user-defined function indeed possesses a neutral element (for the given type). It might be worthwhile for the MPI Forum to consider functionality for this purpose, e.g. a function of the form MPI_Op_set_neutral(op,type,zero) with zero being in C a pointer to the neutral element for user-defined operator op for the (possibly derived) type type, allowing a library internal optimization like the one described here to look up the neutral element also for a user-defined operator. If no neutral element has been defined this way, the MPI library shall internally assume that none exists, and the optimizations described here will have no effect.

7   Concluding Remarks

We described, implemented and evaluated simple support functionality that enables neutral-element elimination to be incorporated into any implementation of the MPI reduction collectives. This consists in three library internal functions SPS_Op_get_neutral, SPS_Send_elim and SPS_Reduce_local_index. Together, these effectively hide the internal data format for vectors with eliminated neutral elements, and can thus readily be plugged in in any (point-to-point) based implementation of any of the MPI reduction collectives. A reduce function called SPS_Reduce with the same interface as MPI_Reduce was implemented by means of binomial tree and linear pipeline algorithms, and demonstrated improvements of a factor of more than 3 with increasing sparsity. Anecdotally, these implementations performed considerably better than the native MPI_Reduce implementation of the `mvapich` library installation on the Nehalem cluster. We suggest that neutral-element elimination can be incorporated as viable optimization in MPI libraries and used as default for the reduction collectives.

Neutral-element elimination was recently given as a student project at the University of Vienna. The students agreed on a common representation of vectors with implicitly represented neutral elements (which turned out to be different from the one presented here), and in small groups implemented either a set of test programs, the send-receive functionality, the local reduction operator application function, and three standard reduction algorithms (binomial and pipelined binary tree, linear pipeline). The resulting implementations gave performance improvements comparable to those presented here. This further shows that neutral-element elimination can be implemented in a modular way with little effort.

Acknowledgments. Thanks to the participating students and the tutor, Martin Wimmer.

References

1. Alpern, B., Carter, L.: Message compression for high performance. In: SIAM Conference on Parallel Processing for Scientific Computing (PPSC), pp. 814–819 (1995)
2. Filgueira, R., Singh, D.E., Calderón, A., Carretero, J.: CoMPI: Enhancing MPI based applications performance and scalability using run-time compression. In: Ropo, M., Westerholm, J., Dongarra, J. (eds.) Recent Advances in Parallel Virtual Machine and Message Passing Interface. LNCS, vol. 5759, pp. 207–218. Springer, Heidelberg (2009)
3. Gropp, W., Lusk, E.: Reproducible measurements of MPI performance characteristics. In: Margalef, T., Dongarra, J., Luque, E. (eds.) PVM/MPI 1999. LNCS, vol. 1697, pp. 11–18. Springer, Heidelberg (1999)
4. Hoefler, T., Siebert, C., Lumsdaine, A.: Scalable communication protocols for dynamic sparse data exchange. In: 15th ACM SIGPLAN Symposium on Principles and Practice of Parallel Programming (PPoPP), pp. 159–168 (2010)
5. Hofmann, M., Rünger, G.: MPI reduction operations for sparse floating-point data. In: Lastovetsky, A., Kechadi, T., Dongarra, J. (eds.) EuroPVM/MPI 2008. LNCS, vol. 5205, pp. 94–101. Springer, Heidelberg (2008)
6. Ke, J., Burtscher, M., Speight, W.E.: Runtime compression of MPI messanes to improve the performance and scalability of parallel applications. In: ACM/IEEE Supercomputing, p. 59 (2004)
7. Lee, H.-J., Park, K.-L., Koh, K.-W., Kwon, O.-Y., Park, H.-W., Kim, S.-D.: Improving the performance of grid-enabled MPI by intelligent message compression. In: International Conference on Internet Computing, pp. 772–780 (2003)
8. MPI Forum. MPI: A Message-Passing Interface Standard. Version 2.2 (September 4, 2009), www.mpi-forum.org
9. Rabenseifner, R., Träff, J.L.: More efficient reduction algorithms for message-passing parallel systems. In: Kranzlmüller, D., Kacsuk, P., Dongarra, J. (eds.) EuroPVM/MPI 2004. LNCS, vol. 3241, pp. 36–46. Springer, Heidelberg (2004)

Use Case Evaluation of the Proposed MPIT Configuration and Performance Interface

Carsten Clauss, Stefan Lankes, and Thomas Bemmerl

Chair for Operating Systems, RWTH Aachen University, Germany
{clauss,lankes,bemmerl}@lfbs.rwth-aachen.de
http://www.lfbs.rwth-aachen.de/content/research

Abstract. In this contribution, we present our experiences gained while prototyping the MPIT Configuration and Performance Interface that is currently under discussion for being integrated into the next MPI standard. The work is based on an API draft that has been recently released by the MPI Tools Working Group [1]. As a use case, we have already developed a simple tuning tool on top of the proposed MPIT interface that can help to optimize protocol thresholds of MPI implementations according to communication characteristics of the respective applications.

Keywords: MPI 3.0, MPIch, Tools Support, Information Interface.

1 Introduction

Currently, the definition of the upcoming MPI 3.0 standard is under active discussion by the respective working groups of the MPI-Forum [2]. One of these working groups deals with the definition of an additional interface that should help to enhance the interaction between MPI implementations and additional tools like debugger and profiler [1]. In this context, it is intended to offer an additional set of tool-oriented functions within a separate namespace, namely the MPIT configuration and performance interface. In contrast to the traditional PMPI profiling interface, that simply offers alternate entry points to MPI functions, the MPIT configuration and performance interface is rather a generic information interface that allows for querying MPI-internal configuration variables and performance counters. While the set of routines of this new interface is to be defined by the new MPI standard, names and intent of the accessible configuration and performance variables are left to the respective MPI implementation. For that reason, variable names and data types have to be retrievable via the MPIT interface, too, as well as their actual meaning. However, the question arises how a certain tool should cope with these library-specific configuration and performance variables.

2 Prototyping the MPIT Interface

In order to evaluate the potential of this new information interface, we have prototyped a major part of the proposed MPIT functions on top of an MPI

R. Keller et al. (Eds.): EuroMPI 2010, LNCS 6305, pp. 285–288, 2010.

library called MP-MPICH [5]. MP-MPICH, which stands for Multi-Platform MPICH, is a modification and extension to the well-known MPICH distribution. It covers several subprojects like SCI-MPICH (support for SCI cluster interconnects), NT-MPICH (support for Windows operating systems) and MetaMPICH (a Grid-enabled MPI library) and conforms to most parts of the MPI-2 standard (including process-spawning and one-sided communication on certain platforms).

Since MP-MPICH is derived from the original MPICH, it also makes use of the three common message transmission protocols *Short*, *Eager* and *Rendezvous*. The decision which of these is to be used is based on the respective message size. Usually, the thresholds between the protocols are set as static configurations before application start, either depending on certain resource restrictions or according to communication characteristics of the MPI applications.

For example, when using SCI-MPICH, the optimal threshold between Short and Eager protocol depends on the size of the largest *atomic* data package within the SCI network (e.g. 64, 128 or 1024 Bytes, depending on the SCI Link Controller used). On the other hand, for an optimal threshold between Eager and Rendezvous protocol, the fraction of *unexpected messages* occurring during an application run can be the decisive factor, as Gropp and Lusk have shown in [3].

By implementing the MPIT interface on top of MP-MPICH, it is now possible to query and even to adjust these protocol thresholds via configuration variables (MPID_EAGER_THRESHOLDS and MPID_RNDV_THRESHOLDS on verbosity level MPIT_VERBOSITY_TUNER_BASIC) for each pair of processes at runtime. For this purpose, an application (or rather an adjustment tool utilizing the MPIT interface) can call the proposed MPIT_Config_get() function, returning an array that contains the respective threshold values for all remote ranks. In turn, via the MPIT_Config_set() function, it can be attempted to modify these values at runtime. However, it's up to the underlying communication device (the so-called *channel device* of the ADI2 layer [4]) to accept such a modification, or to reject it e.g. for resource reasons.

3   Use Case: A Simple Tool on Top of MPIT

In order to evaluate the impact of varying these configuration values, we have developed a small tool that makes use of the MPIT interface by adjusting the protocol thresholds during a simple Ping-Pong benchmark. At first, the tool measures the communication time for messages of a length equal to the considered threshold. Then, the tool shifts the threshold (if possible), so that during a second measurement the alternative protocol is applied for the same message length. By comparing the measurement results and repeating this procedure in a nested manner, the tool tries to optimize the threshold values.

An example for such an optimized protocol transition is shown in Figure 1. Here, the measured communication times are plotted over the message sizes, once for a default threshold between Eager and Rendezvous of 16kByte, and once for an adjusted threshold of 27kByte. Obviously, messages of a size between these two values will benefit from shifting the transition from 16kByte to 27kByte in

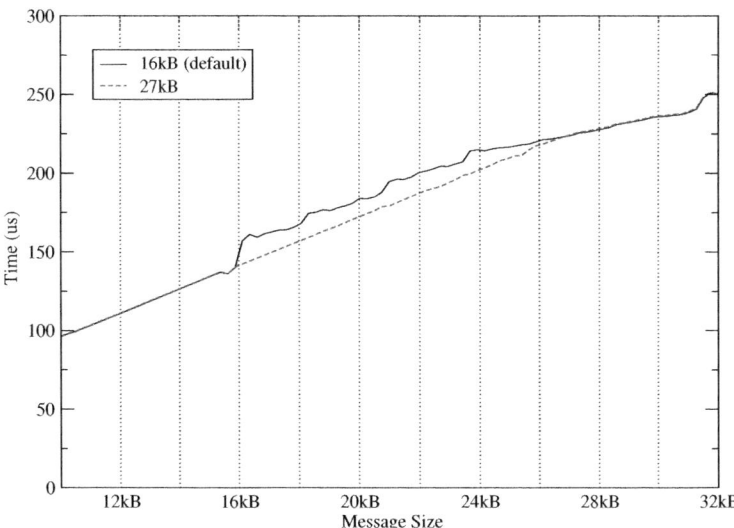

Fig. 1. Roundtrip Time for a Protocol Threshold of 16kByte (default) and 27kByte

this case. Although we have developed this simple threshold adjustment tool on top of MP-MPICH, we have designed the tool independent from the underlying MPI library. For this reason, the actual name of the MPIT configuration variable representing the threshold to be optimized has to be passed to the tool by the user.

However, as we have stated above, the optimal threshold between Eager and Rendezvous protocol also depends on the fraction of unexpected messages within an application run. The case of an unexpected message occurs every time when a message header arrives at receiver side before the final receive buffer is known to the MPI layer. That means that at arriving time of the header, the receiver hasn't yet posted a matching request for the message. The receiving of the actual payload can now either be delayed until the respective request gets posted, or the payload is being received temporarily into an intermediate buffer and copied later on into its final destination.

As shown in [3], the cost for receiving an expected message in Eager mode can be quantified as follows:

$$t_{eager_{expected}} = 2 \cdot t_{lateny} + \frac{h + n}{B}, \quad \text{Communication Bandwidth } B$$

for a message with n bytes of payload and a header size of h bytes.

When the message is unexpected, the payload must be copied into the final buffer later on. Therefore, the cost of an unexpected message in Eager mode can be quantified as:

$$t_{eager_{unexpected}} = t_{eager_{expected}} + \frac{n}{C}, \quad \text{Copy Bandwidth } C$$

In the Rendezvous case, the receiver responses to the initial header likewise with a control packet as soon as the matching receive request is posted. Therefore, the costs for transferring a message via this protocol amount to:

$$t_{rendezvous} = 3 \cdot t_{latency} + \frac{2 \cdot h + n}{B}$$

Finally, if f is the fraction of unexpected messages, then the optimal threshold n between these two protocols can be determined as follows:

$$t_{rendezvous} = t_{eager} \quad \Rightarrow \quad n_{threshold} = \frac{t_{latency} + \frac{h}{B}}{\frac{f}{C}} \approx \frac{t_{latency} \cdot C}{f}$$

In order to take this into account, the user of our tool can specify this fraction f of messages that should arrive during the Ping-Pong measurements before the receiver has posted the respective receive requests.

However, here arises the question how this fraction can be determined for a real-world application in an easy way. For that purpose, we have implemented MPIT-related performance variables within MP-MPICH that count (among others) the occurrence of unexpected messages for each remote rank. By comparing this count with the total number of messages received, the required value f can easily be determined. This appraisal can, for example, be conducted within an overloaded MPI_Finalize() function by just querying the respective MPIT counters (MPID_MSGS_UNEXPECTED and MPID_MSGS_RECEIVED) at the end of the application. By this means, the protocol thresholds of MP-MPICH can be optimized easily with respect to each application.

Announcement. We are not part of the MPI 3.0 Tools Working Group, but we are observing the respective standardization process with great interest.

References

1. MPI 3.0 Tools Support Working Group: Tool Interfaces for MPI (The current draft for the new MPI tools chapter) (April 2010),
   https://svn.mpi-forum.org/trac/mpi-forum-web/wiki/MPI3Tools/draft
2. Message Passing Interface Forum: MPI: A Message-Passing Interface Standard – Version 2.2, High-Perfomance Computing Center Stuttgart (HLRS), Germany (September 2009)
3. Gropp, W., Lusk, E.: MPICH Working Note: The Implementation of the Second-Generation MPICH ADI, Mathematics and Computer Science Division, Argonne National Laboratory, ANL (1996)
4. Gropp, W., Lusk, E.: MPICH Working Note: The Second-Generation ADI for the MPICH Implementation of MPI, Mathematics and Computer Science Division, Argonne National Laboratory, ANL (1996)
5. Bierbaum, B., Clauss, C., Finocchiaro, R., Schuch, S., Pöppe, M., Worringen, J.: MP-MPICH – User Documentation and Technical Notes, Chair for Operating Systems, RWTH-Aachen University, Germany (2009)

Two Algorithms of Irregular Scatter/Gather Operations for Heterogeneous Platforms

Kiril Dichev, Vladimir Rychkov, and Alexey Lastovetsky

UCD School of Computer Science and Informatics,
University College Dublin,
Belfield, Dublin 4, Ireland

Abstract. In this work we present two algorithms of irregular scatter/gather operations based on the binomial tree and Träff algorithms. We use the prediction provided by heterogeneous communication performance models when constructing communication trees for these operations. The experiments show that the model-based algorithms outperform the traditional ones on heterogeneous platforms.

Keywords: Heterogeneous platform, communication performance model, collective communication, scatterv, gatherv.

1 Introduction and Related Work

Algorithms for MPI collective communication operations typically implement them as a combination of point-to-point operations in a tree representing the communication partners and the way messages are exchanged between them. Traditional tree-based implementations target homogeneous platforms, implicitly assuming identical processors and a homogeneous communication layer. When applied to heterogeneous platforms, these implementations may be far from the optimal.

We propose to use heterogeneous communication performance models and their prediction for finding more optimal communication trees for these algorithms. The models take into account the underlying heterogeneous network of computers when constructing communication trees. In this work, we only use $t(i, j, m)$, the prediction of the execution time of sending a message of size m from process i to process j, in the algorithm design.

Optimization of collectives is not a new research topic and a wide range of optimized collective algorithms have been proposed in the past [1,3,5]. Communication performance models [4] have been used for collectives by predicting the runtime of various collective algorithms and switching between them accordingly [6]. In this work, we use model predictions during the dynamic construction of communication trees either by changing the mapping or changing the tree structure altogether. This topic has not been of practical interest to the best of our knowledge, except for [7], where the regular gather collective operation is optimized.

R. Keller et al. (Eds.): EuroMPI 2010, LNCS 6305, pp. 289–293, 2010.

2   Model-Based Algorithms of Irregular Scatter/Gather

The shortcomings of the homogeneous algorithms are also valid for scatter and gather algorithms. While the regular variants of these operations only allow for same-sized chunks of data to be scattered or gathered, their irregular counterparts support different data sizes at each process. Irregular scatter/gather operations are used particularly for heterogeneous algorithms which distribute data according to the different computational capacity of the different processes. In this work, we demonstrate our approach on the example of two existing tree construction algorithms for irregular scatter/gather operations. We integrate the model prediction $t(i, j, m)$ into the algorithms to produce more optimal communication trees.

A model-based algorithm for tree construction derived from an algorithm that does not use models can differ from it by changing the process mapping to nodes or by constructing a tree with a different tree structure. From the two algorithms we present, the first one changes the mapping, while the second one may construct a different tree structure as well.

Model-based binomial tree scatterv/gatherv. In this algorithm we use point-to-point predictions to map processes to a binomial tree. The binomial tree is constructed in a depth-first manner, starting with the lowest-order subtrees. Each new tree node receives the process number i from the set of free processes that has minimal $(minimum - first)$ or maximal $(maximum - first)$ predicted communication time $t(parent, i, m_i)$, where m_i is the message size assigned to process i. A good choice of mapping depends on the runtime platform. For example, on a heterogeneous cluster with a single switch, $maximum - first$ mapping may be better since the subtrees of a parent node will be balanced in their communication costs to it. In a hierarchy of clusters $minimum - first$ mapping may be better because intra-cluster processes are likely to be mapped to the same communication subtree.

Model-based Träff algorithm for scatterv/gatherv. We will significantly modify an algorithm by Träff [2] which targets irregular scatter/gather operations when constructing a communication tree. Träff considers the message size assigned to each process and assumes identical links between all nodes, while we consider both the message size and the characteristics of the links between nodes by using the prediction function $t(i, j, m)$. Even for a fixed node count, the original algorithm can generate different trees depending on the message sizes at the node level. Since our modified algorithm observes the weight of the links instead, both the process mapping as well as the tree structure can differ from the original algorithm.

Given :

- set of nodes S with corresponding sendcounts/recvcounts arrays defining the message size to be sent to/received from each process
- defined predictor $t(i, j, m)$

Result :

− a communication tree

Algorithm :

Starting from a set of processes to build a tree with a given root (Fig. 1a), we sort them decreasingly using $t(i, j, m)$ and partition the sorted set into subsets(Fig. 1b). These subsets are balanced in their total communication cost with the root node - left subtrees have less processes with slower transfer times while the right subtrees have more processes with faster transfer times. Each subset then chooses the process i which would take the least time to transfer all messages of this subset to/from the parent node j, removes it from the subset and an edge (i, j) is created (Fig. 1c). We repeat the algorithm for all further subsets until the tree is fully constructed. The outlined phases have to be repeated in each step since we predict the communication time of new process pairs.

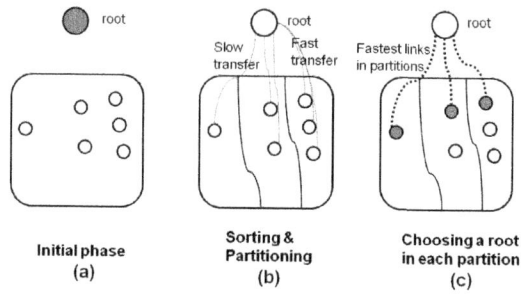

Fig. 1. Main phases in the modified Träff algorithm [2]

3 Experimental Results

We present results for the two irregular scatter/gather algorithms described in section 2. Our experiments used benchmarks implemented in the CPM/MPIBlib framework [4]. We used two different platforms the HCL cluster with a single Gigabit Ethernet switch and a larger and more heterogeneous multi-site cluster of clusters known as Grid5000. On both platforms, we used the Hockney model for our predictions. Since the prediction we use is not model-specific, other communication models can be used as well. An important consideration is that scatterv/gatherv operations can use any distribution of message sizes. We used a setup which assigns each node a message size based on its CPU speed (delivered by a trivial benchmark).

We first tested this message size distribution on the HCL cluster and Open MPI (1.2.8) on 14 nodes, running one process per node. Since this platform did not provide a high level of network heterogeneity, the improvement demonstrated by our algorithms was not significant. We then experimented with a message size distribution with a larger ratio between maximal and minimal message size. The

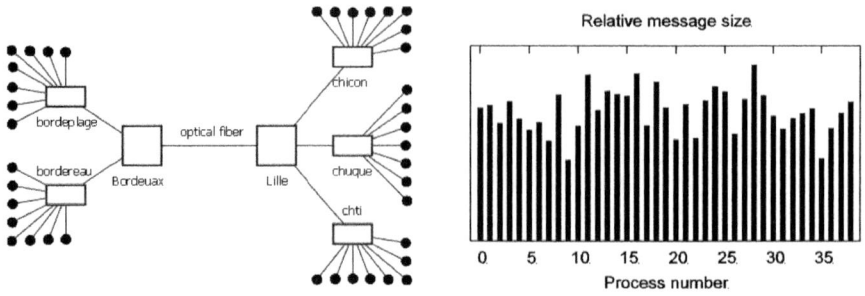

Fig. 2. Experimental setup and message size distribution on Grid5000

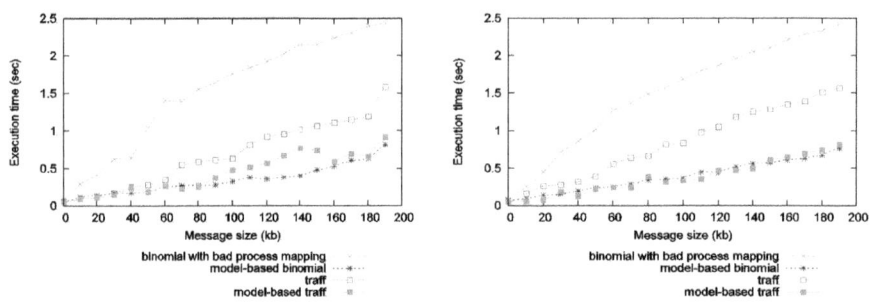

Fig. 3. Benchmarks on Scatterv (a) and Gatherv (b) operations on Grid5000

modified algorithm of Trff had similar timings to the original algorithm. The results were best for the model-based binomial tree algorithm with maximum-first mapping - compared to the original algorithm using an arbitrary binomial tree, it was faster by 25-35% or more for larger messages. In the case of a high message size variation, the prediction still had a positive impact for this cluster.

The experiments on Grid5000 used 39 nodes from 5 clusters located on 2 sites, running one process per node. Fig. 2 displays the experimental setup and the CPU-based message size distribution. MPICH2 (version 1.2.1) was used with TCP/IP as communication layer. The results (Fig. 3) demonstrate that on heterogeneous networks both model-based algorithms clearly outperform their non-model-based counterparts we observed time reductions for scatterv and gatherv of up to 75% for the binomial algorithm (minimum-first mapping) and up to 60% for Träffs algorithm. This confirms that our approach is particularly useful for platforms with high network heterogeneity.

Acknowledgments. This publication has emanated from research conducted with the financial support of Science Foundation Ireland under Grant Number 08/IN.1/I2054.

References

1. Thakur, R., Rabenseifner, R., Gropp, W.: Optimization of Collective Communication Operations in MPICH. Int. J. of High Perf. Comp. App. 19, 49–66 (2005)
2. Träff, J.L.: Hierarchical Gather/Scatter Algorithms with Graceful Degradation. In: IPDPS 2004, vol. 1, pp. 80–89. IEEE, Los Alamitos (2004)
3. Worringen, J.: Pipelining and Overlapping for MPI Collective Operations. In: LCN 2003, pp. 548–557. IEEE, Los Alamitos (2003)
4. Lastovetsky, A., Rychkov, V., OFlynn, M.: Accurate heterogeneous communication models and a software tool for their efficient estimation. Int. J. of High Perf. Comp. App. 24, 34–48 (2010)
5. Chan, E.W., Heimlich, M.F., Purkayastha, A., van de Geijn, R.A.: On optimizing collective communication. In: Cluster 2004, pp. 145–155. IEEE, Los Alamitos (2004)
6. Pjesivac-Grbovic, J., Angskun, T., Bosilca, G., Fagg, G., Gabriel, E., Dongarra, J.: Performance analysis of MPI collective operations. Cluster Comput. 10(2), 127–143 (2007)
7. Hatta, J., Shibusawa, S.: Scheduling algorithms for efficient gather operations in distributed heterogeneous systems. In: WPP 2000, pp. 173–180. IEEE, Los Alamitos (2000)

Measuring Execution Times of Collective Communications in an Empirical Optimization Framework

Katharina Benkert[1] and Edgar Gabriel[2]

[1] High Performance Computing Center Stuttgart (HLRS),
University of Stuttgart, 70550 Stuttgart, Germany
benkert@hlrs.de
[2] Parallel Software Technologies Laboratory,
Department of Computer Science, University of Houston, Houston, TX, USA
gabriel@cs.uh.edu

Abstract. An essential part of an empirical optimization library are the timing procedures with which the performance of different codelets is determined. In this paper, we present for four different timing methods to optimize collective MPI communications and compare their accuracy for the FFT NAS Parallel Benchmarks on a variety of systems with different MPI implementations. We find that timing larger code portions with infrequent synchronizations performs well on all systems.

Keywords: Empirical Optimization, Abstract Data and Communication Library (ADCL), Collective Communication, NAS Parallel Benchmark.

1   Introduction and Motivation

Automatic performance tuning is an area of research defined by one of the fundamental questions in computing: how to obtain for a kernel (computational or communication pattern) platform-independently an equal or superior performance compared to hand-tuned code. Among the pioneers of empirical tuning software in High Performance Computing (HPC) are the Automatically Tuned Linear Algebra Software (ATLAS) [1] which uses pre-runtime tuning. Projects applying runtime tuning include FFTW [2], PhiPAC [3], and STAR-MPI [4].

The Abstract Data and Communication Library (ADCL) [5] is an empirical auto-tuning library targeting, but not restricted to, MPI communications with an own API. It provides predefined sets of codelets for collective operations and for a Cartesian neighborhood communication. Information about the communication operation is encapsulated into ADCL objects and combined to an ADCL_Request object which represents a persistent communication object and similarly to its MPI counterpart for sequential persistent requests can be 'started' using ADCL_Request_start. The latter routine executes the communication and controls the optimization process. During the first calls to ADCL_Request_start,

R. Keller et al. (Eds.): EuroMPI 2010, LNCS 6305, pp. 294–297, 2010.

the search phase, execution times of alternative codelets are measured empirically, i.e. by actually running them multiple times one after another. Thereafter we compute an outlier-aware average on each process as explained in [6] and select the codelet judged best-performing which is then used during the rest of the simulation, the *production phase*. The assessment of the performance of different implementation alternatives is critical as it is the basis for the selection of the optimal codelet.

2   Timing Methods

Embracing the codelet with timing routines *inside* the library (fig. 1 left) is the simplest way to measure the execution time of a codelet and transparent to the user. Including the synchronizations during the search phase creates additional overhead but ensures a common starting point. The timings represent a worst-case scenario since eventually delays would cancel. Even systematic errors may be introduced since applications features (e.g. process arrival patterns) are ignored. We refer to this timing method as *barrier*. Alternatively, the *nobarrier* method has no synchronization but still couples timing and codelet inside the library. It is faster during search phase, closer to reality as it does not add any disturbances, but the timings might influence one another and be difficult to evaluate since the common starting point is lost.

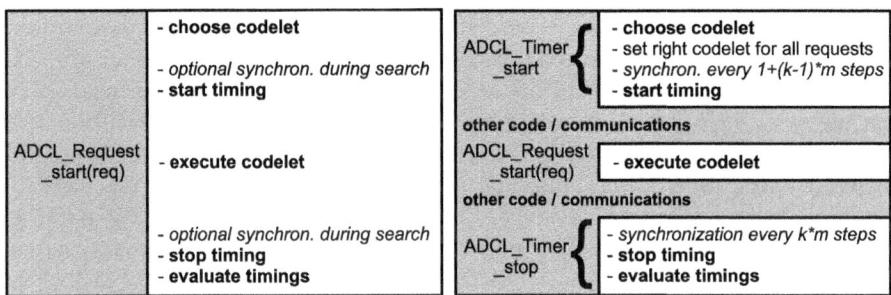

Fig. 1. Timing routines with user code (grey) and ADCL code (white) embrace only the codelet (left) or the codelet plus its environment (right); $m \geq 1, k = 1, \dots$.

To mimic the application behavior, we introduce a new ADCL object, the timer object, along with functions to start and stop the timing. The idea is to not just measure the codelet in isolation, but rather the codelet plus (a part of) its environment (fig. 1 right). This makes it possible to optimize communication operations that are part of larger code portions, such as code sections which overlap communication and computation, as well as the simultaneous optimization of different communications, i.e. multiple requests, at the same time which may interfere with one another. However, it is up to the user to select the right environment to time. For the method *timer* with $m = 1$ the treatment of outliers as explained in [6] does not change whereas for *timer_multistep*, we set $m > 1$ but leave the number of calls to ADCL_Request_start the same ($k * m = const.$). This results in fewer already averaged measurements per codelet, which might be more susceptible to the outlier problem. With well-balanced values for k and m, this method should demonstrate an "undisturbed" execution behavior but also provide enough measurements to allow outlier handling.

3   Timing of Collective Communications

To investigate different timing techniques for collective communications, we use the MPI FFT Benchmark from the NAS Parallel Benchmarks 3.0 [7]. The main loop consists of an evolution step and the computation of the Fast Fourier Transform (FFT) which involves an all-to-all communication and is in this version automatically tuned using ADCL. The test systems used are a Nehalem cluster with InfiniBand interconnect, a Cray XT5m and a NEC-SX8 installation at HLRS, the SGI Altix at LRZ Munich and the BlueGene/P system at the Supercomputing System Jülich. For each system, we executed the FFT benchmark for various classes K with different numbers of processes n (one MPI process per core), denoted in the following as Kn, and eventually with multiple MPI implementations.

Within a single batch job, we execute three set of runs. Each set contains 12 runs, one for each of the 8 codelets for the all-to-all operation (the *verification runs*) and 4 for the different timing methods, with 200 FFT iterations each. Each codelet is measured 20 times, a number found sufficient in the past to determine the winner implementation reliably. For *timer_multistep*, we chose $m = 4$. The execution times over the runs are averaged over the sets, as performance data normally slightly varies. However in some cases deviations in execution time between the three sets of verification runs were equal to or exceeded the possible gains between the codelets, most notably for the SGI Altix with OpenMPI or the Nehalem Cluster with Mvapich. Collectives are known to have the potential to cause network flooding. This fact by itself can produce large deviations and it worsens if resources are shared. Although the reasons for these deviations are worth considering, they lay outside the scope of this paper and are difficult to carry out for closed-source MPI implementations.

The averaged execution times of the verification runs provide some ranking of the 8 codelets. Ideally, the results of the timing procedures should follow this ordering, i.e. be a monotone ascending function, and reproduce qualitatively the slopes of each segment. Nearly vertical slopes signify almost equal performance whereas steep slopes imply a larger performance gap between two codelets.

We found that *barrier* and *nobarrier* have problems on a variety of machines, especially in case of medium deviations. *Timer* has some problems on the SGI Altix. Best results gives the *timer_multistep* procedure, which just fails once in a case where all timing procedures have problems.

4 Summary and Outlook

This paper introduced the timer object within the Abstract Data and Communication Library (ADCL) which allows to time not only a single codelet but also its environment. We showed that for an all-to-all communication pattern the accuracy of performance prediction for performance data generated with this timer object is superior to that when timing just the codelet itself. An enhanced version of the timer object nearly exactly reproduced quantitatively the performance data obtained from longer runs. These results set the stage for further investigations on the simultaneous optimization of multiple, dependent communications and allow adding predefined sets of codelets to optimize overlapping computations and communications.

Acknowledgments. This work was funded by the project STEDG within the BMBF Software Initiative for High Performance Computing and supported by a short-time scholarship of the DAAD. Partial support for this work was provided by NSF under award no. CNS-0846002. Any opinions, findings, and conclusions or recommendations expressed in this material are those of the authors and do not necessarily reflect the views of the National Science Foundation.

References

1. Whaley, R.C., Petite, A.: Minimizing development and maintenance costs in supporting persistently optimized BLAS. Software: Practice and Experience 35(2), 101–121 (2005)
2. Frigo, M., Johnson, S.G.: The Design and Implementation of FFTW3. Proceedings of IEEE 93(2), 216–231 (2005)
3. Bilmes, J., Asanovic, K., Chin, C., Demmel, J.: Optimizing matrix multiply using PHIPAC: a Portable, High-Performance, ANSI C coding methodology. In: Proceedings of the International Conference on Supercomputing, Vienna, Austra (July 1997)
4. Faraj, A., Yuan, X., Lowenthal, D.: STAR-MPI: self tuned adaptive routines for MPI collective operations. In: ICS 2006: Proceedings of the 20th Annual International Conference on Supercomputing, pp. 199–208. ACM Press, New York (2006)
5. Gabriel, E., Feki, S., Benkert, K., Resch, M.M.: Towards Performance Portability through Runtime Adaption for High Performance Computing Applications. Concurrency and Computation — Practice and Experience (2010) (accepted for publication)
6. Benkert, K., Gabriel, E., Resch, M.M.: Outlier Detection in Performance Data of Parallel Applications. In: 9th IEEE International Workshop on Parallel and Distributed Scientific and Engineering Computing (2008)
7. Bailey, D., Barszcz, E., Barton, J., Browning, D., Carter, R., Dagum, L., Fatoohi, R., Fineberg, S., Frederickson, P., Lasinski, T., Schreiber, R., Simon, H., Venkatakrishnan, V., Weeratunga, S.: The NAS Parallel Benchmarks (1994)

Dynamic Verification of Hybrid Programs[*]

Wei-Fan Chiang[1], Grzegorz Szubzda[1],
Ganesh Gopalakrishnan[1], and Rajeev Thakur[2]

[1] School of Computing, Univ. of Utah, Salt Lake City, UT 84112, USA
[2] Math. and Comp. Sci. Div., Argonne Nat. Lab., Argonne, IL 60439, USA

Overview

Hybrid (mixed MPI/thread) programs are extremely important for efficiently programming future HPC systems. In this paper, we report our experience adapting ISP [3,4,5], our dynamic verifier for MPI programs, to verify a large hybrid MPI/Pthread program called Eddy Murphi [1]. ISP is a stateless model checker that works by replaying schedules leading up to previously recorded non-deterministic selection points, and pursuing new behaviors out of these points. The main difficulty we faced was the inability to *deterministically replay* up to these selection points because ISP instruments only the MPI calls issued by an application, whereas thread level scheduling non-determinism may change the course of execution. Instrumenting both MPI and Pthreads API calls requires an invasive modification of ISP which was not favored. The novelty of our solution is to determinize thread schedules using a record/replay daemon and demonstrating that this approach works on a realistic hybrid application: the Eddy Murphi model checker.

Verification Challenge

Figure 1 illustrates the architecture of Eddy Murphi, a parallelized and distributed model checker. It essentially implements a BFS approach algorithm to explore the state space. Each process of Eddy Murphi (the node shown in Figure 1) consists of the worker thread and the communicator thread (CT). The communicator thread issues MPI sends and MPI wildcard receives.

If we are to successfully verify Eddy Murphi using ISP, we must have ISP explore the space of non-deterministic receives of the communicator threads. ISP must not be "confused" by the Pthread schedule changes that may vary the order in which the worker and communicator threads obtain the mutex lock that guards common data structures to these threads.

Determining Solution

ISP's operation is as follows. It collects MPI calls from all processes and waits for the processes to reach their *fence points*. Once all processes have reached fences, ISP chooses a sender process in the set of potential matching senders

[*] Supported in part by Microsoft, NSF CNS-0935858, CCF-0903408, and DOE ASCR DE-AC02-06CH11357.

R. Keller et al. (Eds.): EuroMPI 2010, LNCS 6305, pp. 298–301, 2010.

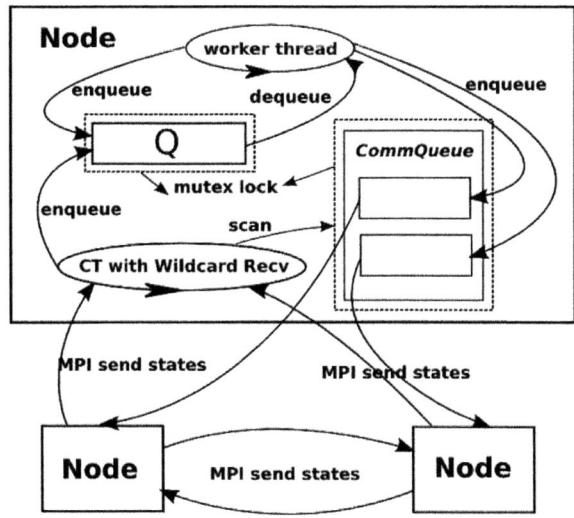

Fig. 1. The Architecture of Eddy Murphi

to a wildcard receive and rewrites the wildcard receive into a specific receive matching this sender, thus determinizing the *MPI schedule* from that point. We call such a step a determinized receive (*DR event*) event.

Figure 2 illustrates how we augment ISP with a daemon that helps record a previous Pthread schedule (of Pthread mutex calls, *etc.*) up to a DR event into a log file and enforces the recorded schedule when we replay up to this DR event. Essentially, the daemon sends positive acknowledgements (ACKs) to threads calling Pthread routines in an order matching the recorded order, and sends negative acknowledgements (NACKs) to threads calling in a non-matching order. The NACKed threads have to re-issue their Pthread calls or be blocked on the calls.

For the very first schedule exhibited, our daemon is in the record mode, recording the entire schedule. When replaying the schedule, the daemon stays in the replay mode till it sees ISP pursue a new behavior (a different DR event is pursued by ISP). The daemon then switches to record mode, extending the partially replayed schedule with a new record sequence. This concatenated sequence forms the 'seed' for the next execution. Such a record/replay mechanism is easily implemented by keeping only two log files: the previous one and the current one.

Assumptions (satisfied by Eddy Murphi):

- All read/writes are protected by mutual exclusion locks.
- The MPI threading level is `MPI_THREAD_FUNNELED`.
- Processes communicate only through MPI calls.
- Other API calls (besides MPI and Pthreads) are not allowed.
- The inputs provided by our test harness are deterministic.

Related Works: The idea of our "record/replay" mechanism is inspired by ODR, output deterministic replay [2]. Our "record/replay" method is similar

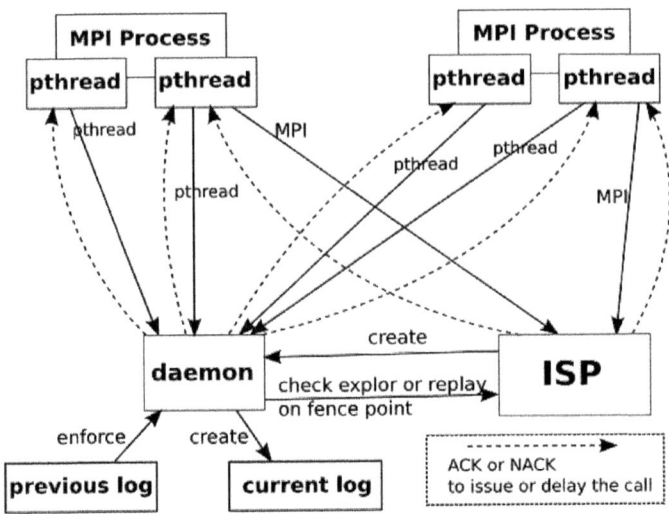

Fig. 2. ISP-Daemon System

to the SI-DRI recording approach to recording the lock order used in ODR. ODR gathers the schedule trace, input trace, and read trace from the original execution and generates symbolic path constraints. These are then solved using a constraint solver to enforce previous schedules. Our mechanisms are much simpler in comparison.

Findings and Concluding Remarks

We chose the *n_peterson* model as the test input of Eddy Murphi and set the depth bound of it to control the scale of exploration. Table 1 presents our experiment results. The *original isp* denotes the old version of ISP which cannot handle thread non-determinism (it crashes when threads change their schedules). The *isp-daemon* denotes the new version of ISP which can enforce Pthread schedules. The column *ISP version & configuration* denotes the version of ISP we used, the

Table 1. Experiment on n_peterson Model

ISP version & configuration	interleaving explored	min./max./ave. DR events
original isp / p3 / d3	11	112 / 112 / 112
original isp / p3 / d5	fail on 2ed	133 / 133 / 133
original isp / p4 / d3	fail on 4th	145 / 149 / 146
isp-daemon / p3 / d3	11	112 / 112 / 112
isp-daemon / p3 / d5	61	133 / 133 / 133
isp-daemon / p3 / d10	over 1500	179 / 179 / 179
isp-daemon / p3 / d20	over 1600	723 / 765 / 727
isp-daemon / p4 / d3	6	141 / 145 / 143
isp-daemon / p4 / d5	over 1097	174 / 174 / 174
isp-daemon / p4 / d10	over 2000	300 / 304 / 303
isp-daemon / p4 / d20	over 2400	898 / 898 / 898

number of processes, and the depth bound. For instance, "original isp / p3 / d5" means the result of running Eddy Murphi on the old version ISP with three processes created and with a 5-level depth bound BFS. The column *interleaving explored* denotes the number of interleavings explored by ISP while verifying Eddy Murphi. The column *min./max./ave. DR events* denotes the minimum, maximum, and average number of DR events we encountered in one execution.

The main limitation of our method is that it does not guarantee coverage of the Pthread non-determinism space. It covers only the MPI non-determinism space for particular determinizations of the Pthread schedule space. We briefly explored trying to iterate through the Pthread schedule space through methods such as preemption bounded search, and re-running ISP for each such altered Pthread schedule. Such an approach will result in an extremely large schedule space to cover – equaling the product of the Pthread and MPI schedule spaces. A better approach may be to conduct a random-walk across the Pthread and MPI schedule spaces. Our main conclusion is that unless hybrid programming is approached with discipline, building a tractable verification approach becomes nearly impossible. Our future work will examine how to make ISP capable of dynamically verifying more types of hybrid programs such as MPI-OpenMP, MPI-CUDA, *etc.*

From a practical point of view, it is important to cover mixed MPI/OpenMP programs. However, this opens up a number of challenges. First, it would be necessary for *OpenMP implementors to expose the underlying OpenMP scheduling points.* Without this, it would be impossible to guarantee any sort of coverage. Since OpenMP implementations differ significantly from each other, ideally one would standardize such an API pertaining to scheduling so that hybrid dynamic verifiers can resort to a single standardized solution in determinizing the OpenMP behaviour while exploring MPI behaviours.

Even with such help, we do need a well articulated set of programming practices around which to build dynamic verifiers for hybrid programs. Without such programming practices, the product schedule spaces of the individual concurrency models will be so huge that it cannot be covered within reasonable amounts of time.

References

1. Eddy murphi. distribution, http://www.cs.utah.edu/formal_verification/mediawiki/index.php/Eddy_Murphi
2. Altekar, G., Stoica, I.: ODR: Output-deterministic replay for multicore debugging. In: 22nd symposium on Operating Systems Principles (SOSP), pp. 193–206 (2009)
3. Vakkalanka, S.: Efficient Dynamic Verification Algorithms for MPI Applications. PhD thesis (2010), http://www.cs.utah.edu/Theses
4. Vakkalanka, S., Gopalakrishnan, G., Kirby, R.M.: Dynamic Verification of MPI Programs with Reductions in Presence of Split Operations and Relaxed Orderings. In: Gupta, A., Malik, S. (eds.) CAV 2008. LNCS, vol. 5123, pp. 66–79. Springer, Heidelberg (2008)
5. Vo, A., Vakkalanka, S., DeLisi, M., Gopalakrishnan, G., Kirby, R.M., Thakur, R.: Formal verification of practical mpi programs. In: PPoPP, pp. 261–269 (2009)

Challenges and Issues of Supporting Task Parallelism in MPI

Márcia C. Cera, João V.F. Lima, Nicolas Maillard, and Philippe O.A. Navaux

Universidade Federal do Rio Grande do Sul, Brazil
{marcia.cera,joao.lima,nicolas,navaux}@inf.ufrgs.br

Abstract. Task parallelism deals with the extraction of the potential parallelism of irregular structures, which vary according to the input data, through a definition of abstract tasks and their dependencies. Shared-memory APIs, such as OpenMP and TBB, support this model and ensure performance thanks to an efficient scheduling of tasks. In this work, we provide arguments favoring the support of task parallelism in MPI. We explain how native MPI can be used to define tasks, their dependencies, and their runtime scheduling. We also discuss performance issues. Our preliminary experiments show that it is possible to implement efficient task-parallel MPI programs and to increase the range of applications covered by the MPI standard.

1 Task Parallelism in MPI Programs

Explicit task parallelism is a simple and elegant programming paradigm that allows to unfold irregular parallelism efficiently. The programmer identifies independent units of work (tasks), dependencies among them, and the runtime takes care of the scheduling. A large set of lightweight tasks are specified, leaving it up to the runtime to unfold parallelism and to decide the mapping of tasks: either they may execute on different units of execution, or run sequentially.

We show how it is possible to develop task-parallel MPI programs that tackle these issues, as well as experimental results of task-parallel MPI programs. which include in their source-code the control of granularity. This control follows an *Adaptive* approach adjusting the size of the grain according to the number of available processing elements.

The development of task-parallel programs involves the definition of tasks, their dependencies, and scheduling decisions [1]. In this Section, we show how native MPI features can be used to treat these issues, after a quick review of classical implementations in shared memory.

Related Work. Many programming interfaces or languages have been proposed to support explicit task parallelism. Cilk [2,3] was the precursor parallel programming interface to deal with explicit task issues. OpenMP 3.0 [4] and Intel© *Threading Building Blocks* (TBB) [5] are also parallel APIs for multithreaded programming. These parallel APIs have some common concepts concerning task-parallel programming. While Cilk represents tasks as procedures, OpenMP uses a block of instructions (defined by a task construct) and TBB uses instances of a task class. Task dependencies are expressed as barriers using keywords in Cilk (sync) and OpenMP (taskwait), and TBB allows

R. Keller et al. (Eds.): EuroMPI 2010, LNCS 6305, pp. 302–305, 2010.

Listing 1. Source-code of Fibonacci calculation using MPI-2 dynamic processes: one parent process computes Fibonacci(n). It spawns two children processes that compute Fibonacci($n - 1$) and Fibonacci($n - 2$). Then it blocks in an `MPI_Recv` to wait for their return. Declarations of variables have been omitted.

```
1   void mpi_fib( int n ) {
2    if ( n < 2 )     MPI_Send( &n, 1, MPI_INT, 0, 1, parent );
3    else {
4      sprintf( argv[0], "%d", (n - 1) );
5      MPI_Comm_spawn( "mpi_fib", argv, 1, info, myrank,
6                             MPI_COMM_SELF, &child[0], err );
7      sprintf( argv[0], "%d", (n - 2) );
8      MPI_Comm_spawn( "mpi_fib", argv, 1, info, myrank,
9                             MPI_COMM_SELF, &child[1], err );
10     MPI_Recv( &x, 1, MPI_INT, MPI_ANY_SOURCE, 1, child[0], &st );
11     MPI_Recv( &y, 1, MPI_INT, MPI_ANY_SOURCE, 1, child[1], &st );
12     fibn= x + y;
13     MPI_Send( &fibn, 1, MPI_INT, 0, 1, parent );
14   }
15  }
```

synchronization with either the *blocking* style (similar to Cilk) or the *continuation* style. Cilk and TBB schedule tasks efficiently using work-stealing, while OpenMP includes several simple strategies but is still under development to include more complex and programmer-friendly strategies.

Defining and Spawning Tasks in MPI — The MPI standard defines tasks as having their own address space, and most MPI distributions map a task to an O.S. process. MPI-2 has added the support for dynamic process management through `MPI_Comm_spawn`.

A simple and well-known example of task-parallel program is a recursive implementation of the Fibonacci calculation. Listing 1 shows this trivial implementation in which the `MPI_Comm_spawn` creates MPI tasks and the exchanges of messages express the dependencies. However, an efficient execution of task-parallel MPI programs requires some control of granularity.

In task-parallel programs, the programmer only identifies the potential parallelism to be unfolded at runtime. Recursive algorithms often include a threshold to indicate the point where sequential computations are more efficient than recursive calls. In task-parallel MPI programs, the threshold may indicate that the spawning of new processes must stop. This naturally increases the granularity of the tasks.

2   Experimental Results

All the results have been obtained on the French Grid'5000. Each node has two Intel© Xeon E5310 Quad Core 1.60 GHz processor (eight cores per node) and 16 GB of memory. We have used GCC 4.3 with OpenMPI. All the presented measures are the speedup

Table 1. Speedups of Fibonacci and Matrix Multiplication with MPI and OpenMP upon 1, 2, 4, and 8 cores of a multicore machine

Applications	API	Number of Cores			
		1	**2**	**4**	**8**
Fibonacci	MPI	1	1	2.62	3.40
	OpenMP	0.55	1.09	2.07	3.89
Matrix Multiplication	MPI	0.76	1.48	1.55	3.94
	OpenMP	1.17	1.77	2.25	2.47

relative to the best sequential running time, and each running time is the mean of 30 executions, with standard deviation always smaller than 3%.

We have developed two test applications, based on recursive algorithms: The Fibonacci calculation and a recursive implementation of the traditional matrix multiplication algorithm (multiplication of rows per columns) for tests with heavier tasks. A recursive threshold is used to determine when sequential executions must replace the dynamic process creation. Furthermore, an *Adaptive* approach has been used to control the granularity of the tasks aiming to provide a better load balancing.

Achieving Parallelism with Tasks in MPI. In the results presented here, the Fibonacci program computes the $53th$ element in the sequence, and the matrix multiplication uses two input matrices of 8192×8192 elements. Thresholds are the $46th$ element in Fibonacci and 256×256 elements for matrix multiplication. Besides, sequential executions took almost $1,240$ seconds for Fibonacci and 80 seconds for matrix multiplication.

Table 1 shows the speedups of the applications using MPI and OpenMP upon a shared-memory processor with 1, 2, 4, and 8 cores.

We have verified that our MPI versions achieves mostly lower speedups than OpenMP, because it creates processes and exchanges messages into a shared-memory environment. On the other hand, the difference is low and the MPI-2 task-parallel programs show speedups close to OpenMP, even without efficient use or shared memory.

Performance of Task-Parallel MPI Programs. This section aims to verify the behavior of task-parallel MPI programs in distributed-memory environments. Here, the input of matrix multiplication has been increased to 16384×16384 elements which took almost 412 seconds in sequential. The Fibonacci input as well as both thresholds are the same as exposed above.

Figures 1a and 1b show respectively the speedups of Fibonacci and matrix multiplication from 1 processor with 8 cores to 8 processors (64 cores). In the Fibonacci case, our MPI implementations achieve speedups mostly lower than the naive version. This means that when tasks involve few computations, the use of a threshold is already enough to ensure good performance. On the other hand, an adaptive granularity control in matrix multiplication allows a considerable gain of performance.

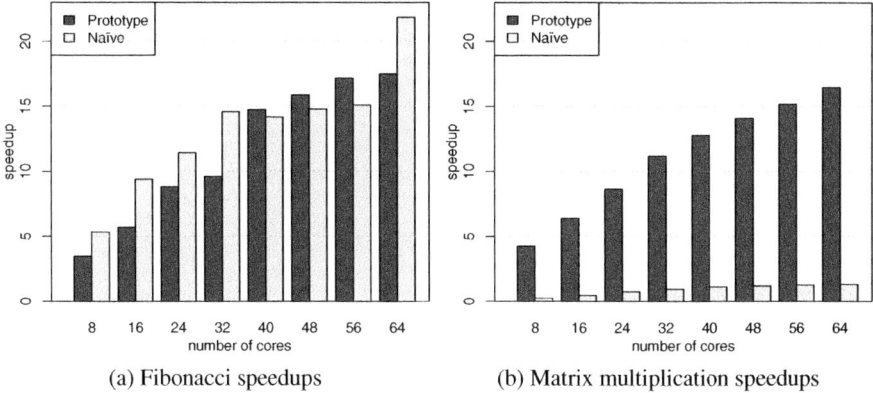

| (a) Fibonacci speedups | (b) Matrix multiplication speedups |

Fig. 1. Performance results with Fibonacci calculation and matrix multiplication comparing our prototype with a naive implementation

3   Conclusion: Blueprint for Task Parallelism in MPI

MPI has not been originally designed to support task parallelism. However, our study provides hints at solutions.

MPI_Comm_spawn is a natural and native way to define and fork new MPI tasks. The original MPI 1.2 norm already defined MPI tasks independently of the notion of process (in the OS sense). However, the current MPI_Comm_spawn is meant to run new images of MPI binaries, *i.e.* to create new processes. An interesting improvement would be to allow the call of functions instead of, or together with MPI binaries. Thus the programmer could easily improve the granularity choosing between processes creation or recursive calls. The choice between spawning functions or binaries could be let to the programmer by the use of a special field of the MPI_Info parameter passed to MPI_Comm_spawn.

Acknowledgments. We would like to thank CAPES for the financial support.

References

1. Mattson, T.G., Sanders, B.A., Massingill, B.L.: Patterns for Parallel Computing. In: Software Patterns Series. Addison-Wesley, Reading (2004)
2. Blumofe, R.D., Joerg, C.F., Kuszmaul, B.C., Leiserson, C.E., Randall, K.H., Zhou, Y.: Cilk: An efficient multithreaded runtime system. J. of Parallel and Dist. Comp. 37(1), 55–69 (1996)
3. Leiserson, C.E.: The Cilk++ concurrency platform. In: Proceedings of the 46th Annual Design Automation Conference, pp. 522–527. ACM, New York (2009)
4. Chapman, B., Jost, G., van der Pas, R.: Using OpenMP: Portable Shared Memory Parallel Programming. In: Scientific and Engineering Computation Series. MIT Press, Cambridge (2008)
5. Reinders, J.: Intel Threading Building Blocks: Outfitting C++ for Multi-core Processor Parallelism. O'Reilly & Associates, Inc., Sebastopol (2007)

Author Index